W9-ATP-765

3 9153 0094223c 5

DISCARD

FINANCING

THE METROPOLIS

Public Policy in Urban Economies

Volume 4, URBAN AFFAIRS ANNUAL REVIEWS

INTERNATIONAL EDITORIAL ADVISORY BOARD

HOWARD S. BECKER
Northwestern University

ASA BRIGGS
University of Sussex

JOHN W. DYCKMAN
University of California (Berkeley)

H, J. DYOS
University of Leicester

MARILYN GITTELL
Editor, Urban Affairs Quarterly

JEAN GOTTMANN
Oxford University

SCOTT GREER
Northwestern University

ROBERT J. HAVIGHURST
University of Chicago

WERNER Z. HIRSCH
University of California (Los Angeles)

EIICHI ISOMURA
Tokyo Metropolitan University

OSCAR LEWIS
University of Illinois

WILLIAM C. LORING
U.S. Public Health Service

MARTIN MEYERSON
State University of New York (Buffalo)

EDUARDO NEIRA
Interamerican Development Bank

HARVEY S. PERLOFF
University of California (Los Angeles)

D. J. ROBERTSON
University of Glasgow

WALLACE SAYRE
Columbia University

P. J. O. SELF
The London School of Political Science and Economics

KENZO TANGE
Tokyo National University

FINANCING THE METROPOLIS

Public Policy
in
Urban Economies

edited by
JOHN P. CRECINE

Volume 4   URBAN AFFAIRS ANNUAL REVIEWS

SAGE PUBLICATIONS / BEVERLY HILLS, CALIFORNIA

Copyright © 1970 by Sage Publications, Inc.

Printed in the United States of America

All rights reserved. No part of this book may be reproduced
or utilized in any form or by any means, electronic or mechanical,
including photocopying, recording, or by any
information storage and retrieval system, without permission in writing
from the Publisher.

For information address:

SAGE PUBLICATIONS, INC.
275 South Beverly Drive
Beverly Hills, California 90212

International Standard Book Number 0-8039-0008-2

Library of Congress Catalog Card No. 72-103479

FIRST PRINTING

Contents

Part II   PUBLIC EXPENDITURES IN THE URBAN ECONOMY

INTRODUCTION   97

PUBLIC EXPENDITURES: THE TOTAL SIZE OF THE PIE

PUBLIC EXPENDITURES: HOW THE PIE GETS DIVIDED

BUDGETARY REFORM AND THE RESTRUCTURING
OF PUBLIC EXPENDITURE DETERMINANTS

Part III SUBECONOMIES IN THE URBAN SETTING

Part IV

URBAN SERVICES AND THEIR DELIVERY

INTRODUCTION

Introduction

Purpose and Focus

JOHN P. CRECINE

☐ PUBLIC POLICY does and can affect the urban condition in two ways:

(1) direct governmental provision of goods and services; and
(2) intervention through regulation, subsidization, taxation, and the like, in the normal affairs of individuals and nongovernmental institutions.

This volume is concerned with both types of public policy, with actual policy outcomes, and with new policies and procedures designed to change outcomes.

Oversimplifying somewhat, local governments through political and administrative mechanisms raise revenues to directly provide particular mixes and quantities of goods and services. An understanding of governmental decision processes and how these processes are affected by the urban environment they are embedded in is required to comprehend this aspect of public policy. Part II of this volume is concerned with the processes and problems associated with the total level of local governmental resources, the allocation of that total among the various goods and services provided by local government, and proposed reforms in the organizational (budgetary) process that generates public resource allocation decisions.

Urban governments induce or force others to provide public goods, eliminate public "bads" (see Buchanan, this volume), and redistribute wealth in the community by their provision of capital goods (for example, vacant land in an urban renewal program, new street and .utility construction, and so forth), subsidization of nongovernmental institutions (for example, welfare payments, grants to local art museum, and the like), and regulation of various activities (for example, zoning laws regulating land use, air pollution ordinances, and so forth). To understand these phenomena, it is necessary first to understand the particular environments or subeconomies within which public sector intervention is taking place. Only by knowing the particular characteristics and structure of the system of behavior that government is trying to affect, is it possible to understand the implications for outcomes of those relatively few inputs controlled by government. Consequently, both the overall structure of this volume and Part III view government policy in the context of the larger whole. Part I, the introductory section, is designed to provide an overall picture of the urban economic system in toto. As Thompson argues in his article, the role of the public sector in the urban economic system is a relatively minor one, at least in the short or intermediate run. This situation makes it all the more important for students and designers of public policy to know the overall characteristics of the objects of policy. Part I also examines the economic rationales for local governmental activity and some of their implications for public policy.

Part III focuses on a particular population group—the urban poor. In most large metropolitan areas, Negroes constitute the major subset of the urban poor. Discussions of the impact of public policy on the urban poor necessarily include detailed descriptions of the several subeconomies or systems of behavior these groups find themselves embedded in and affected by, for example, urban property and housing markets, racial segregation phenomena, ghetto economies, and commercial locational processes. Public policy is examined in the context of the several subeconomies rather than the reverse. This approach is implicitly and explicitly advocated in this volume. Space does not permit an approach similar to that found in Part III for other population groups and policy areas.

Another theme which runs throughout is that the mode of actual delivery of governmental goods and services is an important characteristic of the good or service itself. In Part IV, the relationship of people's perceptions of the levels of urban services to actual service levels (Aberbach and Walker, this volume) and alternative systems for delivering urban services (Eisinger, this volume; Reiss, this volume) are both discussed.

The urban condition is all too often not a happy one. Yet, in spite of benevolent motives of government officials and concerned citizens, considerable public and private resources expended in attempts to eliminate problems, vastly increased public awareness of deplorable conditions, and the like, urban problems persist and multiply. Few generalizations about urban problems are valid; *one* valid generalization, however, is that the abatement of urban problems requires much more than good intentions and financial resources. Public policy cannot be effective, except by dumb luck, unless based on correct notions about the way relevant sectors of the environment work. Consider the following familiar example. A desire to eliminate deplorable living conditions, deteriorating and dilapidated buildings, and slums led to urban renewal legislation (see articles by Davis and by Muth, this volume). Low-quality housing was torn down; the land was then cleared and offered for sale to private developers. The cleared land, after remaining vacant for several years usually ended up being used for construction of high-rise luxury apartments or office buildings. Not surprisingly, the correlation between low-cost and low-quality housing is extremely high. Urban renewal policy, by intervening in urban property markets, has in many cities significantly reduced the supply of low-quality, low-cost housing in a period when the quantities of low-cost housing demanded were rising. As every sophomore economics student knows, under such conditions, prices rise in the short run and incentives to increase the supply of low-cost, low-quality housing are created in the longer run. The results, of course, were that the poor who represent most of the demand for low-cost housing had to absorb the price increases, either through increased dollar payments, greater overcrowding, higher work-trip costs, or even-lower-quality housing. On the other hand, with a price increase, incentives are created to produce more low-cost housing. The most developed technology for producing low-cost housing involves individual landlords who no longer have proper incentives to maintain their property, etc. Thus, supply shortages in low-cost housing are met by increases in the quantity of deterio- rating structures. Stricter building and housing code enforcement should have roughly the same effects as a general price increase for low-quality housing. This example is admittedly oversimplified but illustrates the insufficiency of sincere motives. The *implication* is that public policy should be directed toward reducing the correla- tion between cost and some measures of housing quality, the production of *new,* durable low-cost housing or an emphasis on inexpensive rehabilitation of older units.

Almost all of the articles in this volume are written by members of academia: economists, lawyers, political scientists, and sociol- ogists. By example, a view is expressed of the proper role of the academic as an input to the policy-making process: the uncovering and cumulation of the requisite knowledge of the underlying char- acteristics and structure of the behavioral systems toward which policy is directed. For example, how do urban property markets work? what would be the effects of stricter housing codes and intensified enforcement efforts? how do governmental organiza- tions allocate resources under their control? how might those decision processes be modified?

Collectively, we also hope that the knowledge and insights actually included here will prove directly useful to policy makers and to students and analysts of public policy and the policy- making process.

—J.P.C.

Part I

THE URBAN
ECONOMIC SYSTEM

Part I   THE URBAN ECONOMIC SYSTEM

Introduction

Wilbur Thompson sketches out a theory of urban growth and development, emphasizing the interrelationships between the externally-oriented (basic) and internally-oriented (service) sectors of the urban economy, between public and private institutions, between labor supply and industrial mix, and the like. It is clear from Thompson's overview that in the short run, the public sector has control over relatively few of the mechanisms producing outcomes in an urban area. In the long run, through its provision of a skilled and educated labor supply and other amenities attractive to newly emerging industries, a city may be able to more or less continually ride the crest of successive technological advances, acquiring new economic and cultural activities at rates fast enough to more than upgrade and absorb those activities that are naturally declining and dying out. Thompson sees much of the long-run viability of an urban area as evolving from a continued ability to capture society's new and more rapidly growing enterprises, thus acquiring the intellectual and physical capital and short-run monopolist profits characteristic of industries utilizing new technology. These "excess profits and resources," Thompson suggests, make it possible for a city to deal with many of its long-run problems.

James Buchanan argues that the urbanization process has reached a point where the primary policy needs are shifting from a need to provide public goods and services (for example, streets, fire protection, utilities, and the like) to a need to eliminate "public bads," a need to deal with the negative aspects of urbanization, and away from a primary role of facilitating urbanization. He then traces the implications of this change in perspective for public policy at the local level.

In the final article in Part I, Otto Davis examines the results of public action in the urban renewal area and evaluates this policy in terms of the nominal goals of this program. His piece provides an excellent example of the kind of theorizing and evaluation of policy proposals that should go on prior to the adoption of policy. Some knowledge of the operation of urban property markets is required before government programs in this area can be properly designed or evaluated. Government policy usually operates through nongovernmental institutions.

–J. P. C.

THE URBAN
ECONOMIC SYSTEM

An Empirical
and Theoretical Overview

1

Internal and External Factors in the Development of Urban Economies

WILBUR THOMPSON

☐ URBAN-REGIONAL GROWTH THEORY has subsisted since the uncertain date of initial systematic enquiry, sometime in the early twenties, on one simple but powerful idea, the export base concept. Because the history of this concept has been traced at length elsewhere,[1] as a prelude to the present discussion we need only remind ourselves of the concept's main characteristics and limitations as brought out through the years of debate.

THE LEGACY OF THE EXPORT BASE THEORY

From the beginning, economic geographers, economic historians, urban planners, and urban land economists have all seen the usefulness of distinguishing between those local industries which sell outside the "local economy" (later interpreted more precisely as

EDITOR'S NOTE: *Reprinted from* Issues in Urban Economics, *edited by Harvey S. Perloff and Lowdon Wingo, Jr. Published for Resources for the Future, Inc. by the Johns Hopkins University Press. Copyright © 1968. Reprinted by permission of the author and the publisher.*

the job commuting radius about the urban center) and those which sell within the "local labor market." The export industries clearly generate a net flow of income into the local economy from which the necessary imports can be financed. In this most immediate sense of current money flow, the export sector is basic and the local service sector is derivative in origin.

When, in the postwar period, economists began to take a more sustained interest in regional development, a parallel was seen between the export base concept and the export multiplier concept from Keynesian economics. The contributions of the economists here tended to reinforce the notion of the primacy of the export sector in the performance of the small-area, open economy because usually the economist was thinking in cyclical terms—of the expansion and contraction of output and employment within given export plant facilities. The direction of causation is clear as national cyclical change ebbs and flows through the portal of the local export sector and swirls about the passive stores and offices.

But the attempted marriage of the export multiplier and the export base theory was more than a little forced. Those with deeper and more sustained involvement with the city—urban planners especially—did not have in mind the direction of causation of variations in per capita income of a (nearly) fixed population and labor force over the business cycle. Urban planners care more about the lines of linkage in aggregate growth of total population over decades. Consequently they were interested in a very different kind of "multiplier" that defined locational linkages in the long run where new investment in plant and plant relocations attract other new plants and homes and stores.

Through the postwar period it has not always been clear when a given discussion was addressed to cyclical changes in local output, income, and employment of interest to, say, local public revenue and public assistance officials, or when long-run public investments in streets, utilities, schools, or whatever were at issue. Both the projecting of next year's tax revenue and the pouring of cement are significant social questions, but the time horizons are, of course, quite distinct, as is also the relevant theory.

Understandably, some urban planners rebelled at the invidious comparison suggested by the terms "basic" and "local service," especially as these words implied a higher policy priority to the needs of the export sector. The export firms were more often than not those drab manufacturing plants on the edge of town, or currently moving there, while the local service sector was clustered in the more elegant high-rise buildings that dominated their favored downtown sites and symbolized "city" itself. Usually, the urban planner was not sophisticated enough in economics to defend well his intuition that in his world of the very long run (locational economics) it may well be the local service sector which is enduring (basic) and the manufacturing plant which is transitory.

As we move from local cycles to local development, we shift our interests from money flows, which clearly run from exports to local services, to concern for comparative costs relative to competing urban centers. The lines of causation are then turned inside out to the degree that comparative costs rest on the efficiency of local transportation systems, public utilities, banks and other financial institutions, schools and universities and other training and retraining centers, and a host of other critical supporting services. Such services are, moreover, likely to be more permanent than the particular factories of a given time period.

We begin to see more clearly the relationship between the time period and the lines of causation in urban change and development when we review more precisely the simplistic and somewhat tenuous dichotomy of export and local service. With time and the contributions of the economist, the concept of the export sector has been extended beyond direct exports to other regions to include local firms which supply intermediate products to local exporters. As one moves, however, from the more obvious "linked industries" (tire manufacturers in Detroit, for example) farther and farther back in the stages of production to trucking firms hauling inputs and outputs, to accounting, financial, and promotional services, and even back to technical curricula in local schools, a near continuum of activity emerges. It becomes very difficult to dichotomize local activity, and the most basic work is indeed performed by the very same banks, schools, and utilities that service the households.

These complex lines of linkage among the many local activities can be, and were, brought out crisply and definitively in the many local input-output tables constructed around the country. But, again, the input-output tables quantify the flows among a given set of local industries and are most relevant to short-period variations, although some limited extension to growth paths is possible. We do not, however, have the full development equivalent, a locational matrix that tells us which industries follow a given industry to a locality and when. This is what the urban planners and economic geographers had in mind.

With greater urban size comes a tendency toward greater local self-sufficiency. More than a decade ago, critics of the early, more simplistic form of the export base theory noted the tendency for the base-to-service ratio to fall with increased size. While perhaps none of the early developers of the export base mechanism would have insisted, on being pressed, that the ratio of export to local service workers was fixed, they all too easily slipped into the habit of multiplying the change in the number of, say, manufacturing workers by some customary fixed coefficient to find the change in total employment and then remultiplying that employment figure by the (inverse of the) labor force participation rate to yield the change in total population of a community. Perhaps the number of retail food store clerks or barbers per thousand population is nearly constant across the spectrum of urban areas, but with larger size the threshold of economic local production for the local market is passed for one new activity after another. First come those activities with only modest economies of scale and high transportation costs (medical clinics, for example), and with great size even those activities with very large scale economies and low transportation costs (such as investment banking) come to be produced locally for local sale.

To hold the prime role for an export sector that accounts for a half of all local activity in an urban place of ten thousand population is a far cry from retaining that role when metropolitan areas reach one-half million and over and where the export sector may account for only one-quarter of all activity. (And this applies even in cyclical analysis, where shifts in local investment and in the "local-consumption-of-local-production" function probably come to rival export multipliers in local cyclical impact.) Clearly, then, the relative roles of the export and local service sectors change with the relevant time period and the size of the urban region under analysis.

In the analyses to follow, an all-embracing term will be used: the "local industry-mix." In static or very short-period analysis of a small urban area the distinctive and determining part of the local industry-mix is the current export mix, and the impact of the current local industrial specialties on local economic welfare will be traced through—on the pattern of income as well as the growth rate. But as we move through time and to larger size, the analysis will be directed to the process by which structural change takes place in the local industry-mix. The analyses will, therefore, begin with the export industry-mix and progress toward increasing involvement with the local service sector, especially those parts of it which facilitate adaptation to change and/or produce change itself.

URBAN-REGIONAL INCOME ANALYSIS

The easiest place for the economist to take hold of urban-regional growth and development is in the middle of the story, that is, at the present. As we come to understand better how a given—the current—industrial structure affects the performance of a local economy, even in static terms, we will come to see better the inexorable development of forces leading to change. The most conventional and powerful tools and skills of the economist can, moreover, be most quickly brought to bear in a static regional income analysis. Here we can begin simplistically by casting a local economy as a mere bundle of industries in space: Tell me your industries and I will tell your (immediate) fortune. How could a highly specialized economy fail to reflect in its level, distribution, and stability of income and growth rates its distinctive industry mix? Our analysis will, moreover, retain those traditional dimensions of economic welfare that follow from national income analysis.

The major point of departure will be that no attempt will be made to discuss the general level of income or the general degree of inequality or overall national economic stability, only the pattern of regional deviations about the national average. Our charge in urban-regional economics is to determine why some urban areas are richer or more egalitarian or more unstable—or all three (as in the case of Flint, Michigan). Urban-regional economics, freed in the beginning from explaining the general state of the national economy, becomes the study of variations within the system of cities.[2]

While urban-regional income analysis is more than merely industrial analysis, it is intriguing and surprising to see how far one can go in explaining the level, distribution, and stability of local income as mere extensions of the local industry-mix. But because this has been done elsewhere at considerable length (Thompson, 1965a: Chapter 2-5; or for a revised but briefer version, Thompson, 1966), it should be sufficient here to summarize what seems, deductively, to be the more important lines of interaction between the local industrial structure and the local income pattern before pushing beyond into the much more difficult subject of change.

THE LEVEL OF INCOME

The many subtle and interacting forces that operate to enrich a community may be factored into two generic determinants of the level of income: skill and power. We would expect, typically, to find above-average skill in local labor forces that are wrestling with the production of the newer products. Implicit here is some form of the "learning curve": new tasks tend to be the most demanding of intelligence and creativity, but in time the production process becomes rationalized and relatively routine. Not only are skilled workers more scarce and therefore higher paid, but skilled workers on new products tend to share the monopoly power of the early lead in a new industry; the high profit margins enjoyed, temporarily, by local innovators are passed on in part to local labor. Innovation, then, gives rise to both the skill needed to rationalize the new and strange and the price power of an early lead.

Other local economies producing older, slower-changing products may also profit from the weak competition that originates in pronounced internal economies of scale relative to the size of the market. Large investments in plant and equipment give rise to high fixed costs; the latter leads to substantially lower unit costs at very large outputs; this leads, in turn, to fewer plants (firms) and (implicit) collusion in pricing. Large investments reinforce this market power by impeding the entry of new firms as potential competitors. The price power of oligopoly is not enough, however, to enrich a community, especially if the local plant is absentee-owned. Where there are large, multiplant corporations whose stockholders are scattered, oligopoly power in the local export sector can be translated into high wage rates and/or family incomes only when it is wedded to labor power. The necessary condition would seem to be an aggressive, industrywide union only nominally resisted by a secure oligopoly that is able to support higher wages through higher prices.

Even a moderate degree of price elasticity in the demand for local labor in the oligopolistic export industry is not likely to constrain local income if the adverse employment effects of high wage rates can be worked off in an expanding market (in the form, perhaps, of a new or income-elastic product) or by easy out-migration. If the oligopolistic export industry is a growth industry, the community will enjoy the benefits of economic power to the fullest as the monopoly wage rate of the export sector "rolls-out" into the local service sector, enriching local teachers, bus drivers, and bankers by more than it deflates the money income of the export workers through increased local prices. Wage roll-out assumes that there is some appreciable labor mobility between the export and import sectors and some appreciable immobility of movement into the community from the outside, so that the high export wage largely benefits local residents. The effects of innovation, market power, and unionization on the level of income are summarized in Figure 1.

THE DISTRIBUTION OF INCOME

Unions not only raise local wage rates and average family incomes; they also tend to reduce the degree of inequality. The frequent threat of skilled workers to break away and create their own union is indirect evidence either that "unionism" (ideology) or economic democracy (assuming majority rule and representative union government) is more egalitarian than the free competitive market. The lesser degree of (median family) income inequality exhibited by the more highly industrialized metropolitan areas is probably due in significant measure to the relatively narrow range of skills characteristic of manufacturing, especially branch plant, assembly operations. Still, the labor union would seem, a priori, to be an equalitarian force of comparable magnitude.

Some appreciation of the complexity of the forces determining the distribution of local income is gained by noting that the local labor markets that generate a large number of jobs for women tend to be among the most egalitarian of all. A high female labor force participation rate may come out of light manufacturing (as in a textile town), but probably more often it is the commercial, financial, medical, educational, or governmental center which generates a balanced demand for labor.[3] Implicit here is the assumption, for which there is reasonably good supporting evidence, that working wives and other female second-income earners come more than proportionately from the lower-income house-

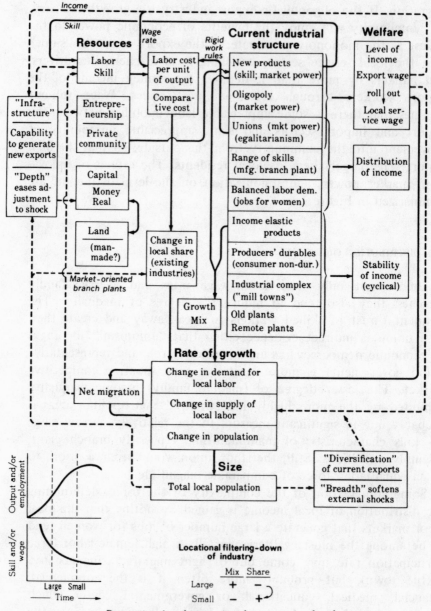

Patterns of urban-regional economic development.

Figure 1

holds. Certainly, second-income earners also raise the average family income but their impact on the distribution of family income would seem to be more significant both quantitatively and socially.

THE STABILITY OF INCOME AND EMPLOYMENT

Two standard concepts from "national economics" seem highly transferable in any attempt to trace through the lines of linkage between the local industrial structure and the relative severity of the local business cycle. Those areas specializing in producers' durable goods should be most unstable and those specializing in consumer nondurables most stable, following from the greater variation in investment expenditures than in consumer expenditures over the cycle and from the simple fact that durable goods can easily be made to last a little longer while the replacement of nondurables is less postponable. The early statistical evidence suggests, however, that an industry-mix explanation of the local cycle will not be sufficient.

Specialization in durables is not, for example, significantly correlated with the degree or rate of decline in employment during the 1957-1958 recession. True, the census data do not permit sure and easy separation of *producers' durables* from consumer durables. Still, the tendency for durable goods to cluster (agglomerate in industrial complexes) while nondurables often stand alone as the sole support of the smaller place (the textile town, for example), may blur the sharp national contrast by posing industrial centers concentrating on *diversified* durable goods against isolated towns concentrating on *specialized* nondurables. Any significant lead or lag in the timing of the industry cycle of any of the local durable industries softens the aggregate local fluctuation as the leading industry begins to fall while other local industries are still climbing, and begins to rise from its trough while the others are still falling.

Further, as we translate national economic concepts into regional economics, we must remember that a local economy is not only an atypical bundle of industries, it tends also to be a nonrandom, or a biased, sample of plants by age. The older industrial areas tend to have more of the older plants, and these tend to be the higher-cost facilities that are cut back first and brought back last over the cycle. To the degree, moreover, that the newer plants are more capital-intensive (automated), the higher ratio of fixed to variable (labor) cost should lower marginal cost and reinforce the tendency for the newer facilities to absorb more than their share of the smaller recession market and lesser share of the larger boom output.

Let us draw out the asymmetry of national and regional cycle analysis a little more by adding that most neglected facet: space. A locality may have relatively high-cost facilities not because of age but because of a more out-of-the-way location. Rarely do we incorporate the cost of transportation of either inputs or outputs into our cost curves. To do so, however, is to see clearly that remoteness from the market raises marginal delivered costs and amplifies output variations in response to demand and price fluctuations. Moreover, "remoteness" may have to be interpreted in the context of differential oscillation of regional markets. Products with high transportation cost are not ordinarily sold across the nation. Thus, a locality's cycle will tend to reflect the regional industry-mix as well as its own mix and its locational position in the region. There is little doubt that a paper-mill town in a region with a state capital and a large university would be in relatively good shape over the business cycle.

GROWTH AND INCOME PATTERNS

By influencing the rate of growth of the local economy, local industrial characteristics also act indirectly to shape the local income pattern. A local export sector which emphasizes either new products or income-elastic products will tend to experience a greater than average expansion in output and in demand for labor. A steadily rising national per capita income acts directly to stimulate the growth of localities producing income-elastic products. Through increasing discretionary income more than proportionately, growing affluence also acts directly to stimulate the demand for new products. The buoyancy of a new-products economy comes not merely from the opportunity to exploit a new market unconstrained by depreciation rates and replacement schedules, but also from the eager search for variety—temporary enchantment with *a* new product becomes a persisting obsession for new products.

A greater than average increase in the demand for local labor interacts with an average increase in the supply of local labor (typical birth, death, and labor force participation rates) to create rising wage rates, overtime, upgrading, part-time jobs, relaxed job requirements, and other reflections of a tight labor market. A greater than average rate of growth in the demand for local labor should then lead to a relatively fast rising and, ultimately, a relatively high median family income, but even more surely should lead to a lessening of the degree of local income inequality. The marginal worker who finds a full-time job will, of course, experience the greatest rate of increase in income. Recent studies of income inequality have shown that it was the tight labor market of the war years which effected the great decrease in inequality and that there has been little or no reduction in interpersonal inequality since the end of World War II (see for example, Miller, 1955).

Again, we distinguish between the national economy, which is a closed economy, and the local economy, which is open to the flow of people as well as goods. Classical economic theory would predict an increase in the supply of labor as the high wage rates, abundant jobs, and high welfare payments of this prosperous community attract the unemployed and unemployable. The seeming paradox of poverty amidst plenty in the booming locality is quite in conformance with classical theory. In the long run, growth industries do not make a locality richer, only bigger.

TOWARD A THEORY OF URBAN-REGIONAL GROWTH

POPULATION SIZE

Since size is simply the cumulation of growth, those places which grow faster tend to get big more quickly and are at any given time larger. Growth creates size and size reacts to restructure the local economy so as virtually to ensure further growth at a near average rate—reacts, that is, to produce growth stability. The simplest, most dramatic and thus most widely appreciated structural change accompanying large size is toward a diversified mix of current exports. A large sample of exports, such as we find in Chicago or Philadelphia, is likely to include some very new, fast-growing industries, some middle-aged slow-growing ones, and even a few that are old and declining. Moreover, a large sample would be likely to mix income-elastic with income-inelastic products. Consequently, we would not expect to find the large, diversified export sector generating unusually rapid growth or very slow growth.[4]

The very large metropolitan area is even more distinctive in its depth than in its breadth. The local social overhead—the infrastructure—that has been amassed is, more than export diversification, the source of local vitality and endurance. Stable growth over short periods of time, say up to a decade, is largely a matter of the number of different current exports on which employment and income are based. But all products wax and wane, and so the long-range viability of any area must rest ultimately on its capacity to invent and/or innovate or otherwise acquire new export bases.

The economic base of the larger metropolitan area is, then, the creativity of its universities and research parks, the sophistication of its engineering firms and financial institutions, the persuasiveness of its public relations and advertising agencies, the flexibility of its transportation networks and utility systems, and all the other dimensions of infrastructure that facilitate the quick and orderly transfer from old dying bases to new growing ones. A diversified set of current exports—"breadth"—softens the shock of exogenous change, while a rich infrastructure—"depth"—facilitates the adjustment to change by providing the socioeconomic institutions and physical facilities needed to initiate new enterprises, transfer capital from old to new forms, and retrain labor.

IMPACT OF SIZE ON LOCAL RESOURCES

Whatever may be the prime source of national economic development, entrepreneurship must be the heart of comparative regional growth in any system of open regions. The large urban area would seem to have a great advantage in the critical functions of invention, innovation, promotion, and rationalization of the new. The stabilization and even institutionalization of entrepreneurship may be the principal strength of the large urban area. In an earlier work, the argument was advanced that a large population operated to ensure a steady flow of gifted persons, native to one area. A population of fifty thousand that gives birth to, say, only one commercial or industrial genius every decade might get caught between geniuses at a time of great economic trial such as the loss of a large employer, but in a population cluster of five million, with an average flow of ten per year, a serious and prolonged crisis in local economic leadership seems highly improbable (Thompson, 1965a).

On further thought, the great viability of the large aggregate in a time of quick and sharp change seems to lie even more in the separate but coordinated institutionalization of the many entrepreneurial functions, especially invention, innovation, and quick adaptation to change. The very large metropolitan area typically hosts a large state university with well-developed programs in basic research and in graduate, professional, and continuing education. The main medical centers grow up to serve the nearby population, and the many advantages of scale draw medical research personnel and funds. As we become more of a service-oriented economy, the city itself becomes the very product that is being redesigned and reengineered—becomes the experiment as well as the laboratory. Small wonder that the largest metropolitan areas can be so little concerned with promoting area industrial development, compared with the frantic activities of this kind conducted by the smaller areas.

When we turn to consider the cooperating factors of production, the case for the large urban area does not suffer. Money capital has long been recognized as the most spatially mobile factor, and is so handled both explicitly and implicitly in the locational analyses of Edgar Hoover and August Lösch. The advent of the modern large corporation, financing internally from depreciation reserves and retained earnings, and externally in a national money market, further weakens the locational influence of local capital supplies for the large, well-established, national-market business. But existing real capital is highly immobile—most often permanently fixed in space. The large urban area has by definition the largest and most varied supply of existing real capital—infrastructure—referred to above as the real economic base of the city. It is, in fact, difficult to define the boundaries between entrepreneurship and capital; consider the research laboratories of the large universities—a case where public capital spins off new products and new businesses.

Conversely, small business is often highly dependent on local capital supplies, both commercial credit for working capital and equity money. But the greatest proportion of small business is in local services, making such activity derivative from, not determinant of, regional growth, at least in the short run. To the extent, moreover, the local service activity does play a significant developmental role in the very long run, the large urban area is probably better able to finance this special group of small business which does depend on local capital supplies.

Turning to labor as a locational factor, the case for the large urban area is again reinforced. With the spread of unionism comes the spatially invariant wage, an indispensable concomitant of labor's bargaining power. An industrywide wage rate in a given industry is a big step toward spatially invariant labor costs, although regional differences in productivity may remain owing to regional differences in skill and motivation. But probably, on net, increasingly automated production processes operate to reduce the opportunity for variations in worker skill or effort to a point where the quantity or quality of output is not significantly altered.

When workers quote the same wage for a given occupation at all places, they give up all influence on the location of work. If blue-collar, middle-income workers should happen to prefer small towns or medium-size cities as places that offer pleasant living and fishing, this is irrelevant as a locational factor. What could be most relevant is that the wives of the corporate managers prefer the theatre. Unionism acts to shift location from a production to a consumption decision and increasingly into the hands of management—urbane professionals.

Finally, let us consider land, which has always been a critical determinant of the locational pattern of economic activity. Most urban areas began by exploiting some natural resource or strategic position on the earth's surface; but for some time now, the share of activity accounted for by the extractive industries and raw-material-oriented processing industries has been declining. However, while on the production side the tie to land has loosened, increasing affluence has tightened it on the consumption side. With high incomes and greater leisure, land in the form of natural beauty and good climate retains significant locational power. But we must also recognize that more and more we live in a "man-made environment" where we fabricate lake-front lots and ski-slope snow, and where we can probably look forward to plastic trees and mountains. That is, capital becomes increasingly a substitute for land; capital is subject to economies of scale and is the magic wand of the entrepreneur-illusionist.

A FILTERING-DOWN THEORY OF INDUSTRY LOCATION

If the larger urban areas are, in fact, more than proportionately places of creative entrepreneurship, then we might hazard a broad hypothesis on the nature of regional growth patterns—one that can be tested with available data (U.S. Department of Commerce, 1965). Following current practice, let us factor regional growth into two major components: mix and share. An area may grow rapidly either because it has blended a mix of fast-growing industries (those of new products as in Los Angeles, or those with income-elastic demands as in Detroit), or because it is acquiring a larger share of the older, slow-growing ones (the movement of the textile industry to various North Carolina towns is such a case). The hypothesis offered is that the larger urban areas tend to combine a fast-growing mix with a steadily declining share of these growth industries. We modify here, then, the earlier position that the large urban area is highly diversified by suggesting that its rich mix is biased at least a little toward growth industries, but that its growth is dampened by the steady spinning-off of these industries. The result is typically a near-average growth rate.

The larger urban area is believed to invent, or at least innovate, to a more than proportionate degree and, therefore, to enjoy the rapid growth rate characteristic of the early stage of an industry's life cycle—one of exploitation of a new market. As the industry matures into a replacement market, the rate of job formation in that industry slows nationally and the local rate of job formation may slow even more if the maturing industry begins to decentralize—a likely development, especially in nonunionized industries, because with maturity the production process becomes rationalized and often routine. The high wage rates of the innovating area, quite consonant with the high skills needed in the beginning stages of the learning process, become excessive when the skill requirements decline and the industry, or parts of it, "filters-down" to the smaller, less industrially sophisticated areas where the cheaper labor is now up to the lesser occupational demands.

A filter-down theory of industrial location would go far toward explaining the southern small towns' lament that they always get the slow-growing industries. Out-of-the-way towns like these find they must run to stand still, because their industrial catches come to them only to die. Meantime, these smaller industrial novices struggle to raise per capita income over the hurdle of industries which pay the lowest wage rates. Clearly, the characteristics of slow growth and low wage rates (low skills) might be viewed as to facets of the aging industry. The smaller, less industrially advanced area struggles to achieve an average rate of growth out of enlarging shares of slow-growth industries, originating as a by-product of the area's low wage rate attraction.

The larger, more sophisticated urban economics can continue to earn high wage rates only by continually performing the more difficult work. Consequently, they must always be prepared to pick up new work in the early stages of the learning curve—inventing, innovating, rationalizing, and then spinning off the work when it becomes routine. In its early stages an industry also generates high local incomes by establishing an early lead on competition. The quasi-rents of an early lead are in part lost to the local economy, as dividends to widely dispersed stockholders, but in part retained as high wage rates, especially if strong unions can exploit the temporarily high ability to pay. It would seem, then, that the larger industrial centers as well as the smaller areas must run to stand still (at the national average growth rate); but the larger areas do run for higher stakes.

In order to develop, it seems that the smaller, less favored urban area must attract each successive industry a little earlier in the industry's life cycle, while it still has substantial job-forming potential and, more important, while higher-skill work is required. Only by upgrading the labor force on the job and generating the higher incomes—hence the fiscal capacity—needed to finance better schools, can the area hope to break out of its underdevelopment trap. By moving up the learning curve to greater challenge and down the growth curve toward higher growth rates for a given industry, an area can encourage the tight and demanding type of local labor market that will keep the better young adults home, lure good new ones in, and upgrade the less able ones.

THE DELICATE BALANCE BETWEEN EQUILIBRATING AND DISEQUILIBRATING FORCES

While we have much to learn from static regional analysis, it is necessary to move toward a more dynamic framework if we are ever to understand how the intricate interactions of development arise. A fully dynamic analysis is still far off, but a form of "stages" analysis provides the means of gaining new insights into the development process. At any given time the industrial structure of a local economy acts directly on the stock of local resources so as to alter their quantity or quality. Each process or stage of regional development leaves a legacy in the form of an altered stock of labor, capital, or entrepreneurship which acts to change the path of development. Only one illustration of a possible development sequence is offered here, but it is one which brings out the delicate balance between equilibrating and disequilibrating forces in regional development.

In the static, industrial-type analysis dealt with in the previous section, it was argued that a locality could parlay an oligopolistic export industry and a strong union into high wage rates and high incomes. A locality that transforms market power into local wage rates that exceed the relative productivity of its labor force will live well but dangerously on overpriced labor. As the relevant industry matures it will slow its rate of job formation, and especially the rate of *local* job formation, if substantial decentralization takes place with maturation, as is usually the case.[5] Confronted by a rate of local job formation inadequate to absorb the natural increase in the local labor force, the search for new industries is greatly hindered by its overpriced labor.

In the very long run local monopoly power may backfire in another way. Strong unions may set rigid work rules that become more binding and expensive as time passes and technology progresses. New processes which, with no greater effort than before, permit a worker to tend more spindles at one time or drill more holes in a given time period cannot be introduced locally. The new process must and will be innovated elsewhere, leaving the locality with a sharply declining share of what might still be a rapidly growing industry. In such a case, moreover, this high-cost labor market will tend also to become burdened with old plants and heavy cycles as well as slow growth. Thus the classic advantages of an early start—a pool of labor where "the skill is in the air" and bankers who know the industry and identify with it—all may be undone by rigid work rules in a world of rapid technological change.

The locality may have to go through a painful wage deflation before it can recover; in short, there may be a day of retribution. Then, again, there may not. If the beneficent oligopoly exhibits an extended stage of strong growth and is slow to decentralize, it could enrich its host area for a full generation or more, as in the case of the automobile industry and Detroit. That locality may come to acquire the education and skill which merits a relatively high local wage rate. Local affluence, however acquired, provides the private means and the tax base needed to build superior social overhead, and thereby to provide a wholly new local economic base to support new export bases. By the time the moment of truth arrives, the local labor force could have risen in education and skill to the point where it is no longer significantly overpriced, and has become, in fact, a scarce factor.

Thus, in this latter case, no equilibrating force would ever come into play. The original nexus of market power could lead instead to a set of disequilibrating forces, as power leads to affluence and affluence leads to education and skill and further affluence. Clearly, whether equilibrating or disequilibrating forces prevail is a matter of timing. The longer a locality holds market power the more likely it is that there will be no day of retribution.

Local rates of growth in employment, labor force, and population substantially different from the national average rate—or the local rate of natural increase—induce net migration flows, and thereby also create a tension between equilibrating and disequilibrating forces. Migration acts most simply to adjust the size of the local labor force and population to the level of economic activity and serves therefore as an equilibrating mechanism in the local labor market, with reference to both employment and per capita income. But migration also acts to change local population profiles to the extent that migrants differ from nonmigrants. If the more mobile (migratory) persons are the younger, more talented, more energetic or aggressive, then migration may serve to reduce the future economic potential of contracting or slowly growing areas, as it drains from them the next generation of professionals and entrepreneurs. Meanwhile, the more rapidly growing areas which receive the young migrants are building up a favorable age distribution of population, ensuring future growth. Migration, then, may begin as an equilibrating force but end as a disequilibrating one, as the rich become richer and the poor become poorer.

NEXT STEPS

Clearly, we are now enmeshed in complex questions of timing. An oligopolistic export industry may die or leave before the community it enriches has time to grow up to its inflated wage, and thus the transition to a new industrial base becomes especially difficult. Or the beneficent industry may set in place an educated second generation before it slowly fades away. Again, net out-migration may relieve the pressure of a redundant labor supply long enough to allow the locality to transfer public funds from, say, public welfare to education and be born again. Or the net out-migration may become a "brain-drain" that leaves the community worse off after each "corrective adjustment" than it was before. We would profit greatly by turning here and now to good urban-regional time-series data to test some of these hypotheses and to suggest new ones. But we lack such data. Still, the available census data do permit some preliminary inquiry that can provide a beginning in what will surely be a long search into the nature and causes of urban-regional economic development. A preliminary empirical inquiry of this nature is presented in the Appendix to this paper.

A NOTE ON THE GROWTH OF THE LARGEST METROPOLITAN AREAS

Any analysis of urban economic development[6] that stresses, as this one has done, the relative cost of doing business in urban areas of various size, is bound to be bullish on city size. The larger places have a clear and sizable advantage in such areas as cheaper and more flexible transportation and utility systems, better research and development facilities, a more skilled and varied labor supply, and better facilities for educating and retraining workers. Further, these economies of scale are captured by private business as lower private costs; at the same time private business is able to slough off on society various social costs that its presence imposes, such as its addition to traffic congestion and air pollution. If, then, the external diseconomies of business-created noise, dirt, congestion, and pollution are some increasing function of city size and/or density, factor market prices are biased in favor of larger urban areas and understate the true marginal costs of production in the metropolis. In the absence of sophisticated public policy and the even more sophisticated public management that would be needed to implement price reform, factor markets so biased promote urban growth and great size.

Effective limits to urban size would seem to have to originate in the household sector. Certainly, alarms of an urban crisis are almost invariably couched in the color words of amenities: congestion, pollution, and other aspects of bigness, or at least poorly managed bigness. It is, in fact, not at all clear from this largely impressionistic (and frequently impassioned) literature whether the hypothesized rising costs and/or deteriorating quality of urban life with greater scale is due to some naturally scarce factor, such as fresh air or clean water, or due instead to the probable or demonstrated failure of urban public policy and management to apply the best technology and/or arrange the best combination of factors. It could be argued that urban public management is the scarce factor in the latter case, just as management has always been seen as the ultimate limit on firm size in neoclassical price theory. But the probabilities and strategies of relaxing natural resource constraints versus managerial constraints may be quite different. None of this is meant to minimize the difficulties involved in achieving better urban policy and management in our largest, most complex urban agglomerations.

To date, the most that can be made of these popular, if not classic, "problems" of great city size is that they have slowed slightly the growth of the largest urban areas: probably New York and less clearly Chicago. But, paradoxically, the loss of amenities with great size may redound to support the growth of the second echelon of metropolitan areas. Metropolitan areas with over a million population but less than Chicago's eight million offer substantial infrastructure in support of modern business, although they do not rival New York and Chicago on this score. But then neither are their "problems" quite as big either—it is easier, if not quite as exciting, to live in these second echelon metropolitan areas.

We could argue that New York must wrestle with a somewhat more advanced form of each of the classic urban problems, or we could argue alternatively the equivalent: that New York must face each new problem in urban management first. Thus, New York was the first to have to learn how to handle ten million people, and must soon be first to master the problems posed by twenty million. Each successively smaller city, roughly in its rank order, has one more example from which to profit, whether the examples be good or bad. If there is a downward sloping learning curve in urban public management and the challenges of the field are a function of size, Chicago finds the path a little easier because New

York has gone before, and Detroit profits from the pioneering of both. Detroit, that is, should be able to offer in 1970 a better organized version of the five-million population cluster, as a partial offset to the disadvantages it suffers living in the shadow of the greater choice and urbanity of Chicago.

A recent fifty-year projection of the Detroit metropolitan area, which assumed a zero rate of net migration, was criticized as being too optimistic because "the automobile industry is a maturing industry and will, of course, experience the typical slowing rate of growth." Certainly, the automobile industry will probably trace out its own variant of the classic growth curve. But to project the *long run* growth of an urban area from its current industry-mix would be to project the slowing growth and ultimate decline of all areas, as all existing industries are slowly dying. Who, in retrospect, would correlate the growth patterns from 1900 to 1950 of citrus fruits and Los Angeles, or buggies and Detroit? Why, then, must we link automobiles and Detroit for the next fifty years? Ten-year regional growth projections based on the local industry-mix may be defended, but for periods of two or more decades a static, fixed industry-mix technique is heroic if not naive.

Given the embryonic state of urban-regional growth theory, what fifty-year projection might we make for the Detroit metropolitan area? I would argue that Detroit is big enough to assemble the infrastructure necessary to: (a) support a rapidly advancing technology, (b) employ productively and creatively an increasingly educated labor force, and (c) serve an increasingly affluent and sophisticated set of households. From such a base, Detroit should grow at about an average ($\approx$ zero net migration) rate. At one end of the spectrum, the smaller metropolitan areas should about hold their current share of population in that the population of rural areas and small towns will soon be reduced to a bare minimum and the rural to urban migration cannot much longer feed these cities. At the other end, few would argue that New York and/or Chicago is likely to grow faster than its natural rate of increase and drain off growth from the second echelon areas. A zero net migration projection would, in fact, be an optimistic projection for either New York or Chicago, and even this high projection would not encroach on the future population growth of the Detroit economy.

Detroit's competition in the national system of cities would seem, then, to come largely from its peer group: Cleveland, Pittsburg, Cincinnati, Milwaukee, and St. Louis. The question can be turned around from "Why project a zero net migration rate for Detroit?" to "Why not?" What cities will suck growth from these second echelon areas: the bigger ones with more problems? the smaller ones with less infrastructure? A good case can be made that each of the two dozen or so urban areas with a million to five million population will net out to about an average growth rate over the next fifty years, and more than double in size; New York and Chicago will pave the way, perhaps at a slowing rate (with therefore some convergence in size at the top). All this assumes, of course, the absence of national policy that would restrict the continued growth of big cities. And at this time and vantage point, it seems likely that our national policy will be directed more toward mastering the management of large population clusters than toward preventing their growth.

NOTES

1. See especially the series of nine articles by R. B. Andrews, in *Land Economics,* Volume 29, (May, 1953); Volume 32, (February, 1956); and the writings of other early proponents and critics of the export base theory of regional growth, all reprinted in Pfouts (1960).

2. Worthy of attention at some other time and place is a careful investigation of the reverse side of the coin: the way in which a nonoptimal system of cities depresses the national income.

3. Since these nonmanufacturing places tend to create more income inequality out of their wider ranges of occupations, their higher propensity to produce jobs for women tends only to make them less unequal than they would otherwise be, but still not as egalitarian as factory towns.

4. The coefficient of variation $(\sigma/\overline{\chi})$ of the percentage change in population, 1950-1960, for the twenty-two metropolitan areas with a million population or more was about one-half that of the smaller metropolitan areas. See Thompson (1965a: 192), or for a more extended treatment of the effect of great size on stability, see Thompson (1965b).

5. Even if we assume that the industry faces a nationwide union and a spatially invariant wage and, thereby, rule out relocation into lower-wage areas as operations become more routine, as argued under the filtering hypothesis above, decentralization to reduce transportation costs is a common pattern in maturing industries.

6. This section results from comments made by Harold J. Barnett on an earlier version of this paper.

REFERENCES

MILLER, H. P. (1955) Income of the American People. New York: John Wiley.

PFOUTS, R. W. (1960) The Techniques of Urban Economic Analysis. Trenton: Chandler-Davis.

THOMPSON, W. R. (1966) "Urban economic development." In W. Z. Hirsch (ed.) Regional Accounts for Policy Decisions. Baltimore: Johns Hopkins Press for Resources for the Future, Inc.

––– (1965a) A Preface to Urban Economics. Baltimore: Johns Hopkins Press for Resources for the Future, Inc.

––– (1965b) "The future of the Detroit metropolitan area." In W. Haber, W. A. Spivey and M. R. Warshaw (eds.) Michigan in the 1970's: An Economic Forecast. Ann Arbor: University of Michigan Press.

U.S. Department of Commerce (1965) Growth Patterns in Employment by County, 1940-50 and 1950-60. Volumes 1-8. Washington, D.C.: Government Printing Office.

2

Public Goods and Public Bads

JAMES M. BUCHANAN

☐ IN ANY BEHAVIORALLY relevant sense, "goods" and "bads" are reciprocal. Supplying a good eliminates a bad, and destroying a bad produces a good. Conversely, supplying a bad eliminates a good, and destroying a good produces a bad. Until and unless the evaluating and/or choosing agent is specified, however, goods and bads have little or no meaning. Who decides what is good or bad? And for whom? By whom is the choice to be made? In a world where scarcity cannot be ignored, choices must be faced which means, quite simply, that things which are goods for some persons may simultaneously be bads for others. In the particular context to be examined in this paper, behavior aimed at producing good for the individual, as a privately acting independent agent, may be bad for the same individual when he considers himself as a part of a defined collectivity, and vice versa. This elementary fact of life provides the starting point for almost any meaningful discussion of collective or group action, including that of individuals who organize themselves into political units.

The reciprocal relationship between goods and bads has been obscured in much of the modern discussion. A developing literature in a subdiscipline of economic theory concerns itself with "public goods," while the word "public bad" scarcely appears at all. The approach which has become standard emphasizes the failure of voluntary individual behavior in organized markets to produce public goods, more correctly described by the term collective-consumption goods. The failure to supply such goods in efficient or optimal quantities is, of course, equivalent to a failure to eliminate existing public bads in efficient quantities, but the mere change in wording may have explanatory significance. More importantly, by indirection the standard approach has concentrated almost exclusively on a presumed once-and-for-all institutional change, the shift of activity from the market to the governmental sector. Almost no attention has been given to the problem of preventing the *erosion* of public goods once these are supplied, an erosion that is generated by the individual behavior of persons producing public bads. A redirection of emphasis is a central objective of this paper. In the development of the argument, many of the problems that are confronted continuously by modern municipal governments will be recognized. It seems clear, for example, that progressive urbanization and economic development reduce the need for collective or governmental financing of public goods in the more traditional context while, at the same time, these increase, often dramatically, the need for collective or governmental action designed to eliminate multiplying public bads. A shorthand, if somewhat inaccurate way of stating this is to say that municipalities could probably become and remain viable if existing public bads could be eliminated, even if no additional public goods of the standard sort should be provided. In practical policy application, this implies that relatively less emphasis be placed on the financing problem and relatively more emphasis be placed on the problems of enacting and enforcing effective rules for personal and institutional behavior.

PRIVATE GOODS AND PUBLIC GOODS

Bernard Mandeville, in his famous poem, *Fable of the Bees,* was one of the first social philosophers to demonstrate that the results emerging from the interaction of many persons need not be those intended or planned by any one person or group of persons. Under some situations, he argued, qualities of private individual behavior that might seem vicious or self-seeking may be precisely those required to produce desirable social results when persons interact in a complex environment. Adam Smith, in his *Wealth of Nations*, further developed and elaborated this essential vision, which allows us to predict the emergent results of an interaction process independently of the specific overall plans or objectives of the individual participants. To Smith, social or collective welfare is generated, not by the benevolence of the butcher but out of his regard for his own self-interest in supplying meat for sale. Indeed Smith's genius, and his legitimate claim to be the founder of scientific economic theory, lies precisely in his convincing demonstration that the self-interested, independent behavior of individuals in a competitive market process produces socially desired results. Socially desired in this sense is defined strictly in terms of the mutually attainable benefits for all members of the community.

Public good is, therefore, produced by the workings of the competitive market process. Aggregate values, derived from individual choices, are generated and distributed in such a way that all participants secure gains. The predominant share of national produce, even in the mixed economies of the West in the 1970s, is made up of values derived from the production of goods and services through the market, through the so-called private sector. These values are public or social and the increase in the value of national product and in its rate of growth through time are widely shared objectives for economic policy. These values derived from the market are not, however, those to which the specific term public good is normally applied in the postwar discussion. This term, in its current technical and professionally used sense, tends to be reserved for those goods and services that are financed and provided through the governmental-political process, through the so-called public sector. The subdiscipline of economic theory that is now sometimes called the modern theory of public goods is largely devoted to an analysis of why governmental provision of these goods and services may be required, and of how such provision should and does take place.

PRIVATE GOODS AND PUBLIC BADS

If we 'take an economic approach to an analysis of individual behavior, we assume that all persons act so as to maximize utility, whether they act as participants in a market or in a nonmarket setting. In attempting to maximize utility, an individual is, of course, trying to achieve those goods that he values most highly. Indeed, this is merely another way of stating the same thing. The distinguishing feature of the market process is that individuals are allowed, independently and privately, to seek their own private goods, without direct and explicit consideration of the effects of their behavior on other persons in the community. This is the same as saying that in the market process individuals act independently rather than collectively. Decisions are made privately, by the acting individual alone, and he is not required to seek the agreement of anyone other than his partner in the two-way trading that characterizes the market. As suggested above, such utility-maximizing behavior on the part of many persons produces an efficient allocation of economic resources and a maximally valued bundle of goods and services, provided that several important side conditions are either fully satisfied or approximated. Among these conditions, and the only one that is relevant for this paper, must be the *divisibility* of goods and services among separate persons. When it is efficient to share goods and services among several or many persons, the need for cooperative or group action arises. Under some conditions, efficient sharing arrangements may be worked out through an extension of the market process without difficulty (for example, when the sharing groups are relatively small, or when the costs of excluding users are negligible). Under other conditions, however, efficient sharing arrangements may require cooperative or group action organized through the governmental-political process.

When divisibility of goods and services among potential users is not present, and when extended market arrangements are not efficient, the independent or private utility-maximizing behavior of individuals in the market, the pursuit of private goods, yields aggregate results that may be relatively undesirable to all members of the group. The pursuit of private goods may produce public bads in the evaluation of all persons. An inherent conflict emerges between private, utility-maximizing behavior on the part of individuals and the overall results that these same individuals would desire to see achieved. That is to say, individuals may value the results of their private-independent behavior, when viewed in an aggregative sense, less than they would value the results of some alternative overall arrangement. When conflict of this sort arises, the competitive market process, by definition, does not produce a maximally valued bundle of goods and services. Care must be taken in all such discussions as this, however, to insure that public or social valuation refers only to that which is derived from individual utility functions. There is no public or collectivity that exists apart from the individuals, at least in the economist's standard model. Hence, by referring to a conflict between the pursuit of private goods and the achievement of public bads, we mean only that the individuals themselves may be behaving in an institutional framework that produces outcomes that they do not themselves value so highly as possible alternatives. They may be locked in a many-person analogue to the familiar prisoner's dilemma of modern game thoery.

The phenomenon has long been recognized and, of course, it has been discussed in many forms. It provides the basis for the contract theory of the state as well as for the modern theory of the public sector. As Hobbes told us long ago, in an anarchic society the life of man is "nasty, brutish, and short." From this, he derived his central argument concerning the need for a voluntary relinquishment of power to a sovereign governmental authority.

The phenomenon is omnipresent in the conurbations of the world in 1970. Traffic, smog, water pollution, urban blight—these are only a few of its familiar manifestations. In each of these familiar examples and in many more (all of which we may summarize under the common rubric, "congestion"), the aggregate results of the behavior of many persons who act privately and independently—each one seeking to maximize his own utility given the environmental constraints that he faces, each seeking to secure his own private goods—may be less desirable potentially to all members of the community than alternative institutional arrangements. The untrammelled pursuit of private goods may generate public bads.

A simple nonurban example will be helpful, one that retains considerable relevance in many underdeveloped areas of the world. Assume that separate owners of cattle share grazing rights to common territory. Land is commonly owned, and all cattle owners have grazing rights. In this situation, each cattle owner may have a private incentive to overgraze the common land. Each cattle owner, acting in his own interest, may find it advantageous to behave in such a fashion that all owners are in a worse position than they might be under other forms of behavior or under other institutional arrangements. In economic terms, the result of private utility-maximizing behavior is nonoptimal in the Pareto sense.

General recognition of the inherent conflict between the pursuit of private goods and the generation of public bads has led to many proposals for change designed to eliminate or to surmount the problem. These may be separately discussed.

PROPOSALS FOR BEHAVIORAL CHANGES

Although economists have only rarely discussed the problem in such terms, the way toward elimination of the conflict that perhaps first comes to mind lies in some modification of the private behavior of individuals. The goods that enter as arguments in individuals' preferences or utility functions are not chosen from on high, and different individuals surely value different sets of goods. If utility functions could be so modified that persons would be led to treat as private goods only those activities that would result in an absence of the conflict, no further action would be required. The conflict between the pursuit of private goods and the generation of public bads simply would not emerge.

Various ethical systems and principles can be interpreted as attempts to formulate behavioral solutions to the problem here. Christian ethics, with its emphasis on love as a guiding principle of human behavior, may be interpreted as an effort in this direction, one that probably must be judged as ultimately unsatisfactory. More germane to the particular problem (because it is less specific) is the system of Kantian ethics. Kant attempted to formulate a fundamental principle for personal human behavior that would eliminate the conflict. His generalization principle, sometimes called the categorical imperative, provides a proposed solution. If, in fact, each person could be induced to act in such a manner that his behavior, if generalized to all members of the community, would produce results that he desires, the pursuit of private good would guarantee the simultaneous provision of public good. Almost by definition, in a purely Kantian world, public bads could not be observed.

At a considerably less sophisticated level, political scientists have often proposed reforms in this same spirit through their emphasis on the political obligation. Upon this principle political leaders are held to be obliged to behave in such a way as to promote the public interest as opposed to their own private interest.

PROPOSALS FOR INSTITUTIONAL ORGANIZATIONAL CHANGES

A second approach toward elimination of the conflict is one associated with Adam Smith and the classical economists generally, although its origins extend back to Aristotle's defense of private property. Here an attempt is made to modify the structure of property rights so as to reproduce the harmony between the pursuit of private goods and the generation of overall public good as a result, the potential harmony that Mandeville and Smith both stressed. The simple example noted above can illustrate the approach. An institutional reform that would surely eliminate the overgrazing on the land is the granting of private ownership rights to land, to convert commonly owned property into private property. Once this step is taken, the private owners of the land have an incentive to restrict the usage of the scarce resource to socially optimal levels. Land will be used efficiently; total social product will tend to be maximized by the private utility-maximizing behavior of individuals.

The differences between the behavioral and the institutional approaches to eliminating the problem of conflict should be noted. The first concentrates on modifying the behavior of the private individuals in such a way as to make them more conscious of the social or public good. The second accepts the utility functions of individuals more or less as given and attempts to modify institutional arrangements so as to produce the natural harmony that can emerge. In either or both of these approaches, the result, if achieved, is the best of all possible worlds. In either case, the world is transformed so that the pursuit of private goods generates public good.

PROPOSALS FOR SPECIFIC COLLECTIVE ACTION

A third approach comes into play only in those situations where neither of the first two approaches is considered to be practicable. Even under the most plausible assumptions about behavior, and even under an ideal set of private property arrangements, there may still remain inherent indivisibilities such that the conflict between private good and public good will emerge. In such cases, public bads may be observed. It may, for example, be difficult to think either of human behavior being sufficiently reformed or of property rights being so defined as to eliminate such a conflict in the usage of "atmosphere" (generally described). To an extent that economists perhaps do not sufficiently recognize, however, both of the first two approaches are significantly productive of social harmony in any civilized society. Respect for the needs and desires of others surely is a strong motive in individual behavior patterns, and unless this be indeed present to a major degree it is difficult to even imagine how civil life could proceed at all. Also, the traditional arrangements for private property rights allow many potential areas of the conflict to be dissolved without notice. The important areas of disharmony arise only where traditional standards of behavior along with traditional and long-accepted, property-rights structures allow evident public bads to emerge. Hence, proposals for the production-supply of public goods (the destruction of public bads) arise, proposals that incorporate explicit action on the part of the community acting as a collectivity. This action may take several forms; these will be discussed subsequently. At this point, one first clarifying step must be taken in the argument. The failure of private, independent behavior to produce socially desired results can arise equally from action or inaction. Analytically, these are equivalent. If individuals do not provide, say, an efficient quantity of internal defense through their private utility-maximizing behavior, they create a public bad in precisely the same manner that they do when their private independent behavior fouls the air with smog. Modern developments in the theory of public goods have concentrated too much on the failure of individuals to act positively in financing or supplying common consumption goods and services, and too little on the excessive negative action of individuals in generating or producing public bads. This analytical bias has been present despite the fact that any given situation can be treated in strictly reciprocal terms. For example, consider once again the overgrazing of common lands. This can be examined as a situation where independent private behavior generates a public bad. Or, conversely, it can be looked at as embodying a failure to create a public good, in this case defined as the effective policing of the usage of the common property. Concentration on the strictly reciprocal nature of all collective-consumption relationships is helpful in clarifying several ambiguities in the analysis.

IMPLICIT OWNERSHIP UNDER COLLECTIVE ACTION

In a celebrated paper, R. H. Coase (1960: 1-44) demonstrated that, under a regime of private ownership and effectively working markets, the precise structure of ownership rights need not affect the final allocation of economic resources. So long as separate private owners are allowed freely to negotiate trades, one with another, under conditions where transaction costs, as such, are not significant, we should predict that roughly the same allocative results will emerge under any initial assignment of ownership rights. Coase's example was that of two adjacent farms, one of which raised cattle, the other of which raised wheat. The interaction arises because of the likelihood that the cattle will stray and trample the wheat crop. Coase's central point was one of demonstrating that roughly the same allocative outcome emerges regardless of the legal assignment of liability for damages. For economic resource allocation, the assignment of property rights tends, in this model, to be irrelevant.

We may, by analogy if not directly, extend this analysis to what we may call the implicit ownership rights to common facilities and to collective or governmental action generally. Let us assume that all structural changes in property rights which might remove the conflict between the pursuit of private goods and the public good have been made, and that individual behavior is taken as given. Explicit collective or governmental action is indicated. This, in itself, tells us nothing about the starting point, the position from which collective action commences, the structure of the status quo. With the existence of legal titles to private property, the definition of the status quo in terms of rights to exclude others from the usage of defined physical assets is usually not difficult. When we come to discuss common facilities, or those which are potentially common, mere definition of the status quo becomes difficult. There must exist an implicit structure of rights to use commonly shared facilities, but this structure may take several forms.

Who owns the common property? To answer this by saying, "the whole community," becomes operationally meaningless until and unless we specify precisely the rights of individuals to use the property held in common ownership. These rights of action with respect to commonly owned facilities and services define an implicit pattern of ownership, and it is from this pattern that collective action aimed at changing the results must start. Such rights may not, of course, be described specifically in statutes; they may be traditional rights based on the evolution of historical experience. Understanding this structure may, nonetheless, be highly relevant in any discussion of explicit collective action. The reciprocal nature of public goods and public bads can clarify some of the issues here in terms of a familiar municipal example, smog.

There are two implicit ownership extremes, two polar cases. The first is that which has, until quite recently, prevailed in most urban centers, or indeed everywhere else. All persons may possess rights to use the commonly shared resource, air, with no restrictions at all on this usage. At the other extreme, we might think of an implicit ownership pattern in which individuals' rights to use the common resource are severely restricted by outright legal prohibition. This solution seems to be that which is proposed by many of the advocates of pollution control, e.g., Ralph Nader. In this polar extreme, individuals might simply be prevented, by law, from discharging pollutants into the atmosphere.

In the first pattern of rights to usage of the commonly shared resource, we refer to the generation of a public bad in the absence of specific collective action. In the second pattern of rights, we might perhaps be tempted to refer, at least initially, to a public good as the resultant outcome. Surely, it may be argued, since clean air is a good, valued by all members of the community, and since it is commonly available to all, this is a reasonable usage of terms. The emotive strength of this rather simple extension of the standard analysis lends great strength to antipollution campaigns. However, what is overlooked in this facile extension is the necessary presence of private bads in the extreme no-pollution pattern of rights. Individuals are, or may be, prohibited from behaving in ways that they, independently and voluntarily, value more highly than that which the collectively imposed antipollution rules require. The private goods—public bads conflict has merely been replaced by the private bads—public goods conflict. Since individuals place values on both the goods and services through ordinary market processes, the so-called private goods, and on the goods and services generated, directly or indirectly, through the political process, the so-called public goods, it should be clear that the overall public good or public interest or social interest, in its more general sense, must involve some balancing off of these two sets of values. If we are limited to a comparison of the two extreme patterns of rights to usage, no restrictions on usage on the one hand and strictly controlled usage on the other, there is simply no a priori means of determining which of the two is the most productive of economic efficiency. In terms of the values that individuals place on results, the smog-laden result of uncontrolled private behavior may well be preferable to that result which would emerge should Ralph Nader be allowed to impose antipollution controls.

The two extreme patterns discussed here are analogous to the different assignments of liability in the Coase problem. In the latter, liability may be assigned to the cattle raiser, on the one hand, or to the wheat grower on the other. *If no trades were allowed to take place between these two parties, the allocation of resources would be different under the two ownership patterns, and there would be no way of determining,* in the absence of such trade, *which of the two patterns would be most efficient. If,* however, *trade is allowed to take place,* roughly the same results will emerge under either assignment of liability. And, importantly, it seems evident that *the results would not normally be expected to be equivalent to those that would emerge* in either assignment *under the no-trade possibility.*

GEOMETRICAL ILLUSTRATION

The analysis can be illustrated easily in a simple diagram. In Figure 1, at the left-hand origin we place completely uncontrolled usage of the air. At the right-hand origin we place strictly controlled usage, with outright prohibition on any discharge of pollutants. Along the abcissa between these limits we measure either air purity or impurity, depending on the starting point and direction of measurement. Along the ordinate we measure values in dollar terms.

In the polluted or smog situation, let us say that the whole collectivity of persons places some positive value on the good or activity that we may call depollution. The aggregate *marginal*

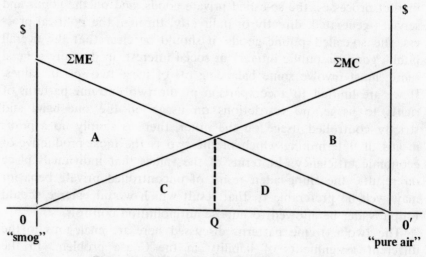

Figure 1. PRIVATE GOODS—PUBLIC BADS
BALANCED AGAINST PRIVATE BADS—PUBLIC GOODS

evaluation curve is shown by ΣMC. This curve is derived by vertically summing the marginal *evaluation* curves of all persons in the collectivity.[1] In the polluted or smoggy extreme, some measures could be taken which might reduce pollution levels significantly without great cost. This cost could be expected to increase as depollution is extended. The *marginal cost* of cleaning up the air is shown by the rising curve ΣMC in Figure 1. As the diagram is drawn, the most efficient or optimal level of depollution (or pollution) is that shown at Q, where the marginal evaluation placed on depollution, summed over all persons in the group, equals the marginal costs of the same activity, also summed over all persons as required. If we start from the full pollution position, from 0, the net social gain secured from production of the public good (from reducing the public bad) in the quantity 0Q is represented by the triangular area, A. Individuals, in their capacities as collective consumers of air, place a value of A plus C on the depollution activity. However, these same individuals, in their capacities as consumers of other goods and services, place a cost of C on the activity. The net gain is limited to A.

If we ignore the feedback effects that may result from different schemes for sharing the net gains,[1] it is relatively easy to see that the same solution emerges if we assume that collective activity commences at the other polar extreme. This equivalence in solution here is the precise analogue to that developed by Coase in his private property-market structure problem. The great difference between the two models is, of course, that the efficient solution emerges in the Coase model from the interaction of traders. In the collective-action model, by contrast, the efficient solution, regardless of the implicit rights structure assumed as the starting point, must be reached through some collective decision-making process. Suppose that, at the outset, no person is allowed to discharge pollutants into the air. We commence at the right-hand extreme. As drawn on Figure 1, this situation is inefficient; the air is too pure, as determined by individuals' evaluations. The marginal evaluation of additional cleanliness is below marginal cost, or, reciprocally, the marginal evaluation of dirtiness is greater than the marginal cost. As we move leftward from the right-hand ordinate in Figure 1, the analysis is fully symmetrical with the rightward movement previously traced. The ΣMC curve can now be treated as a ΣME curve for the creation of the public bad, whereas the ΣME curve can be treated as the ΣMC curve for the same activity. Once again, moving leftward, the efficient or optimal position is attained at Q, or when an amount $0'Q$ of the public bad—in this case, air pollution—is produced.

This somewhat unusual manner of looking at the familiar pollution problem prompts the question: Why should individuals place any positive value on the production of a demonstrably public bad? The answer here is that they do so because such production must involve the complementing production of a private good which may outweigh the costs of the public bad as evaluated by the same persons. The inherent conflict between the achievement of public good and the independent utility-maximizing pursuit by individuals of their own private goods is not eliminated at any position along the left-right spectrum of Figure 1. So long as the conflict exists, what is required for efficiency is not the total elimination of public bads (total satiation of demand for public goods), but, instead, a proper balancing off, *at the margin,* between public and private values. It must be kept in mind that, in either case, the values are those of the individual members of the community and of no external agent. This indicates that the optimal solution is achieved only at Q. If the collective action commences from the rights structure described by the right-hand origin, a net gain from pollution is secured as measured by the triangular area B. The gross gains from pollution are measured by B plus D, but D

is the cost that must be subtracted to get the net surplus generated.

A SCHEMATIC MATRIX REPRESENTATION

The geometrical presentation makes clear that efficiency is
secured only when collective action incorporates the inherent conflict between private goods and public goods. This is implicit in the
standard norms for public goods efficiency, but the reciprocal
nature of the problem has tended to be obscured because of the
implicit assumption that costs are technologically defined and
hence objectively measurable. The standard treatment involves the
statement of the necessary conditions for optimality, the equation
of summed marginal evaluation with marginal cost. This locates the
efficient position, but it *tends to imply,* even if indirectly, *that the
conflict is somehow resolved once a financial outlay is made. This
causes a major problem in the provision and supply of public
facilities and services to be neglected, namely the usage of a
publicly provided facility or service once it is financed and constructed.* The erosion of existing and potential public services and
facilities has hardly been discussed. In terms of a familiar example,
the standard treatment proceeds as if a once-and-for-all installation
of smog prevention devices will resolve the problem of pollution.

The importance of recognizing the continuing nature of the
inherent conflict can be represented in a simple matrix scheme. In
the two-by-two matrix of Figure 2 we place the alternatives for
individual, independent behavior in the two rows. The individual,
considered here to be a representative for the group, may behave in
only two ways. He may act so as to maximize utility, given the
constraints that he confronts, that is, he may seek his own private
goods. Or, by contrast, he may behave so as not to maximize his
own utility. To the economist, this would be irrational behavior if
it were voluntary, but constraints may be externally imposed on
the individual by the collectivity. In the columns of the matrix of
Figure 2, we place the alternative results for the collectivity of
individuals, as evaluated by a representative individual in the group.
These can be either good or bad, as evaluated by the individuals
themselves.

The matrix can perhaps be best understood by looking first at
the cell numbered 1. This cell is the best of all possible worlds.

COLLECTIVITY

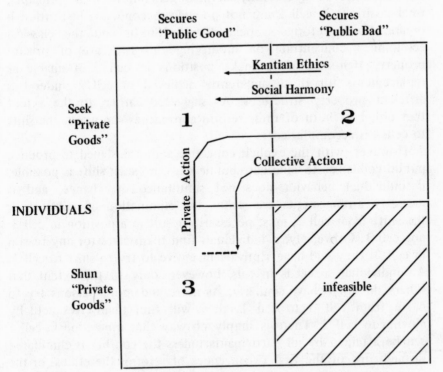

Figure 2. RESULTS OF STRATEGIES FOR SHIFTING BALANCE
BETWEEN PRIVATE AND PUBLIC "GOODS" AND "BADS"

Through individual behavior aimed at maximizing individual utilities, at seeking private goods, an overall social result is generated
that individuals optimally prefer. Public good is secured. The classical economists demonstrated that the working of competitive
markets tends to produce results in this cell. In the cell numbered
2, individuals are behaving so as to maximize their own utilities,
but the aggregate result is one they do not prefer. Public bad
results from the pursuit of private good. This is, of course, the
position that we have been discussing in this paper.

Consider, now, the various avenues for reform when this conflict is recognized. First, we mentioned the system of Kantian
ethics. The emphasis is on modifying individuals' preferences, or
more precisely, the arguments in individuals' utility functions so
that individuals will behave differently. The attempt here is to
make individuals seek as their own good only those activities that
produce outcomes within the first cell of the matrix. Under univer-

sal adherence to a carefully defined generalization in principle, final positions in cell 2 are not possible. Second, the institutional-organizational reforms associated with Smith and the classical economists concentrate on modifying the structure of private property rights so as to make positions in cell 1 attainable as replacements for those apparently achieved in cell 2 under less efficient property structures. As suggested earlier, to the extent that either or both of these reform approaches can work, the shift to cell 1 is accomplished.

However, with the explicit collective action designed to produce public goods, or to eliminate public bads, no such shift is possible. if individual behavior does not simultaneously change, and if property arrangements are stable, the elimination of public bad, the shift from cell 2, must necessarily result in a position in cell 3, not cell 1. Conversely, if individuals find themselves for any reason in cell 3, they will have a private incentive to try to shift to cell 1. An individual succeeds in this, however, only to the extent that others do not behave similarly. As more and more persons try to move from cell 3 to cell 1, they will find themselves actually moving to cell 2. There is simply no way that a position in cell 1 can be attained by all participants unless the conflict is eliminated through one of the first two avenues of reform, the ethical or the institutional.

The predominant emphasis in the modern theory of the public sector has been on the demonstration that individuals acting voluntarily through markets will not escape the dilemma position of cell 2. Hence, the argument is made for collective action, for the levy of compulsory taxation to finance publicly supplied goods as a means of overcoming the failures of voluntary market processes. This emphasis is, of course, important and, within its limits, fully appropriate. Where the modern theory of the public sector has been remiss is in its failure to follow up the analysis, and to stress the private incentives for individuals to escape from cell 3. Instead of a successful shift to cell 1, which may have been implied in some of the more naïve presentations, the resultant position in cell 3 should have been more fully recognized. Once this is done, the problem of public goods erosion will have emerged as a natural extension of the analysis. Individuals finding themselves in cell 3 have precisely the same incentives for allowing a public facility or service to erode, to run down, to be destroyed, as they do for failure to provide the goods or services in efficient quantities when they find themselves in cell 2. The failure of individuals, through voluntary channels, to shift from cell 2 is no different from the positive action of individuals in shifting from cell 3. The argument for the levy of compulsory taxes is on all fours with the argument for the imposition of compulsory rules concerning the usage of publicly provided facilities, goods, and services.

The argument of this section may be summarized by reference to the arrows on the matrix diagram in Figure 2. If some Kantian-type Ethical principle is effective, the shift may be made from cell 2 to cell 1, as indicated by the labeled arrow. Similarly, if re-arrangements in property rights can convert potential conflict into Social Harmony, a shift from cell 2 to cell 1 may move along the parallel arrow. If neither of these approaches can be effective, collective action to finance-provide public goods may be under-taken in an attempt to move along the arrow marked "Collective Action." However, it is impossible to shift from cell 2 and to remain, as might be desired, in cell 1. Inexorably, the process winds up in cell 3. Once in cell 3, however, individuals will necessarily attempt—when and where possible—to move to cell 1 along the arrow indicated by "Private Action." If and as they succeed, however, and as others adopt similar patterns of behavior, instead of finding themselves in cell 1, as optimally desired by all, they wind up in cell 2.[2]

MUNICIPAL EXAMPLES

The argument developed here has important applications for many municipal governments. Examples where erosion can take place are familiar. The mere financing and construction of a munic-ipal park is only a first step in guaranteeing that individuals will, in fact, be able to enjoy the services of the park through time. If the collectivity stops short at adequate provision and maintenance of the facility, and if it does not impose and effectively enforce rules on individual usage, erosion will rapidly convert the public good into a public bad. Experience in the financing and provision of housing by governmental units lends empirical support to the argument here. The investment of capital in slum clearance may lead to very limited long-range improvement value to the commu-nity unless the rate of erosion or deterioration is modified. Even more familiar, and empirically observable by everyone, are the effects of the construction of urban freeways. Individual users simply expand traffic volumes to generate additional congestion, producing grossly inefficient usage of the facilities. Traffic congestion can be interpreted in our terms as an erosion of scarce highway-street facilities. Air and water pollution have already been discussed.

A less familiar, but increasingly urgent, example is the erosion and disruption of educational facilities and processes, from lower schools to higher. Vandalism and destruction of physical facilities, violent disruption of traditional classroom instruction, unlawful invasion of public property, and terroristic threats followed by acts against personnel have become characteristics of education in the late 1960s. The argument sketched above concerning the erosion of publicly supplied facilities and services does not fully explain the actions of the terrorists and the lawbreakers, but it does explain the failure of the nonviolent and lawful users to defend their common property. Individually and privately, they have no incentive to organize themselves in defense against the militants. This remains true although one and all may recognize that the situation, without effective defense, becomes progressively worse through time.

In all of these cases, and in many others that could be added, individuals behave as the economist predicts. They act so as to maximize their own utility in all of their choices, private and public. The net effect is that publicly or collectively supplied facilities will tend to erode so long as this is (1) technologically possible, and (2) overt constraints against eroding behavior are not embodied in a structure of legally binding rules about usage. It is important to recognize that a behavior pattern equivalent to the "free rider" dilemma is present here. The user of a publicly supplied, commonly used facility or service has no privately valued incentive to protect and preserve the facility or service from deterioration, or at least this incentive tends to be grossly undervalued. The conservation of publicly supplied facilities may require the coercive implementation of rules on usage, and all members of the group, at least conceptually, may approve the adoption of such rules.

PUBLIC FINANCING AND PUBLIC ORDER

Cities need financial resources; nothing in this paper is intended to reject this fundamental proposition. Collective-consumption goods must be provided, and the effective range of indivisibility for many of these is limited to the boundaries of single metropolitan units. But all will not be well with mere financing, and the implication that cities could become utopian if only there were (federal) taxpayers willing to provide the billions needed had best be quickly forgotten. The man from Mars who looks at some of the blossoming literature on the urban crisis would surely think that such utopias could sprout everywhere if our cities could only escape their financial straitjackets.

My central objective in this paper has been to show that an alternative and necessarily complementary approach to urban improvement requires more attention, research, and application. In terms of economic value received, the current yield at the margin is surely greater from enforcing more effective usage of facilities than in enlarging the quantities of the facilities themselves. It should also be noted that the continuing process of expanding public financing and provision, without the accompanying imposition of effective rules concerning use and/or abuse, creates a continuously mounting problem of maintenance. Is it not predictable that many of the ills of the modern city arise precisely because of the shift from private to public property? If this is accepted, the suggested avenue for reform need not be a return to private ownership, although this may, in certain instances, be efficient. The argument of this paper is that an effective avenue for reform may be the enactment and enforcement of stricter rules, preventing the erosion of public goods and services. Public schools require public order; public parks require public policing of usage. The modern theory of public goods demonstrates why it is necessary that the single taxpayer be compelled to pay his taxes. *The theory's extension developed in this paper demonstrates why it is also and equally necessary that the single user of any public good, public service, or public facility be compelled to behave in a manner that will insure that the good, service, or facility so supplied shall be maintained.* The conservation of private property can normally be left to the workings of the market process. The conservation of public property requires explicit public policy.

NOTES

1. These individual marginal evaluation curves, and hence their summation, will not be uniquely determinate until and unless a cost-sharing scheme among individuals is specified or unless income-effect feedbacks are assumed to be absent. This complexity must be recognized, but it need not eliminate the usefulness of the geometrical illustration. For purposes of simplicity, assume that income effects are absent.

2. Critical readers may raise questions at this point concerning the reasons why changes in property rights are indicated as successfully accomplishing a shift from cell 2 to cell 1, whereas the provision of collective goods or facilities embodies a necessary shift to cell 3. Under any set of private property rights, it may be urged, individuals must adhere to general standards, whether legally or ethically imposed, of respect for the rights of others. In one sense, therefore, they remain in cell 3. If they can secure gains from violation of private property rights (if they have an opportunity to steal), they will do so under strictly defined utility maximization. Why, it may be asked, is the overt violation of private property rights treated differently in this whole discussion than the failure of individuals to preserve public properties?

These are complex issues that have not been sufficiently discussed. To the extent that individuals do not respect the private property rights of others, all property becomes common property when defined behaviorally. When we refer to institutional-organizational changes designed to shift from cell 2 to cell 1, changes in the structure of private ownership rights, we must assume implicitly that these will be accompanied by general acceptance of the new structure in the behavioral standards of persons. It would, of course, do little or no good to modify the property rights structure if individuals refuse to acknowledge it. In this situation, the change in property rights, as a means of moving from cell 2, would generate a situation in cell 3, just as is the case with collectively provided goods and services. Conversely, if the provision of collective goods and services should be accompanied by a shift in behavioral standards that would secure the maintenance and preservation of these, a situation in cell 1 would be achieved. The analysis of the paper, therefore, assumes that the mere financing and provision of goods, services, and facilities collectively or governmentally will not be accompanied by any change in behavioral standards, whereas it is assumed that private property rights, old and new, will be respected. These assumptions seem to be descriptive behaviorally although, of course, they should be tested empirically.

REFERENCES

ARROW, K. J. (1963) Social Choice and Individual Values. New York: John Wiley.

BAUMOL, W. J. (1965) Welfare Economics and the Theory of the State. Cambridge: Harvard University Press.

BLACK, D. (1958) The Theory of Committees and Elections. Cambridge: Cambridge University Press.

BOWEN, H. (1948) Toward Social Economy. New York: Holt, Rinehart & Winston.

BUCHANAN, J. M. (1968) Demand and Supply of Public Goods. Chicago: Rand McNally.

——— (1967) Public Finance in Democratic Process. Chapel Hill: University of North Carolina Press.

——— and G. TULLOCK (1962) The Calculus of Consent. Ann Arbor: University of Michigan Press.

COASE, R. H. (1960) "The problem of social cost." Journal of Law and Economics III (October): 1-44.

MUSGRAVE, R. A. (1959) The Theory of Public Finance. New York: McGraw-Hill.

OLSON, M. (1965) The Logic of Collective Action. Cambridge: Harvard University Press.

SAMUELSON, P. A. (1955) "Diagrammatic exposition of a theory of public expenditure." Review of Economics and Statistics, XXXVII (November): 350-356.

——— (1954) "The pure theory of public expenditure." Review of Economics and Statistics, XXXVI (November): 387-389.

3

Economics and Urban Renewal: Market Intervention

OTTO A. DAVIS

☐ FOR MORE THAN twenty years now, the cities of America have witnessed the turmoil inherently associated with our federal Urban Renewal Program. Although over the same period of time greater aggregate federal expenditures have been devoted to, say, efforts to maintain the incomes of farmers, the Urban Renewal Program has sparked much more controversy than the Farm Program. Part of the reason for the controversy lies in the fact that urban renewal is a very visible activity. One can hardly help but observe that buildings are torn down and that new structures rise to replace them. In addition, urban renewal is a program which raises so many of the questions which intellectuals find relevant about cities that almost every project is likely to require an examination of first principles. What is the purpose of a city? Whose interests and values are to be served? What sacrifices should be made for the common good, however defined?

In view of the evidence of turmoil, controversy, and just plain confusion, it is hardly surprising to find that the purposes of the urban renewal program are not clearly defined. Partly for this reason, partly because the initiative in the actual conduct of the program is left largely in the hands of local communities, and partly because urban renewal resembles a tool more than a goal, one finds that the program has served a great variety of ends. In some places, renewal has meant erecting civic monuments in a revived downtown plaza. In others, it has meant getting "undesirables" out of "desirable" neighborhoods to save portions of the city for the middle class. In still others, it has meant clearing and developing land that might attract new businesses into the community or even removing unpopular businesses. It has meant "saving" urban universities from being engulfed by the tide of encroaching slums. It has meant assembling tracts of land upon which expensive apartment houses could rise. It has also meant assembling tracts on which subsidized low- or middle-income housing might be built. It has been used to provide otherwise unavailable land for the construction of a sports stadium thereby encouraging athletic teams to move into, or not depart from, some given city. It has been used to provide a tract for the construction of a civic arena and, incidentally, to block off the downtown area from a menacing Black ghetto. It has been used to assemble tracts of land for the construction of public housing projects, the residents of which are mostly Black. Given this welter of achievements and implied aims, some good and some bad, it is understandable that urban renewal should mean very different things to different people. It is also understandable that the program stirs deep-seated emotions.

In view of the above, it is probably best to begin by examining some of the bare-boned details of how the program operates and then to move on to consider the rationale for the program, the criticisms of it, and how it might be improved.

HOW THE PROGRAM OPERATES

The Urban Renewal Program began with the passage of the Housing Act of 1949. Although later acts have changed the program in important ways, the fundamental operating procedures and the main thrust of the program remain basically unchanged.

Title I of the Housing Act of 1949, as amended, authorizes advances, loans and grants to "local public agencies" for renewal. Thus urban renewal is a local program. Projects must be locally conceived, locally plannned and locally carried out. The federal role is primarily one of providing financial assistance, leadership, and general program direction.

Since Title I uses the simple generic term, "local public agency," in referring to the local body authorized to carry out urban renewal projects, one finds, as might be expected, that there is considerable variation in the nature of local agencies responsible for renewal. In some areas, the city government itself is designated as the local public agency and carries out the program through one of its departments. In others, a separate agency is authorized, usually with a governing board appointed by the mayor and subject to approval by the city council.

Although there is diversity in organizational forms of the local agencies which administer renewal programs, the ultimate responsibility for making the basic decisions in renewal always rests with the elected governing body of the locality. In addition to requirements imposed by the various states, federal laws and policies establish several key checkpoints for action by the local governing body. First, as a prerequisite to any federal financial assistance, the local governing body must adopt a "workable program" for community improvement; this workable program is supposed to establish a framework for the component renewal projects. Second, the application for federal project planning funds must be accompanied by a resolution of the local governing body authorizing the application and also designating the urban renewal area. Third, there must be a public hearing and the local governing body must officially adopt the urban renewal plan before the Renewal Assistance Administration of the Housing and Urban Development Department of the federal government can approve any component project for execution.

Before any renewal project can be launched, the local governing body must determine that the selected area is a slum, a blighted area, or a deteriorating area. After such a finding, the local public agency must conduct detailed surveys and studies which aim toward the production of the plans and estimates of what is required to undertake the project. The costs of producing these plans are a part of the expenses associated with a renewal project.

The most visible part of an urban renewal project comes when the local public agency acquires buildings and arranges for their demolition. Acquisition of properties is carried out either through negotiation with owners, or if necessary, by court condemnations under the power of eminent domain. The actual removal of structures is normally done under a contract with a private contractor, as is the work involved in preparing the site and executing such improvements as new streets.

After the site is cleared and the improvements completed, the land is sold to private developers. These developers, of course, must follow the plan and as private construction takes place, it is inspected to assure that it is in accordance with controls specified in the plan.

Land acquisition and clearing obviously require money; a major part of the federal involvement is concerned with the financing of this phase of the project. This financial assistance takes several forms. First, there are *advances* to cover the preparation of project plans. The cost of planning, at least if the project is executed, is counted as part of the overall cost of the project. Second, there are *temporary loans* which serve as the working capital for the project. While the Renewal Assistance Administration itself can offer temporary loans, these may also be secured from private lending institutions with the benefit of a federal guarantee of payment for security. Third, there are *definitive loans* which can be made to the local public agency when the cleared land is leased rather than sold. These loans are for a maximum of forty years. Fourth, there are *capital grants* to cover the federal portion of the net project cost. These capital grants can be for fractions of two-thirds or three-fourths of the net costs, depending upon the population of the locality and the choice of the local public agency. The particular public agency chosen determines differences in inclusions allowable in the cost calculations, making one fraction more favorable than the other for a given project. Finally, there are *relocation grants* to provide assistance to individuals, families, business concerns, and nonprofit organizations which are displaced by the project. Urban renewal is only one of several activities which can make a locality eligible for relocation grants. The local public agency does not have to contribute at all to relocation expenditures (which are not considered a cost of the project).

The heart of the program is obviously the capital grant. An example based upon congressional testimony by William L. Slayton (1966), then Commissioner of the Urban Renewal Administration, may make this point clear. Consider Exhibit 1 (Slayton, 1966: 201). Line 1 gives the "gross project cost" for this imaginary project. These costs consist partly of the direct project expenditures (line 2) incurred by the local planning agency for such items as planning, acquisition of properties, clearing, and improvements, if directly financed by the local public agency. Temporary loans (line 3) may be used to cover these expenditures as well as possible cash grants-in-aid provided by the locality (line 4). The other part of gross project cost consists of noncash grants-in-aid provided locally (line 5). These noncash grants-in-aid generally include expenditures associated with the project but not made by the local public urban renewal agency. Examples are the provision of such site improvements as streets and sidewalks which may be paid for by some other agency of the local or state government, the construction of public facilities in the project area such as schools and parks, the provision of land for low-rent housing, and certain statutory credits such as expenditures by or in behalf of an eligible college, university or hospital.

Since cleared land is sold to private developers, the project obviously has certain revenues (line 6). These are offset against the gross project cost to give net project cost (line 7). Under the usual formula (but with the exceptions noted above) the federal contribution consists of a capital grant for two-thirds of the net project costs (line 8). The remaining one-third of net project cost is covered by local grants-in-aid (line 9) which are the sum of the local cash grant-in-aid (line 4) and noncash local grants-in-aid (line 5).

EXHIBIT 1

EXAMPLE OF THE SHARING OF PROJECT COSTS (2/3 basis)

1. Gross project cost $1,500,000
   Total cost in carrying out project except relocation payments.
   Made up of 2 components:
2.   Project expenditures $1,300,000
       Direct expenditures by project, financed
       from—
3.   Maximum temporary loan (loan funds
           provided to the project either as
           direct Federal loan or by private loan
           under Federal guarantee) 1,150,000
4.   Local cash grant-in-aid (cash provided
           by locality) 150,000
5.   Noncash local grant-in-aid (land donations
         and public works provided by the locality) 200,000
6. Disposition proceeds (funds received from sale of project land
   for redevelopment) .... -450,000
7.   Net project cost (necssary public cost of undertaking the
       project) 1,050,000
       This is met by—
8.   Federal capital grant (2/3 of net project cost) 700,000
9.   Local grants-in-aid 350,000

The above example should make clear the fact that the major ingredient of the program is the capital grant. Although the two-thirds formula is the one usually used, communities having a population of 50,000 or less, or municipalities in areas designated by the Secretary of Commerce as "redevelopment areas," may obtain capital grants for three-fourths of the project costs. In addition, any municipality may elect the three-fourths formula deleting from its cost calculations those expenditures associated with administration, overhead, and legal fees, all of which may be incurred during both survey and planning and project execution.

THE ECONOMIC RATIONALE

Having seen how the Urban Renewal Program works, it is appropriate to attempt to gain some impression of its size and importance. From its inception in the Housing Act of 1949 through fiscal year 1966, Congress has authorized $5.4 billions for this program. The Demonstration Cities and Metropolitan Development Act of 1966 authorized $250 millions in additional urban renewal granting authority, to be available after July 1967. This latter additional authority is limited to renewal projects which are included within an approved comprehensive city demonstration program. The same act increased the authorization for projects in model cities areas by $350 millions.

Appropriations for this program were at the levels of $725 millions for fiscal year 1967 and $750 millions for 1968. The 1968 Independent Offices and Department of Housing and Urban Development Appropriation Act appropriated $750 millions as an advance appropriation for fiscal year 1969 for the main urban renewal program and an additional $100 millions for projects associated with model cities, both of which are to remain available until expended.

Clearly, despite its prominence and visibility, the Urban Renewal Program is not nearly as large as many of our other domestic programs. Yet, the expenditures are not trivial, but sizable and significant. One might expect a program of this size to have a well developed rationale. Such a rationale is not spelled out in the law containing the general aims of the "elimination and prevention" of slums and blighted areas. In this section we shall attempt to provide an economic rationale for the program. We should then be better able to point out the implications of that rationale for the operation of the program.

It is probably best to begin with basics. If urban renewal is aimed at the "elimination and prevention" of slums and blighted areas, as the law states, then it is appropriate to begin by inquiring why slums and blighted areas develop and persist. Of course, one might attempt a simple answer by simply stating poverty, but such an answer is clearly inadequate. While there is an obvious relation between a family's income and the quality of the housing unit which it can afford, it would appear to be blatantly silly to enact a program which, at least under this argument, got at only the symptoms (the slum housing) without consideration of the cause (the poverty). Since the Urban Renewal Program began some years before we declared war on poverty, it seems only fair to give our lawmakers the benefit of the doubt and presume that they did believe that there was something fundamentally wrong, aside from poverty, which might cause the creation of slums and blighted areas.

Any look at our urban areas quickly convinces an objective observer that all but an infinitesimal part, that small portion of our housing stock erected under the public housing program, was provided by the private sector. Hence, it may be helpful to view the problem of slums and blighted areas from an economic perspective.

For the sake of argument, let us imagine that we own a parcel of property which has reached a stage in its life where it must either undergo repairs or begin to deteriorate. What kinds of thoughts might go through our minds? Obviously, we would have to consider the alternatives, repairing the property or allowing it to deteriorate. (For the sake of argument, we omit the alternative of selling the property since the new owner would have to go through the same calculations.) Clearly if our individual returns—monetary or physical, depending upon whether the property is rented or not—exceed the costs of repair, then we should select that alternative. Only if the costs were greater than our individual benefits would it make any sense to allow the property to deteriorate.

The above argument indicates that, at least when we consider the problem of deterioration and blight within the context of a single property, deterioration will only occur if it is not individually "worthwhile" to maintain the property. Hence, *if* there is an economic problem of blight and slums caused by something other than mere poverty, we must go beyond simple individual calculations and try to determine whether there is something in the larger environment—the entire private enterprise market system which allocates our real property resources—which might, in some particular instances, cause some divergence between private and social benefit or cost.

One possible difficulty with urban property markets is fairly obvious. The value of any one property may depend, at least in part, upon the characteristics of the neighborhood in which it is located. Economists have long believed that "neighborhood effects" exist in the market for urban property. Consider the following quote from Alfred Marshall (1920: 445) whose place in the history of economic thought probably stands second to that of Adam Smith.

> But the general rule holds that the amount and character of the building put upon each plot of land is, in the main (subject to the local building by laws), that from which the most profitable results are anticipated, with little or no reference to its reaction on the situation value of the neighborhood. In other words, the site value of the plot is governed by causes which are mostly beyond the control of him who determines what buildings shall be put on it.

We shall now attempt to demonstrate how these neighborhood effects, given certain circumstances, can cause rational property owners individually to decide to allow their buildings to deteriorate and thus allow blight to appear and persist.

In order to explain what is at issue here, it is appropriate to introduce a simple example which is drawn from the theory of games.[1] This example, which belongs to the general class of games known in the relevant literature as "Prisoner's Dilemma games," appears to contain the important points at issue here.[2] For the sake of simplicity in the argument, let us consider only two adjacent properties. Suppose that these two properties have two different owners who can be denoted conveniently here as Owner I and Owner II. Assume further that each owner has made an initial investment in his property from which he is reaping a return, and is now trying to determine whether to make an additional invest-

ment for redevelopment of his property. The additional investment, of course, will alter the return which he receives, and so will
the decision of the other owner.

Let us presume that the situation can be summarized in the
following game matrix.

OWNER II

	Invest	Not Invest
Invest	(.07, .07)	(.03, .10)
Not Invest	(.10, .03)	(.04, .04)

OWNER I

We now interpret and explain this game.

For simplicity let us presume that each owner has made his
initial investment in his property and that he has an additional
sum, which is precisely the amount which is required if he decides
to redevelop his property, which is invested in, say, corporate
bonds. Presume that, at present, the average return on both of
these investments, the property and the corporate bonds considered together, is four percent for each owner. Therefore, if neither
owner makes the decision to sell his corporate bonds and invest
the resulting money in the redevelopment of his property, both
will continue to get the four percent average return. This situation
is represented by the entries within the small parentheses in the
lower right of the matrix where each owner has made the decision
"not invest." The left hand number within the small parentheses
always refers to the average return which Owner I (the row player
in the game) receives, and the right hand number indicates the
return for Owner II (the column player in the game). For the "not
invest, not invest" decisions, the situation is characterized by the
pair of numbers within the small parentheses in the lower right
side of the matrix where both continue to get their four percent
return.

The assumed climate for redevelopment in this example is illustrated by the entries in the upper left of the game matrix which
presumes that both owners choose the "invest, invest" possibility.
Hence, each owner decides to sell his corporate bonds and invest
the resulting money in the redevelopment of his property. In this
possibility, each owner would get a seven percent return on his
capital.

The other entries in the matrix represent the situation when one owner invests in redevelopment but the other does not do so. Consequently, these possibilities require more detailed explanation.

The basic assumption here is that "neighborhood effects" are very strong. Consider for the moment the numbers within the small parentheses in the lower left of the matrix. These entries reflect a situation in which Owner I decided not to invest in the redevelopment of his property while Owner II decided to sell his bonds and invest in redevelopment. The very concept of neighborhood effects implies that Owner I must gain some of the benefits of Owner II's investment. For example, if the two properties happened to be apartment houses and the decision of Owner II meant that he would demolish his outdated structure and construct a new one complete with off-street parking and other amenities, it is easy to see how Owner I might benefit. New facilities for off-street parking implies that the tenants of Owner I can have an easier time finding street parking. Tenants of Owner I may find that their children, by forming appropriate friendships with their peers in the new apartments or by simply violating the rules, may be able to enjoy facilities such as a pool. All these benefits mean that, as soon as leases allow, Owner I can edge up his rents. Hence, his return increases without additional investment. Assume for sake of argument that his return increases to ten percent, which is appropriately entered into the matrix.

Owner II, on the other hand, should find that, since prospective renters also consider the "neighborhood" (including the ill effects of Owner I's "outdated" structure) in selecting an apartment, his level of rents must be less than would be the case if his apartment building were in a location with another new structure nearby. Assume that the return on his total investment (the investment in the now-demolished structure plus the investment in the new structure) falls to three percent. This percentage also is indicated appropriately within the small parentheses at the lower left of the matrix. For simplicity, a symmetric situation in regard to return on investment is assumed for the remaining situation where Owner I decided to invest and Owner II decided not to invest. Thus, the reverse situation is reflected in the numbers in the small parentheses at the upper right of the matrix.

Having described the set of situations which the two owners face, let us now consider the decision process. Presume that both owners are aware of the returns which are available in the hypothesized situations. Consider Owner I first. While he must decide whether to invest or not invest, he should realize that any action taken by Owner II will affect the outcome. Hence, he might use the following line of reasoning. Suppose that he asks himself whether he should invest or not if he tentatively assumes that Owner II does invest. Owner I should see that in the assumed situation if he did invest he would obtain a seven percent return on his capital whereas a decision not to invest would yield a return of ten percent on his capital. Hence, if Owner II were to decide to invest, it would obviously be in Owner I's best interest to refrain from investing. Now tentatively assume that Owner II does not invest. What should Owner I do? If he does invest he will obtain only a three percent return on his capital whereas if he does not invest he will obtain a four percent return. Under these conditions, too, it is advantageous to Owner I not to invest. Hence, no matter what Owner II does, it is clearly in Owner I's best interest to refrain from investing.

Consider Owner II. He should go through a similar thought process. If he assumes that Owner I invests, then he can gain a ten percent return if he does not invest and only a seven percent return if he does invest. If he assumes that Owner I does not invest, then there is a four percent return for not investing compared to a three percent return for investing. The obvious conclusion is that, no matter what Owner I does, it is individually rational for Owner II to refrain from investing.

The situation described above means, of course, that neither Owner I nor Owner II will invest in the redevelopment of their properties. Thus, the kind of neighborhood effects which are summarized in the Prisoner's Dilemma can help explain why blighted areas can develop and persist.

The simple example does not, of course, constitute a full explanation. One might notice, for example, that the entire analysis in the Prisoner's Dilemma example assumes that there is no communication between the persons playing the game. Thus, one might suggest that it should be easy for Owner I and Owner II to get together for a chat and mutually agree that it would be advantageous for both of them to invest in redevelopment so that they might obtain the higher rate of return of seven percent. Not only would such action be socially desirable, but it also would be mutually advantageous. In this simple example, of course, it should be rather easy for the two owners to get together and coordinate their decisions and actions. However, it is obvious that as the number of individual or owners increases, the difficulty of obtaining coordinated action also increases. If any one property owner decided not to invest while all others redeveloped, the former might make a considerable gain. Some real situations, of course, involve many property owners and this fact helps to make the assumption of uncoordinated action reasonable. Of course, there is incentive for one of the two owners to purchase the property of the other, or for some third party to purchase both properties, so that both could be developed and the seven percent return obtained. Certainly, it cannot be denied that this often occurs in reality. It would be a mistake to underestimate the strength of the profit motive. On the other hand, certain institutional realities of the urban property market mean that there is no assurance that such a result will always take place. We might explain this point by means of an example.

Consider an area composed of many small holdings. Suppose that renewal or redevelopment of the entire area would be feasible if coordination could be achieved, but that individual action by the small property owners alone cannot result in redevelopment or renewal due to strong neighborhood effects, such as those incorporated into the previous example. Incentive exists for some entrepreneur to purchase the entire area and invest in redevelopment or renewal. Imagine that some entrepreneur does have such plans and sets out to assemble the properties. If one or more of the owners of the small plots becomes aware of the entrepreneur's intentions, and if their plots are so situated that they are important for the success of the project, these owners are likely to realize that it is possible to expropriate part of the entrepreneur's expected profits by demanding an exceedingly high price for their property. Alternatively, they might refuse to sell at all in anticipation of the neighborhood effects which redevelopment might provide for them. In any event, it is entirely possible for several of the small owners to try to expropriate as much of the entrepreneur's expected gains as possible with the result that, without collusion or communication between themselves, a Prisoner's Dilemma type of situation is created for them and their individually rational action results in the entire project being abandoned.

Those familiar with transactions in real urban property know the difficulties associated with site assembly, the real possibility of price-gouging, and the importance of keeping secret one's plans and intentions. They are probably also aware of the difficulties associated with maintaining secrecy.

The above analysis means that situations can exist where individually rational action may preclude socially desirable investment in the redevelopment of urban properties. Now it is clear that the necessary ingredients for the existence of this kind of situation depend crucially upon: (a) the neighborhood effects being very strong and (b) uncoordinated actions of various property owners who know of or sense plans for redevelopment and who try to expropriate the entrepreneur's profits by demanding more than the fair market value for their property. In general these two conditions probably are not satisfied for the vast majority of urban property. All cities evidence a great deal of private renewal, which indicates that at least one of the conditions is not satisfied. The point here, however, is that these conditions can be satisfied for particular areas so that blight can develop and persist. Thus, there is at least one reason, aside from poverty, for the existence of blight.

CRITERIA FOR URBAN RENEWAL

Having looked at the economic rationale for urban renewal and having found that the allocative device for urban property, the market mechanism, need not always produce the proper solution, we must now turn to the task of determining whether the previous analysis might not contain clues to proper rules for the conduct of the program. In so doing, we are implicitly saying that the program should be aimed toward the attainment of an optimal allocation of our real physical property. Hence, we are ruling out the program's being used to reduce poverty, or even to provide subsidized housing for the poor, and insisting that these objectives should be served by alternative programs specially designed with those ends in mind. One might observe, as will be seen later, that there is good reason to adopt this view since the program certainly has not promoted ends related to items other than physical property.

It is clear from our Prisoner's Dilemma example that the essence of the problem of *preventing* the development of blight is the provision of some mechanism whereby decisions concerning upkeep and repair can be coordinated in such a manner that it will not be individually rational to make only individualistic calculations. It is obvious from the frame that an owner should find it advantageous, and in his best interest, to constrain himself to invest if at the same time he can also constrain his neighbors to do likewise. In other words, it is rational to impose constraints which get one out of the dilemma created by the existence of neighborhood effects. One possibility is a special type of code which might specify such items as minimal levels of repair. Such a code would constitute a preventive strategy.

On the other hand, our task here is to provide some guidance or criteria for the existing urban renewal program. One such criterion can be devised through the utilization of notions from the benefit-cost approach of economics. The real clue here is derived from a consideration of what is accomplished by an urban renewal project. When a local public agency designates an area for renewal, it acquires properties through negotiation with present owners insofar as is possible; but it has behind it the power of eminent domain which should give it the ability to prevent price gouging. This power removes the obstacles to the assembly of sufficiently large sites and, at least if used properly, should mean that the agency has to pay only "fair" prices. When the local public agency clears the assembled properties and begins selling sites to private developers, each entrepreneur is assured that other sites in the renewal area will also be developed. Thus, at least if the plan for the development of the area is properly specified, the very fact that this part of the renewal process has taken place should both provide the desired coordination and eliminate the difficulties caused by the neighborhood effects. In other words, the very process of renewal should "internalize" at least a large part of the neighborhood effects for the local public agency.

One fact needs great emphasis here. The internalization or alienation of neighborhood effects causes social and private products to be equated. This insight allows us to propose an approximate measure for the theoretical benefit-cost rule of economics. This rule states that projects are worthwhile if the benefits are greater than the costs. The usual difficulty in the application of the rule is that it is practically impossible to measure the benefits. In this instance, an approximate measure is available due to the fact that at least a part of the neighborhood effects can be internalized by the local public agency. This internalization allows us to use revenues, augmented with some other incommensurables, as an approximation for benefits. One should emphasize the approximatory nature of the measure. One should also emphasize that even an approximate measure is better than no measure at all.

The costs of the project are easily enumerated. These include costs of planning, acquisition, demolition of existing structures, improvement of sites, aiding in the relocation of displaced families, and interest expenses. The correct enumeration of the relevant revenues is more difficult. The major revenue comes from the sale of the cleared sites to the private developers. According to the usual type of economic argument, if the sites are sold competitively, prices should reflect the "value" of the sites and, in the absence of distorting effects, should be a measure of social benefits. Hence, revenues from the sale of sites are the first component of this approximation of the social benefit of an urban renewal project.

Obviously, these prices are imperfect measures. One difficulty is that the tax on property may bias the prices. Economists have long held that taxes are likely to be shifted to immobile resources, and few resources are more immobile than land. One way to correct for this bias is to take into account the net addition to revenues from property taxes, including a "fictional" component of this net addition which reflects the fact that many projects include tax exempt properties. This projected net addition should be discounted to a present value and the present value of these added revenues should be added to the revenue income of the local public agency, for the project under consideration, and the resulting total is a closer approximation to the true benefits of the project.

A second difficulty with this measure of benefits is that part of the land may not be sold. It is customary in many renewal projects to alter the street plans. Since the local public agency does not actually pay to acquire the original streets, there is little difficulty here, aside from some accounting, unless there are streets which are devoted to through traffic and thus are not primarily for the use of residents of the area. In the latter instance, neither the cost nor the benefit of the street should be assigned to the project. In addition, renewal plans often call for playgrounds and parks. Part of the value of these public facilities will already be included due to the fact that land near playgrounds and parks is likely to sell at a premium; but if persons not residing in the project area use these facilities, then an additional increment of value should also be added. It is only honest to admit that no one has been able to determine the best way of placing a value on a park or playground, or even being certain that a value assigned is in the ballpark; but there are various methods which could be used. One is to try to determine the number of persons which might use the facility per unit of time and then assign values derived from studies where economists have attempted to determine peoples' willingness to pay for unpriced recreation. Another is to follow the simple expedient of valuing the facility at its opportunity cost—the price the land would command if it were sold for some other purpose.

The summation of the above components gives an approximation for the benefits derived from an urban renewal project. The approximation is not perfect, but it should be a usable figure. In addition, there may be, in specific instances, external benefits of the project which extend beyond the area; but if these are present they are likely to be difficult to identify and measure. The principle of including all benefits, but not double-counting, cannot be challenged. On the other hand, the benefits should be identifiable.

The above criteria does not eliminate the need for some judgment in determining whether specific urban renewal projects are worthwhile. Some judgments do have to be used in determining whether external benefits extend beyond the project, and there is some arbitrariness in the above methods of determining benefits. Nevertheless, the general approach of attempting to determine, with the aid of our best judgment, whether the benefits are greater than the costs has basic appeal and should be used whenever possible in governmental decision-making. Of course, benefit-cost considerations need not, and should not, be the sole considerations in determining whether programs should be adopted or continued. For example, one should also consider how the benefits are distributed; other points probably merit examination as well. Some attention will be devoted to these other criteria later. The argument here is that in any program which aims to help achieve a better allocation of resources, benefit-cost should always be one of the criteria by which a program's performance is measured.

PROGRAM PROCEDURES AND THE BENEFIT-COST CRITERION

It is now appropriate to review the procedures of the federal Urban Renewal Program and determine whether they are compatible with the use of a cost-benefit criterion. Obviously, some of the financial tools of the program, although crucial for the conduct of projects, are neutral with respect to this criterion. Thus, we need not be concerned here with items such as advances and temporary loans for working capital.

Some of the details of the costing procedures are relevant. Some of the entries allowed for noncash local grants-in-aid simply are not a part of project costs. Communitywide or general benefit supporting facilities, such as schools and universities, can, under certain conditions, be counted as part of the noncash local grants-in-aid even if the facilities existed prior to the project and even if they are not within, but are within one-quarter mile of, the project area. These kinds of grants-in-aid, at least for the purpose of using the cost-benefit criterion, should not be counted as a part of the cost of the project.

Except for the above class of costs, the procedures of the program on the cost side are largely in accordance with the methods that would be used to measure costs if one were making calculations in order to apply a cost-benefit criterion.

On the benefit side, the procedure does specify that revenues be deducted from costs so that the first step of the benefit calculation is actually made. However, there is no provision for other steps in the efforts to approximate the true benefits of a project. Since these steps are not taken, we cannot, of course, say with certainty on an a priori basis whether the sum of the federal capital grant and the portion of the local grants-in-aid which is relevant, even approximate the additional benefits which would be necessary to justify the program on the basis of a benefit-cost criterion. We can be certain, however, that the rule of two-thirds (or the alternative three-fourths) of the net project cost being the amount of the federal capital grant is entirely too arbitrary, even if the fraction were to be viewed as a mechanism for the local public agency to be compensated for those benefits which imply specific costs but which by their nature cannot be a source of revenue.

Indeed, the very structure of the procedures of the Urban Renewal Program, while they obviously could be adopted to a benefit-cost framework, probably have acted to cause the program to misallocate resources. From our previous discussion it is possible to express the judgment, but only a judgment, that the largest item in the approximation of a project's benefits would be the revenues from the sale of sites. Since these revenues are typically smaller than the contribution of the federal capital grant, one may express the judgment that the present program acts to encourage too much renewal.

URBAN RENEWAL AND OTHER CRITERIA

While one of the stated goals of the Urban Renewal Program is to "prevent and eliminate slums and blighted areas," any casual observer of the present urban scene knows that after twenty years of the program we still have slums and blighted areas. That objective of the program certainly has not been accomplished.

There are other standard criteria that economists generally suggest as applicable to public programs. One of these is a distributional question. Is it the rich, the poor, or the middle class that benefit from a program?

Probably the most careful analysis of the distributional consequences of the Urban Renewal Program is contained in a book by Martin Anderson (1964). Basically, the argument can be briefly and simply stated. The program is aimed at slums and blighted areas. Consequently, the residences which tend to be eliminated are those which house persons at the lower end of the income distribution. Partly because renewal has sometimes been used to preserve "decent" neighborhoods, there has been a tendency for the new housing which is built in renewal areas to be designed for persons with higher incomes than those who previously lived in the area. Insofar as renewal projects expand the supply of housing for these higher income groups and contract the supply of the less attractive housing, which is usually all the poorer persons can afford, the price of the former must fall in relation to what it would have otherwise been while the price of the latter must rise in relation to what it would have otherwise been. Hence, the urban renewal program has historically tended to subsidize the higher income classes and penalize those who are at the lower end of the income distribution. Given this dimension to the provision of housing, the program has not been a contributor to the goal of subsidizing the poor—quite the opposite.

Criticisms such as those indicated above have resulted in some alterations in some of the priorities of the program. For example, the Department of Housing and Urban Development has stated that it will give priority to those projects which contribute to certain national goals. One of these goals is the conservation and expansion of the housing supply for low and moderate income groups.[3] Thus, projects are more likely to be approved if they are planned so as to provide for more than fifty percent of the average being devoted to housing for low and moderate income groups.

CONCLUDING REMARKS

This paper has aimed at explaining how the Urban Renewal Program works, developing an economic rationale for the program, and ascertaining how well the program performs in relation to that rationale. Given the fact that the program has not contributed toward such social goals as the redistribution of income and the expansion of the supply of housing for the poor, and the good possibility that even with altered priorities for the approval of projects these goals will not be served, it would seem that the simple cost-benefit criterion should be emphasized as the major objective of the program. If such a sensible procedure is to be followed, the law must be altered so that the basic ingredient of the present program, the provision of the two-thirds federal capital grant, can be abolished and replaced by a formula which relates to those benefits which are not revenues to the local public agency.

NOTES

1. This particular analysis, and the example, is drawn from Davis and Whinston (1961).

2. For an explanation of the "game theoretic" points of interest in Prisoner's Dilemma games, see Luce and Raiffa (1957: 94-102). The reason for the intriguing title of this type of game is interesting in itself. The name is derived from a popular interpretation. The district attorney takes two suspects into custody and keeps them separated so that they cannot communicate with each other. He is certain that they are guilty of a specific crime but does not have adequate evidence for a conviction. He talks to each separately and tells them that they can confess or not confess. If neither confesses, then he will book them on some minor charge, and both will receive minor punishment. If both confess, then they will be prosecuted, but he will recommend less than the most severe sentence. If either one confess and the other does not, then the confessor will receive lenient treatment for turning state's evidence, whereas the latter will get "the book" slapped at him. The prisoner's dilemma is that without communication and collusion between them, the individually rational action for each is to confess.

3. See the statement of priorities as stated in 3105A of *Urban Affairs Reporter*, Washington, D.C.: Commerce Clearing House.

REFERENCES

ANDERSON, M. (1964) The Federal Bulldozer: A Critical Analysis of Urban Renewal. Cambridge: MIT Press.

DAVIS, O. A. and A. B. WHINSTON (1961) "The economics of urban renewal." Law and Contemporary Problems, 26 (Winter): 105-117. Reprinted without revision, pp. 50-67 in J. Q. Wilson (ed.) Urban Renewal: The Record and the Controversy. Cambridge: MIT Press.

LUCE, R. D. and H. RAIFFA (1957) Games and Decisions. New York: John Wiley.

MARSHALL, A. (1920) Principles of Economics. Eighth edition. New York: Macmillan.

SLAYTON, W. L. (1966) "The operation and achievements of the Urban Renewal Program." Pp. 189-229 in J. Q. Wilson (ed.) Urban Renewal: The Record and the Controversy. Cambridge: MIT Press.

Part II

PUBLIC
EXPENDITURES
IN THE
URBAN ECONOMY

Part II PUBLIC EXPENDITURES IN THE URBAN ECONOMY

INTRODUCTION   97

PUBLIC EXPENDITURES: THE TOTAL SIZE OF THE PIE

PUBLIC EXPENDITURES: HOW THE PIE GETS DIVIDED

BUDGETARY REFORM AND THE RESTRUCTURING OF PUBLIC EXPENDITURE DETERMINANTS

Introduction

Most observers of the urban condition agree on one point: local government needs more money. Although there are many valid reasons for believing this is not always the real problem (see articles by Buchanan, Sandalow, Fitch, Davis, Muth, Netzer, Aberbach and Walker, and Reiss, this volume, for descriptions of some conditions under which increased public expenditures at the local level are undesirable), in general more resources would probably help. This section is concerned with three aspects of public policy: how local government raises revenues to *directly* provide goods and services, how local government allocates resources at its disposal, and some suggested reforms in this public resource allocation process.

Meltsner examines the overall revenue and tax decisions made by public officials. In particular, he examines the political constraints on such decisions, explores various ways for officials to discover their *real* constraints as opposed to perceived constraints, and discusses ways to modify or eliminate constraints. Meltsner contends that public officials are overly pessimistic about tax burdens the public is willing to bear. Riew and Fitch examine some of the advantages and disadvantages of what is generally regarded as the largest potential source of new revenues for cities: a federal revenue-sharing program. Sandalow argues that in assessing the impact of federal grants to local governments, we should not ignore their effect on local government structure. Grants-in-aid aimed at removing or relaxing financial constraints also have influences on local government structure; these effects are not all desirable. Riew, Fitch, and Sandalow all point out many of the limitations on revenue sharing or grants-in-aid as a primary policy lever of the federal government, as well as the obvious benefits of increasing overall local government revenues.

The Wilensky and the Whitelaw articles focus on the determinants of resource allocation decisions made by local governments. Most such decisions are represented by municipal capital and operating budgets. The Wilensky article reviews the many empirical studies done by economists, political scientists, and sociologists on the determinants of per capita public expenditures. These studies largely attempt to test the notion that economic, political, and social characteristics of a community determine, with trivial exceptions, the pattern of public sector resource allocation in a community. That is, population tastes are more or less directly translated into goods and services provided by the public sector. She concludes that these studies fail to provide policy makers with useful advice on such important things as future government revenue needs, likely effects of federal or state subsidies, or questions concerning metropolitan consolidation (economies of scale). She recommends that students of the determinants of public expenditures at the local level direct more of their attention to the political and bureaucratic mechanisms that generate these resource allocation decisions if they wish to increase the policy relevance of their work.

I have argued elsewhere (Crecine, 1968; 1969) that the principal determinants of resource allocation decisions in large municipal governments are the total resources available (that is, those factors which influence tax rates and yields) and the (standard) organizational responses to the annual budget balancing problem. Whitelaw examines both the environmental-determinants view (see Wilensky, this volume) and the revenue-bureaucratic-process view (see also Crecine, 1968; 1969; the Mushkin; the Meltsner; the Wildavsky articles, this volume) in the context of his own empirical study of municipal budgeting in Worchester, Massachusetts, involving an extensive time series. Unlike most public expenditure studies, he included capital items. Whitelaw argues that, particularly on capital items, public officials do take the changing needs and preferences of the electorate into account when making expenditure decisions. Several of his need-and-preference variables could be interpreted as surrogates for total revenues, however.

Regardless of the current determinants of public expenditure decisions by local government, there is a great deal of sentiment for the reform of municipal budgeting procedures. Most current proposals involve some variant of the planning-programming-budgeting system (PPBS) originally introduced in the Department of Defense. The nature of a PPB system, modified for use by local government, is discussed in the Mushkin article. Also found is a discussion of the current state of the movement to adopt a PPB system in cities and her views of the likely next phase of this movement. Donald Borut, responsible for introducing PPBS in the Ann Arbor, Michigan city government, discusses some of the problems of implementation. In particular, he argues that in order to successfully implement a system based almost exclusively on economic-efficiency criteria, it is necessary to explicitly recognize organizational constraints. Borut contends that it is foolish to attempt to introduce PPBS all at once; one should introduce the reform a department at a time, building support for the system by demonstrating its utility to department heads. The first section of the Whitelaw article contains an excellent description of the complexities in the objective reality faced by public officials when making budget decisions; these complexities account for much of the difficulty involved in implementing a reform like PPBS. Finally, Meltsner and Wildavsky argue at length for leaving the current manner of formulating municipal budgets alone. They contend that the political and administrative process that generates municipal budgets is functioning reasonably well, and, in any case, is performing some vital political functions that PPBS cannot and should not be applied to. Meltsner and Wildavsky cite current budgeting practice as a form of "revenue behavior" and as a control mechanism rather than a resource allocation device.

Public expenditure decisions and public expenditures are hardly the entire story. Their *actual* effects are likely to be primarily determined by nongovernmental factors (as will be discussed in Part III) and are influenced in important ways by the mode of service delivery (see Part IV).

–J. P. C.

REFERENCES

CRECINE, J. P. (1969) Governmental Problem Solving: A Computer Simulation of Municipal Budgeting. Chicago: Rand McNally.
––– (1968) "A simulation of municipal budgeting: the impact of problem environment." In W. D. Coplin (ed.) Simulation in the Study of Politics. Chicago: Markham.

PUBLIC EXPENDITURES

The Total Size of the Pie

4

Local Revenue:
A Political Problem

ARNOLD J. MELTSNER

☐ LOCAL OFFICIALS throughout the nation believe that there are insufficient available resources to meet their problems, that the public is hostile to increased levels of governmental taxation and spending, and that they have few choices left except to appeal, with futile expectations, to the federal government. In short, there is a fiscal crisis. I accept the reality of this crisis, but maintain that local officials, despite the difficulties, must accept the responsibility of getting more money as their primary objective. The officials' hard choice is either to get more or to do less, and our cities cannot afford to do less.

The current fiscal crisis of our cities is a political problem, not an economic one. Although there may be some poor states, some poor cities, and some undernourished rural areas, there is no long-run scarcity of resources. In fact, some observers can make our cities' fiscal problems, viewed in the aggregate, disappear. The Tax Foundation (1966), in making fiscal projections for state and local government to 1975, concluded that revenues would grow sufficiently to create a surplus, and that only a minority of jurisdictions would have difficulty meeting future revenue require-

AUTHOR'S NOTE: *The National Aeronautics and Space Administration and the Urban Institute provided financial assistance for my research. I wish to express my appreciation to them and to acknowledge that my opinions are my own.*

ments. We all agree that such projections have to be taken with caution and skepticism; tax yields may not rise automatically with rising incomes and assessments may never catch up with market values. On the bright side, however, wars, both domestic and foreign, have certainly raised and displaced the level of public spending (Peacock and Wiseman, 1961: 24-28). And if foreign wars end, cities may be able to dig deeper into the federal fiscal barrel through some ingenious method of spreading the country's wealth. Once the federal government's largesse is distributed, city managers and mayors can stop thinking about a tipplers' tax where barroom habitués are penalized because they do not like to drink at home. In the aggregate and in the long run, there is no major fiscal problem that rising incomes and economic growth cannot take care of. In the short run, however, the fiscal crisis is very real to local elected and appointed officials. Distant goals such as the implementation of tax sharing may be worthwhile pursuing, but they do not help to balance this year's budget. How to get more revenue now is an important political problem for local officials. What can social science tell these frustrated public servants?

Public finance theorists tell us what is desirable: taxes should be equitable and neutral; they should contribute to fiscal stabilization, encourage productivity and growth, and be simple and easy to administer. Economic, income, and administrative effects should all be considered. If a contemplated tax does not meet our diverse set of criteria, or if it is no better than existing taxes, then we would prefer increasing the rates of the old tax rather than adopting the new tax. Suppose the property tax were the ideal tax and the income tax the worst fiscal option. The local official, in a condition of fiscal extremis, would be told to raise the property tax rate rather than adopt that inelastic, regressive, and far from neutral local income tax. The official would respond that he could not raise the rate because there would be a taxpayers' revolt. Instead he would insist that the tax base be broadened to include the awful income tax, by which he means that he wants some way to get the same taxpayer to pay more for the same public goods and services. Should this local official be castigated because he ignores what is theoretically desirable in doing what is politically possible?

Local officials live with some psychological guilt because they have to manipulate the environment and, at the same time, ignore some of the standards that they have been taught. For example, equity, tax justice, and similar notions are not realistic guides to action, but just sources of bureaucratic anxiety. Myrdal (1954: 185), after examining the principles of public finance, concludes that "it is vain to attempt to isolate a purely economic problem from its political setting." And Myrdal's description (1954: 187) of the behavior of officials is consistent with my own observations: "I do not blame the politicians and tax experts who legislate and administer as best as they can. They look abroad to see how similar matters are ordered there, they quote the literature whenever convenient; they form opinions on the nature and on the practical aspects of their problems and then come to some sort of conclusions. The reasons which they advance bear the unmistakable doctrinal imprint of the high principles. Occasionally a problem of tax incidence is touched upon but hardly ever fully discussed."

Too often, when we say the problems of local finance are political, we mean that public decision-making is inadequate, that local governments are Balkanized so that they provide miniscule services and have minimum support, and that local officials are inept and stupid in their fiscal behavior. Political becomes a garbage-pail word to explain what seems to be irrational behavior. Political, however, can be used in a positive sense to convey the how of doing things. How many political scientists in the country can determine a priori the political limits of a proposed tax? How many could develop a strategy to get a new tax accepted by the public?

If we are going to develop policies that can help solve the financial dilemmas of our cities, we need a multiple approach. We need federal policies that will share the wealth in the long run, and at the same time we need realistic policies and tactics for our state and local officials to pursue in the short run. But it is not enough to urge officials to take a multiple approach, to get money from a variety of sources. What is also required is a new orientation toward the reality of taxation. While social scientists worry about economic elasticity and local officials ignore it, no one is very much concerned about political elasticity. (The need for developing political elasticity coefficients was pointed out to me by Robert Biller.) How does public tax support or resistance change when tax rates are changed? Would it be possible to develop a sensitive measure that could determine the political feasibility of new taxes? Can the range of local options be extended by better understanding of the political conditions of fiscal reality? In the following pages, I want to adopt the viewpoint of the local official. At the same time I want to argue with his definition of fiscal reality. Regardless of how hostile reality seems to the official, he must take action to insure that his resources are equal to his problems. Otherwise our cities will continue to drift into decay and our citizens into disillusionment.

PANACEAS

The problems and solutions of local finance are old hat. The refrain that we need money is well known, and I imagine that people are tired of hearing about the famine of money (Tax Institute, 1955): strengthen the property tax; find new sources of nonproperty tax revenue; states should not fiscally constrain and choke local governments; be efficient with expenditures. These nostrums have been repeated so often that their message is likely to be ignored by local officials.

Instead, the local official is likely to call in a tax consultant who, after reviewing the situation, usually suggests that city officials adopt a new tax. If the particular city does not have a sales tax, then the sales tax will be recommended. If the city has a sales tax, then the income tax will be recommended. The rule for this game is: always recommend a new tax. However, it is unlikely that a novel operational suggestion will come out of this procedure. Either the revenue source has already been tapped or the suggestion has previously been considered and rejected as being politically infeasible. The list of available and practical revenue sources is really quite short, and a quick, comprehensive solution to the local finance problem just does not exist. Yet the advice mill goes on: cities continue to have experts review their fiscal situations (Research Advisory Committee, 1966); professional finance organizations conduct conferences and issue policy guides (Municipal Finance Officers, 1967: 93-96); tax associations and business organizations state their tax preferences and hope for efficiencies (Committee for Economic Development, 1966: 17); and governmental interest groups testify before legislatures and state principles of tax reform (League of California Cities, 1966). However, the payoff for the frustrated local official is difficult to discern.

Nor is the federal government much help. The local official is at the extremities of a Rube Goldberg system known as fiscal federalism, which is a euphemism that masks local poverty and fiscal discretion, and at the same time holds out a fiscal carrot for a better world to come. Local officials who wait for the federal and state governments to pick up the tab are bound to be disappointed. Some local officials do, however, expect the federal government to rescue their cities from fiscal ruin. They appear to believe there is no doubt, in the long run, that some form of fiscal first aid such as tax credits, categorical grants, or revenue sharing will balance problems and resources (Break, 1967; Brazer, 1967: 155-164).

Local officials, however, should not hold their breath waiting (Berle, 1960). A large part of such aid will still be for extras, for national problems on local sites, and not for minor street repairs and new hoses for the fire department. Moreover, someone is bound to tie some strings to make conditional, so-called unconditional, grants; somebody else's money is never quite as free as we would like. Of course some version of the Heller-Pechman plan would provide state and local officials with some discretion, but revenue sharing is probably intended as a not-too-large supplement for local efforts. Using Heller's figures (1968), which would distribute $6 billions or about $30 per capita, California would get about $560 millions. Assuming that the cities would get at least half of the grant and that the money would be distributed by population, Oakland, for example, might get roughly $6 millions, and the city would share some of the grant with the school district and the county. If the city received one-third of the $6 millions, then this $2 millions would be welcomed by city officials, but it would not even cover one year's wage increases for municipal employees.

As Heller views it (1968: 18-27), tax sharing would provide "financial elbow-room" and would slow down state and local tax increases. However, Pechman (1965:78) believes that the "amount of assistance should be large enough to make possible a significant increase in the level of state-local services." But in any case, given other demands for federal expenditures, the assistance, if it comes at all, will not be large enough for service expansion and probably will nourish only a few hungry cities—no small chore. Even if the grants were large enough for city officials to rely on the funds for expanded and usual city services, what happens when the level of national income goes down and there is less to share? The services provided out of federal funds would not be easy to cut. Is it so out of the question that our economy can have a few bad years in which income goes down or are recessions and depressions truly a thing of the past? Income elastic revenue sources are fine as long as income increases. Local officials still require stable revenue sources; otherwise, local budgeting and municipal services may be somewhat uneven and chaotic.

The message is clear; local officials will still have to increase tax rates and search for new revenue sources. They will not find the tax that everybody is happy to pay. But there are ways to cut down the pain of taxation and perhaps enhance the official's revenue efforts.

ASSESSING THE ENVIRONMENT

Local officials do not relish going to the public on money matters because, in their view, it is a risky business. The officials might not get what they want. Instead of looking for revenue, they could be looking for jobs. Of course, there is always the risk of an unfavorable outcome, but this risk can be reduced if officials will invest some of their resources to find out about their own communities before putting a proposed bond issue on the ballot. When bond and tax issues fail to pass, the officials blame it on a hostile public and keep their sanity by not reflecting on their own performances. In talking with local officials, I have urged that they get out of their offices and find out about the environment in which their organization operates. The usual response is that there is no time for excursions; they are too busy fire-fighting the details of internal administration. But if mayors and managers will not gather the basic intelligence to insure an adequate resource base, then who will? Finding out what people want, what they will pay for, and what the traffic will bear may possibly provide clues to building a winning tax coalition, and at the very least such intelligence can dispel some of the fiscal fictions that local officials cherish.

Officials are sometimes reluctant to raise taxes because they believe that taxes have reached a political limit. "How do you know, Mr. Mayor, that the property tax has reached a political limit?" Answer: "I do not know; I just feel it." A political limit is a fuzzy constraint, perhaps fictitious, that local officials worry about, but have difficulty predicting. Even social scientists cannot tell when a political limit is about to be reached. Unlike budget constraints, political limits are not finite. One problem in identifying political limits is that they keep changing over time. For the past thirty years the property tax has been just about to reach the limit of taxpayer rebellion. Nevertheless, the tax limit tends to be raised when people want more, when operating costs go up, and when local officials are in fiscal extremis and feel there is no other way to go except to raise the rate. It would be a mistake for officials to assume that there is a political limit on any revenue source without first doing some checking. If local officials know their community and keep their intelligence current, then it is unlikely that they will be trapped into thinking that they are more constrained than they really are. A political limit is a question of political feasibility, something which the local official should assess for himself.

Some officials will react to my "know your community" advice by saying they are well informed. After all, they talk to a head of the chamber of commerce, a real estate board president, a tax association official, and perhaps a League of Women Voters representative. But can a local official know his community when he restricts his contacts to those who seek him out? The local official should worry about how representative his conversations are of community sentiment. My impression is that local officials generally talk to one another. At professional meetings where the like-minded gather, there are splendid opportunities to reinforce each other's distortions about the public. When everyone around the official is saying that the property tax cannot be raised, it is likely that the official will also believe the same thing.

However, local officials should not fall into the generalization trap, when social scientists find it difficult to generalize about our diverse cities. For example, Harvey Brazer (1959) found wide differences in city expenditure patterns which varied by the state and type of city; a core city of a metropolitan area is different from an industrial suburban city which, in turn, is different from a resort city. Because citizens have different wants and needs, and because state and local officials have come to different agreements on who will do what, cities do have different responsibilities and resources. Cities are not all the same. Although a certain situation is a nightmare for the academic social scientist or the national policy maker, it is an opportunity for the local official. He has a chance to tailor local resources and support to match local desires and demands. This is not to say that local officials should not look at the big picture, on occasion, but the global generalizations, particularly on money matters, should not stultify local action and initiative. The world of local finance is filled with enough "Chicken Littles" who expect that any day our cities will collapse in fiscal ruin.

THE HOSTILE ENVIRONMENT:
REALITY OR FICTION

Local city officials tend to perceive the environment as hostile. It is not that they are paranoid but that there is a fear of private citizens and other public officials. At the local level, special districts and the county can encroach on their tax base. The no-money situation at the local level also tends to discourage cooperation between local governments; what is mine is mine and, if I do have anything for you, you have to pay. Reimbursement rather than cooperation describes local intergovernmental relations. At the state level, the legislature may preempt new revenue sources or take away existing ones. Whatever fiscal meaning is left to notions of independent home rule, the local official feels he is competing with other jurisdictions, not to provide service but to get money. Home rule does not preserve local resources for local problems. Instead, city officials, like other officials, have a restricted hunting license to scramble for revenue.

Thus the city official views other officials with some suspicion and with some sympathy; all of them are very much in the same fiscal rowboat. But the public is another matter. Whether correct or not, the official sees the public more in terms of tax resistance than in terms of expenditure support. When the local school district has trouble raising its tax ceiling and passing bonds, the city official often feels that the public is hostile to tax increases. The city official reasons that the public always supported the schools and, if there is little support for schools, there certainly will not be much support for sewers.

No doubt there are cities where the public is so frustrated and alienated from the local government that direct appeals for tax increases are not feasible. However, there are probably many cities where taxpayers' attitudes are not as hostile as officials think. There is not much empirical evidence that the public is more benign than hostile, but there are some indications that an assumption of public hostility should be questioned. Bollens and his colleagues (1961: 277) found that 37% of St. Louis city residents agreed that taxes were too high when service levels were considered. The poor and less educated citizens had the greatest resentment to local taxes. In a Detroit Area study, Janowitz, Wright and Delaney (1958: 36-44) found a similar dissatisfaction with tax burdens among the lower class. In their total sample, "forty-one percent thought taxes were too high, 'considering what they got from the government.' " For federal fiscal policies, Mueller (1963: 221) found from a national survey that "half of the people interviewed said that they were prepared to pay additional taxes in order to make possible larger outlays on two or more government programs." However, she also found (1963: 222-228) that people's fiscal attitudes are incongruent: they favor increased spending, while disliking increased taxes.

Of course we would like to find that everyone wants and is willing to pay for more public services. Local officials, however, can be encouraged because these numbers are not more negative than they are. Suppose Bollens had found 80% of the people of St. Louis fed up with taxes instead of his 37% figure; then the situation would be as dark as officials describe it. However, roughly 50 to 60 percent of the public may not feel that taxes are too high, so there may be some maneuvering room for increased local taxation.

The cited studies are a little dated, and since public fiscal attitudes vacillate, it would be desirable to have additional evidence before local officials modify the hostility assumption. One positive finding of a recent Advisory Commission on Intergovernmental Relations study (1968) was that over the 1950-1967 period, over 50% of the growth in state tax revenue was a result of the political process. In other words, increases in state revenue are not just a matter of economic growth but also reflect that state legislatures can and do raise rates and increase taxes. In my own state, California, the governor was able to increase taxes and his popularity at the same time. In Cleveland, political leaders were able to double the city income tax, and voters passed a $100 million bond issue to handle water pollution problems (*Newsweek,* 1969: 68). In Michigan, despite growing discontent over increased assessment, "85% of the school districts that asked voters to approve higher millages eventually won the increases" (*Wall Street Journal,* 1969: 1).

One can also find many examples of city voters rejecting bonds and tax referendums. Rejection, however, should not be taken as conclusive evidence of the hostility of the environment. Voter rejection may just reflect the inadequacies of the tactics and assumptions of the local officials' tax campaign. Consider the local official who bases his campaign on the assumption that elections reveal preferences, or that citizens vote only on the merits of the issue, or, specifically, that they want a museum or they don't want a museum. This official may have lost the election before the voting. Why some issues pass and others are rejected is not a simple question to be explained away by assuming that voters are hostile toward better schools, sewers, or a zoo.

Why people vote the way they do in local elections is a complex business. Perhaps in more pastoral times at the turn of the century, local voting was related to local issues. Prior to the Depression of the thirties, for example, local governments were our chief tax collectors (Maxwell, 1965: 15-22, charts 1-2). Today there are many hands in the taxpayer's pocket, and local tax voting may be just a convenient means of getting back at the federal government's Internal Revenue Service.

Some authors have interpreted negative voting on local bond and tax issues as expressing the political alienation of the voter in which the vote becomes a means of protest, not only against the local political system but also against the citizen's generally deprived condition (for example, see Horton and Thompson, 1962: 485-493; Templeton, 1966: 249-261). Usually, local elections do not attract much of a turnout, but when the turnout is high, there is a greater chance for defeat of tax and bond issues (Boskoff and Zeigler, 1964: 16-17; Carter and Savard, 1961; Stone, 1965, for contrary views). One explanation for this phenomenon is that the citizen who is deprived or disinterested in the community usually stays home, while the citizen who is satisfied or interested in the community usually votes in favor of the issue. In a high turnout election more unhappy citizens are drawn into the electorate and thus increase the chances of defeat.

Nor is short-term economic interest, where the voter seems to maximize his income, a reliable indicator of the outcomes of local referendums. Short-term economic interest, such as the property owner without children who opposes school tax issues, does not always prevail because some people may support tax issues out of a sense of civic responsibility or pride. Wilson and Banfield (1964:

876-887) present some evidence that suggests that winning coalitions can be built from the poor who have nothing to lose and those upper-income citizens who have civic responsibility. In a study of bond elections near Atlanta, the authors found a positive relation between income level and approval of bond issues. Both low- and high-income voters supported the bond issues but support was greater among the upper-income groups (Boskoff and Zeigler, 1964: 45-47).

Certainly many citizens consider issues such as the particular service which the bonds will finance, but local officials should not assume that the merits of an issue will automatically sell it. Voting decisions may involve concepts of self-interest. or civic responsibility, or feelings of the worth and performance of the local government, or outlets of protest against national taxes. Since there are many reasons for voting against something, it is encouraging to find examples of taxpayer support. Evidently the taxpayer revolt virus has not infected us all. Finding evidence one way or the other, however, will not disclose the hostility of a particular community. Local elected and appointed leaders should not assume but ascertain the reality of community hostility.

TAX PUBLICS: MAKING LITTLE ONES OUT OF A BIG ONE

Part of the hostility assumption is probably due to the local official's tendency to perceive taxpayers as an undifferentiated group. The term "taxpayer revolt" is a clue to this perception. How often does the local councilman get a nasty letter complaining about high taxes? How often is the assessor's office filled with pickets, signs and all, complaining about increases in assessment? Of course these events do occur but not often; in fact, most people pay their taxes most of the time, and a low delinquency rate is one indicator that taxpayer resistance is not particularly intense.

Moreover, it is my impression that most taxpayers are indifferent to taxation. One reason the public is indifferent to taxes is that taxation, whether by design or not, is complex and difficult to comprehend. It is a policy area where one has to be an expert to understand the arithmetic of rates, assessments, the secured roll, and equalization. Most citizens are not motivated to master the rudiments of tax knowledge. They do not have to understand the whys and wherefores of tax paying nor do they probably know how much they pay or when they pay. The ordinary citizen simply knows that he is a taxpayer.

For a citizen to care about taxes, taxes have to be high enough so that he is conscious of them. The citizen has to know he is paying, and what he is paying has to be significant relative to his income or the profit position of his business. With withholding at the federal level, property taxes prorated as part of a monthly mortgage payment, and the use of indirect taxation by state and local authorities, our tax system has been designed to reduce the public's tax consciousness. One study, for example, found that 31% of high-income taxpayers were unaware of their marginal income tax rates (Gensemer, Lean, and Neenan, 1965: 258-267). Their tax bite was high enough so that these citizens should have been aware, but a large minority were not. Therefore it is reasonable to expect that as the tax bite gets smaller, the proportion of tax-conscious citizens will be considerably reduced.

To say that most taxpayers do not care about taxes does not exclude the fact that some taxpayers do care, which brings us to the subject of tax publics. Clustered around each proposed or existing local tax is a tax public composed of attentive citizens who have some reason to care. One reason for their concern is that these citizens pay much more in relation to most citizens; for example, a charge for city sewer service may only involve $1 or $2 for a residence while a large cannery may pay $2,000 a month. For industry, business, and real estate operators, taxes are a cost element that has to be watched. Being a large payer, however, is not the only reason for concern. Some citizens become members of a tax public because the government has involved them in the administration of a tax. The wholesaler or distributor of tobacco products may get involved, depending on who is tapped to collect a cigarette tax. Sometimes the involvement is strictly to settle administrative matters such as billing procedures to be used by the telephone company for a utilities consumption tax. Sometimes the involvement is of more direct concern because the proposed tax may affect the demand for the goods or services that are going to be taxed. Sales and excise taxes, such as a transient occupancy tax, can generate concern, especially when rates are not uniform at the local level. Sometimes concern turns into stiff resistance which local officials would find hard to anticipate. For example, the city manager of a small community in the San Francisco Bay Area recently proposed a gross receipts basis for a business license. His proposal was defeated by the small businesses in his community; there was no resistance from the large businesses. As he explains the outcome, the small businessman was afraid to let the competition know too much about his business. The license fee was trivial; it was the intelligence opportunity that generated concern.

In addition to the citizens who pay attention to specific taxes, there are the professional tax watchers who, by definition, are members of all the tax publics. The full-time employee of a tax association has a difficult job since his own resources are quite limited in trying to follow the activities of the many agencies within his jurisdiction; he is generally ineffective in terms of support or resistance to tax measures. Nevertheless, as the apostle of efficiency in government, he will show up at city budget and tax meetings. To round out the picture of a tax public, I would also include the representatives of the local chamber of commerce and similar organizations.

To summarize, a tax public is that small group of citizens who care enough about a specific tax to bargain with local officials over rates or perhaps work to influence the broader, usually indifferent public to defeat a tax referendum. Therefore, the tax public is the primary target for official persuasion, action, and perhaps, cooptation in the supportive coalition for a tax increase.

TAX COALITIONS: MAKING A BIG ONE OUT OF LITTLE ONES

Because most people do not care about taxes, the local official, by default, must provide political leadership in building a winning tax coalition. With the exception of the few friends of the museum or zoo, most cities do not have groups of people who advocate increased taxes. At the same time, there are groups that would like lower taxes but only agitate for them sporadically. The political arena of taxation is relatively quiet and it is the local official who must stir things up if he wants adequate resources to work with. In my mind, the question is not whether to stir, but how to stir and how much.

The tactics that the local official can use depend on his assessment of the hostility of the environment. These tactics are bracketed by the official's willingness to confront the community. At one end of the spectrum of his options, he can avoid the public when he believes the environment is hostile. At the other end, he can contact the public and negotiate and work with its attentive segments when he believes the environment is relatively benign. If he does not need money next fiscal year, he can wait for the federal or state donation. If he needs a little money to meet increased operating costs, he can use public avoidance tactics. If he needs a great deal of money, he will have to go to the public whether he likes it or not.

Depending on how much money the official needs, the tax coalition that he will build and work with will vary. If a proposed tax is small and will not be felt by a large number of people, the official need not form a large coalition but only accommodate the small tax public that cares. But if a proposed tax or bond issue will have a widespread effect, requiring at least a majority vote or a large payment by a large number of voters, then coalition formation becomes difficult and complex. It is this latter, more difficult, community coalition which will test the political leadership of the local official.

Although much of American politics is coalition politics, I find it embarrassing to admit, as a political scientist, that I have not learned a secret blueprint for building a winning coalition. However, there are some clues that are suggestive. First, the local official should develop a support nucleus composed of citizen leaders who will work and donate resources for developing communitywide support. The composition of the nucleus should reflect the diversity in the community; it should not be just a group of the official's friends. It should contain nonresidents and residents, Democrats and Republicans, the poor and the rich, the selfish and civic types, and taxpayers as well as beneficiaries. Even if the proposal is to finance a zoo, the Friends of the Zoo are not enough, and there usually are not Friends of the Sewer System to rely upon. Second, the local official should not assume that the citizens in the support nucleus are aware of his particular financial and service problems. Most of these leaders will make the tax-service nexus in a general way; they will understand that costs of government are going up just like the cost of automobiles, and they will appreciate that taxes have to keep pace. It is important, however, that the local official make a specific connection between the proposed revenue source and community benefit. The connection must not be just that the rest rooms in the park need remodeling and so a bond issue is required. Rather, making the tax-service nexus for the support nucleus involves stressing output: less crime, less flooding, fewer accidents, more business, better health, higher incomes and employment skills. The official who uses a capital improvement program that contains only a shopping list of buildings, lands, and improvements is bound to have trouble. The appeal should be tailored to each individual of the support nucleus. For example, the same museum bond issue can attract tourist business for the merchant, provide greater educational opportunities for children and cultural activities for parents, and improve a specific neighborhood. Once this group of leaders sees some personal payoff in the official's proposal, rather than relying on a vague sense of civic do-gooding, the official can turn his attention to campaign tactics to win broad community support.

Although the local official should try to make the tax-service nexus for each leader, there is less reason for doing this when designing tactics to win public voting support. Social scientists are still at the top of the iceberg when it comes to understanding and predicting the public's preferences for taxes and services (Buchanan, 1967: 181-210). The local official, for example, can fall into the trap of activating a segment of the public that has latent opinions hostile to tax increases. The problem is that it is difficult to predict who will support or who will resist when the salience of a particular issue is low (Key, 1961: 263-287; Bauer, Pool, and Dexter, 1968: 94-96). However, there are some premises that the local official should take into account in his decisions. First, he should not assume that voters are aware of existing taxes (their cost burden) or existing services (their benefits). Second, he should not assume that even if voters are aware of taxes or services or both, they will make the tax-service nexus.

Perhaps democratic theorists would prefer a fully informed voter who, after calculating his costs and benefits, would reveal his preferences in an expenditure or revenue referendum (Birdsall, 1965: 280 ff.). Also, local officials would probably like to have referendums that are mandates; they would know where they stand. But this is not the case, and short of doing away with referendums, officials should take advantage of what others see as disadvantages in the system.

The citizen's separation of taxing decisions from spending decisions provides officials with an opportunity to stress the benefits of a proposal rather than its costs. In most situations when a resource constraint is introduced, expenditure suggestions will diminish. This is true in planning activities and capital improvement programming, and it is likely to be true in voting on referendums. There is a higher probability of voter approval when a spending measure is not connected to its means of financing (Buchanan, 1967: 202-204). Awareness of one's tax burden is not necessarily conducive to increased expenditure support. Consider property owners, for example. They are more likely to know their property tax burden than renters, and from statistical studies we know that the greater the proportion of homeowners in a jurisdiction, the lower the level of taxes and expenditure (Lineberry and Fowler, 1967: 712; Davis and Haines, 1966). The voter who is motivated by an economic calculus cannot be solely relied upon to increase the public budget. While he may support bonds for fire stations because of lower ultimate costs to him, he may not support bonds for lighting some other part of the city. If the choices are made clear, he can vote down those proposals where his costs outstrip his benefits. The mix of public services is such that he can always find something by which to express a negative preference.

Ironically, the local official has to rely on voters who do not make individual cost-benefit calculations or who perhaps are just inconsistent. In our cities, the poor want more services, and do not vote; if they did, it might be a vote of protest. But if the local official encourages their turnout, on the basis of benefit, he may receive a favorable vote although implicit in the vote is the expectation that somebody else will pay. Furthermore, the knowledgeable homeowner is not necessarily consistent in his attitudes toward the financing of public goods. He can also be a source of support for some things not in his apparent self-interest. For example, one might think that homeowners without children would tend to dislike the property tax because of its use for schools. However, in one study of tax preferences, homeowners with or without children in the public schools expressed much the same attitudes towards the property tax (David, 1967: 89-90). In a statistical analysis of educational taxes and expenditures, the negative relation to the proportion of homeowners also disappears (Campbell and Sacks, 1967: 150-154).

Stressing benefits by the local official is a sensible counter-weight to the public's tendency for small, "incorrect" budgets because, as Downs states (1960: 544): "rational ignorance among the citizenry leads governments to omit certain specific types of expenditures from their budgets which would be there if citizens were not ignorant." We probably do tend to underestimate and be ignorant of the possible benefits of city programs. At the same time, citizens are inclined to believe that there is waste and bureaucratic nonsense at city hall so that budgets could be cut. Frequently, opposition to tax rate increases will be stated by prescribing efficiency: "They don't really need more money if they would wisely use the money they have." Stressing benefits is a way of making visible the hidden benefits of usually invisible city programs. If Downs (1960: 550) is correct in his observation that "every citizen believes that the actual government budget is too large in relation to the benefits he himself is deriving from it," then local officials should not be squeamish in consciously manipulating the benefit side of a proposal.

Thus, stressing benefits and not costs and leaving the tax-service nexus obscure are tactics that local officials might consider. Those voters who do not know their individual burden and want the particular public good will support the referendum. Those who do not want the benefit but have nothing to lose may stay home, but if they turn out there is less reason for negative voting. Those few citizens who ferret out the cost implications and make their own calculations will vote their preferences. While an economic calculus approach may be suitable for gaining the cooperation of a small tax public, such an appeal is probably not suitable for building voter support. Although a city's financial imperative motivates the official, there is no reason for him to assume that segments of the public will have similar feelings. Community coalitions are not likely to be built where the payoff appears only as increased taxes. Benefits are the cement of the community coalition.

SELLING PUBLIC SERVICES AND GOODS

One way to avoid the public and still get revenue is to allow citizens to buy (only) the public goods and services they desire. The public usually buys some portion of city output through devices such as permits, licenses, fees, and service charges. These revenue sources have different names but much the same character- istics: (1) the citizen has some discretion in his purchases, (2) he gets some benefit which others who do not pay do not get, (3) he pays part of the city's cost of production, and (4) the amount he pays is nominal and painless. For convenience, I will call these revenue sources "user charges." The point I want to make is that if the local official manages his user charges with some finesse, these charges can become a lucrative source of hidden taxation. But there are limitations, as we will explore.

Many economists see user charges as an attractive source of financing due to their obvious parallel to prices in the private market. Some economists, such as Brownlee (1961: 421-432), see limited opportunity for user charges, but others, such as Netzer (1968: 457), believe that there is a "significant potential for greater and more sophisticated application of user-charge-type financing." Along with economists' support for user charges come their concerns about allocating resources efficiently, income distri- bution effects, and the benefit principle. All of these concerns converge in the choice of the government good for user-charge financing and in the determination of its price.

At first glance the choice of which good to sell is easy. At the local level practically all services and goods could be financed from user charges because much of the local government's output satis- fies "merit wants" (Musgrave, 1959: 13). Public education is a good example of a merit want where the individual benefits and could pay for that benefit. However, once a community decides that education is important, not just to the individual but to the community as well, then it becomes legitimate that the property tax pay for the individual's education. Public education is the outcome of perceived social benefits. (Some might argue that with current student unrest, social costs outweigh benefits.) We often create these social benefits by our own change of attitudes. There is nothing inherent in the nature of the good, education, which prevents a simple identification of benefits and charges with the individual.

There are, of course, some goods that technically involve collec- tive benefits and costs but these are not as prevalent as one might think. Many of the services that local governments provide, such as recreation, police protection, and free books have their analogous counterparts in the private sector. Parks can be fenced off, and fire protection can revert to an earlier system of individual payment. Theoretically, most of the goods and services that local governments produce could be sold to the public. I suppose if a survey were taken of the provision of ambulance services in our cities, the investigators would find a wide variety of financing and a mix of public, business, and charitable or voluntary operators. The point is that the service is similar but the means of payment is very much a function of community attitudes.

The citizens of a community define what is public and what is private, and within the public sector they also define what may be sold and what is free. Regardless of theorizing on the nature of public goods or the devising of methods to allocate costs, it is what people expect of their local political system which limits the official's choice of a good for user-charge financing. The official can influence these expectations; he does not have to supinely accept them, but he must recognize that they exist. In any event, the case for user charges is too often viewed as a problem of connecting burdens and benefits, as a technical question of allocating costs and dividing benefits. The "perfect" local user charge is not one where the payer gets the benefit, or where resources are properly rationed, or where there are no income distribution effects. For the local official the "perfect user charge" may have these features, but of overriding importance to him is whether the public will accept it. "Perfect user charges" are those that are paid and bring revenue to the city.

After the official decides on the good or service, he must also decide what price to charge. Many of the problems and opportunities of user charges converge when price is determined. Although it is tempting to suggest that local officials should study price elasticity of demand, and marginal costs of contemplated user charges, it is unlikely that they will or can (Thompson, 1965: 280-283). Most local officials are indeed cost conscious (they like to cut costs), but they do not know what their output costs. The local official believes that the city should be reimbursed for its costs, but in practice he can estimate only a fraction of them. It is not unusual to find local charges that have not been revised for thirty years, and the charge no longer covers the cost of administering it. This lag is a result of the thrust of local public accounting which is directed toward corruption control and not cost information. But I would not suggest completely changing this accounting orientation. If a city is in the utility business, then it will probably pay to install cost information systems because the revenue potential is considerable (Stockfisch, 1968: 89; Scheffer, 1962: 528). On the other hand, I do not see the value of an elaborate cost accounting system to determine that the entrance fee to a museum should be twenty-five cents. For most of the minor fees, price can be and usually is determined by a slight extension of the prevailing local practice.

Officials can also set prices by checking with other cities and professional organizations. Such practices are useful shortcuts to price determination but there is some danger that the price will become a fixed standard unresponsive to changing cost conditions.

Furthermore, the current set of practices contributes to an illusion of full payment. When he pays, the user assumes he is paying for the total costs of the services; he has no way of taking into account the subsidy from other tax sources. If the public golf course green fees are only a fraction of the private courses', the user is more likely to interpret the difference as a matter of lesser quality of service for the public course and not as subsidization by the public for the same quality. As long as local officials are going to incur some political and economic costs because they institute a user charge, it would be preferable from a revenue perspective to get full payment rather than just the illusion of it.

Full cost reimbursement, however, is not enough. In some cases, local officials should attempt to set the prices so that they make a profit. I agree with Vickrey's statement (1963: 64): "If any specific charges are to be made, they should in nearly all cases be designed in part to contribute to the public treasury over and above the amount that would flow in on the basis of charges strictly reflecting marginal costs." It is seldom that one will find a charge that makes a profit, and when a city makes a profit it is more a matter of accident than design. Users generally do not like the idea of paying more just to enhance the city's bank account. They would prefer that their cost be commensurate with the benefit they receive because such a profit would only be a hidden tax. True enough; if taxes are hard to come by, why not have charges that are taxes?

Furthermore, user charges provide an opportunity to get the rich to pay for services that the public subsidizes. Generally it is considered desirable if local governments concern themselves with providing goods and services, and allow the federal government to worry about income distribution and stabilization. For local governments we should, as Brazer says (1964: 128-129): "regard effects on income distribution and stabilization as being incidental and unsought, to be avoided as much as possible." When this criterion of neutrality with respect to income distribution obstructs the local official in his gathering of revenue, however, he should ignore neutrality and let the federal government compensate for his actions. Usually officials are not neutral anyway, as they affect income distribution, in the name of equity, without knowing who is subsidizing whom (Maxwell, 1965: 15-22, charts 1-2). In recreational activities, for example, they set prices low enough (below costs) to accommodate children and low-income families, but this practice often favors middle- and upper-income users. Officials could correct this practice by having price discrimination by income grouping. This may be administratively difficult but is worth trying. At the very least, officials could discriminate by the type of service or good. Marinas, golf courses, and sailing lessons could bear a higher charge because their users are mainly from the middle- and upper-income groups (Thompson, 1968: 31). Before leaving this point, I want to make it clear that I am not making an argument for economic equality, but agree, for the most part, with Galbraith (1958: 72) that "inequality has ceased to preoccupy men's minds." What I am saying is that the rich can and probably are willing to pay more than the poor and that local officials should, in this situation, increase fee schedules at least until the subsidy from general taxation is removed. The problem with the criterion of neutrality is that local officials cannot be neutral when setting prices for the poor and it certainly does not pay to be neutral when setting prices for the rich.

My attitude toward revenue maximization is not constrained by principles, but I offer the following principle: charge the highest amount possible so long as the original purpose of the public activity is not compromised or destroyed. This is more than saying, "charge what the traffic will bear" because it implies that there are limitations related to prices which the official has to consider. Price has to be compromised when it conflicts with the purpose of the public activity. If the city wants to perform its recreational objective for children (regardless of the distribution effects) and have a cool summer for the community, the price for swimming should be low enough to keep the children in the pools and off the streets. Similarly, I do not agree with those who recommend that charges should cover the full cost of regulation (Break, 1967:

219). For example, Stockfisch (1960: 100-101) suggests that in Los Angeles fees be increased to cover the deficit in animal regulation and control. He does not suggest just how he would make sure the dogs would be controlled, given the higher fees. Since many people in our cities are poor, the high license fee plus the cost of a rabies shot will discourage poor citizens from licensing their dogs. If animal control has some known relation to public health, then, perhaps, such charges should be lowered or eliminated. There are times when regulation takes precedence over revenue maximization.

Some goods have to have a nominal or zero price because the charges may alienate voter support for general tax and bond issues. When contemplating revisions to local fee structures, the official restricts his attention to a particular charge and service. The problems of running a local government are complex enough and it is understandable why officials ignore possible second-order effects, such as voter support. But the cumulative effects on the voter of small charges may be serious. After being "nickled and dimed" to death, voters may get the notion that their taxes pay nothing, and their ties to the local political system may be shaken. Thus officials should avoid nuisance charges, not just because of administrative convenience but because such action will avoid undermining voter support in the community. Highly visible charges tend to cause lasting irritation. Nor is it particularly easy to start charging for something that has been free. If there is a tradition of free summer band concerts in the park, it is probably not worthwhile to think of ingenious ticket-selling devices. On the other hand, if the particular activity is attended mainly by nonresidents, then even ongoing activities are eligible for fees. Suppose the park department sponsors an elaborate garden show that attracts flower fanciers from all over the state; an admissions charge is feasible (once the fears of local merchants are assuaged) because nonresidents will be paying the freight.

To minimize the number of people unhappy at the same time, a gradual, selective approach to price increases can be a useful tactic. One major advantage of user charges would be lost if the local official, in effect, also increased the salience of the charge. Ideally, the payment for a service should not linger in the citizen's memory. However, if after an assessment by the official that a particular charge can no longer be raised, then charging for something new is a good way to proceed. I do not completely agree with

Vandermeulen (1964: 402) when she says, "new fees can easily be devised, but they should be justified on the ground that they are either a more equitable or more efficient way of raising revenue than increasing the rates of present levies." The option of increasing existing rates is not always available, and new fees may have to be adopted even if they are less efficient and equitable. Local officials do not have available a large number of alternative revenue sources. Sometimes the importance of the uses for the additional revenue overrides important criteria such as efficiency and equity. Local officials, as I have said before, should not feel guilty just because they have to cope with an imperfect world.

Thus, user charges can be a painless way to increase revenue if the local official knows his costs, tries to achieve revenue increases by way of relatively small charges on services lacking great saliency, avoids conflict with the basic purposes of his organization, and also appreciates the interdependence between user charges and general tax support by the voters in a community.

GIVING THE JOB TO SOMEONE ELSE

City officials can give up part of their public business to meet revenue exigencies. For the local official, one of the opportunities of our federal system and fragmented local government is the existence of other public and private organizations that can assume functions traditionally municipal in nature. Another opportunity is that officials facilitate changes in expectations about what these organizations should do. The recipe for the governmental marble cake keeps changing. For decades now, school districts have been consolidating to provide more specialized services and achieve other economies of scale. Now there are pressures for decentralization of school operations as people express demands for neighborhood control. At the same time, there are those who would like increased centralization (by the federal government) of the financing for such programs as education. Why share taxes when the federal government can pay directly for the services? (Ruggles, 1968: 62-72.) Yesterday's local financing problem becomes today's national problem. Who performs which service is not a static issue. Certainly what citizens expect and what officials do can determine what a public organization does. There is no absolute and unchanging definition of local responsibility.

Local officials can shift the locus of financial responsibility by encouraging the adoption of different boundaries for the performance of local functions; public health is a county problem and, with the increasing mobility of criminals, why not consider police protection a national problem? Oakland, California, for example, did not always have a public library. It may be that as the revenue straitjacket gets tighter, local officials will see some virtue in having the county or perhaps a private organization provide library service. With the fluidity of "who is responsible for what," local officials have an opportunity to give away a few jobs, creating some budgetary slack.

There are basically two ways to transfer public services: (1) create a new governmental unit to do the job, or (2) persuade an existing governmental unit to take over the function. Among those who are concerned about local government, I do not think there is much agreement on the wisdom of creating many new units. The Committee for Economic Development (1966: 17), for example, recommended that: "The number of local governments in the United States, now about 80,000, should be reduced by at least 80 percent." No doubt there are inefficiencies in so many small governments. No doubt the existing Balkanization and overlapping of local government is beyond the citizen's comprehension. Indeed, it is with some temerity that I recommend more governments rather than less. Yet the same financial imperatives which, in part, created the present maze are still with us. One proponent (Alderfer, 1955: 225-226) thought authority financing was the answer over a decade ago, and his remarks are applicable today: "The municipal authority can step into the breach in the walls of municipal finance where constitutional and statutory debt and tax levy limitations and difficult-to-raise property assessments, make it impossible to carry on expanded and new functions through regular municipal financing. . . . The municipal authority . . . is a perfectly legal attempt to get out of the present financial straitjacket which states have fashioned for municipalities."

More recently, Robert Wood (1961: 72-73) suggested the creation of new governments as one political option for financing local government. Wood also suggested adjusting assessments and finding new sources of revenue, but the creation of new governments had a special appeal to him because the same revenue base could be tapped without political repercussions. The special district can, in effect, raise the property tax rate, something that local officials are afraid to do. Wood (1961: 75) also pointed out quite correctly that: "The creation of special districts cannot be extended indefinitely, tapping and retapping the same revenue source to the point of confiscation."

Another problem with creating new governments is not that taxpayers will be unhappy because they are paying too much, but that taxpayers will become indifferent and not pay enough. There is a danger of fiscal undernourishment for all the public organizations in an area when there are too many governments. When mosquito abatement, transit, schools, flood control, air pollution, parks, the county, and municipalities all feed off the same revenue source, citizens can easily get confused as to who is doing what, and withhold their support and interest. Municipalities, in this fragmented situation, may have to face an indifferent and confused public for tax support. Some special districts may be better off; they will be in a position to run themselves by keeping general taxes down, by not alerting the taxpayers, and by resorting to public-avoidance tactics such as employing user charges and revenue bonds. Other districts, such as schools, where public awareness is keen, may have trouble. Margolis, for example (1961: 262), suggests that a single-purpose government will have more trouble gathering voter support than a government which can present to the voters a multipurpose package. As he says, ". . . the single-purpose special districts may not find themselves with sufficient fiscal strength to fulfill their goals." Moreover, in a metropolitan area where there are many municipalities and special districts, citizen preferences are clouded, benefits and costs spill over, and the general conclusion is that resources are underallocated to the local public sector (Brazer, 1964: 142-146). There is also statistical evidence that for some local functions, the larger the number of jurisdictions (not including special districts) in a county area, the lower the per capita expenditures (Adams, 1967: 9-45).

Creating new governmental units does have its problems. The municipality may free resources for the short term by an arrangement that may come back to haunt it in the future. Therefore, in situations where there is already a multiplicity of governments in an area, it would be preferable for local officials to adopt the second tactic and try transferring the job to one of the existing units. Such transfers may involve piecemeal consolidation with other local units, a surrender to the private sector, an emphasis on contractual and franchise devices, or perhaps an assumption of responsibility by neighborhood groups.

Opportunism is important in effecting transfers. Suppose a large regional or state park desires land to expand its service area, and the city has been maintaining an expensive recreational area in the proximity of the state park. What better time is there to get out of the park business, get some cash from the real estate sale, and get rid of an operating headache? Suppose the county is in the health business, but the city is still inspecting kitchens for permit purposes. Why not pay the county to do the inspections and get the city out of the health business? Perhaps the school district could be persuaded to run the city's swimming pools. The civic auditorium might also become a private theater. Garbage collection, rather than incurring city costs, could become a revenue device when a franchise is sold to a private contractor. A neighborhood group may prefer to undertake the sponsorship of a branch library rather than see its doors shut to their children. In all these cases, the recipient feels he is benefiting from the transfer. One man's white elephant is another man's transportation.

I am not suggesting that the city give away all of its jobs and go out of business. On the contrary, I would prefer that cities have sufficient resources to do more, not less. But since they are fiscally constrained, local officials could get rid of those jobs that they do out of habit and that do not reflect any citizen demand for city provision of the particular service. Why should a city maintain a large park system for regional use when it does not have the money for tot lots which its citizens want? In any event, the following would be a good exercise for local officials to go through: how relevant are the city's activities to its citizens? Perhaps another organizational unit is in a better position to meet citizen demands than is the city. Revenue needs, however, should not be the only consideration. When contemplating piecemeal consolidation, officials should consider political and economic scale questions as well. They should consider the effect of giving away a job on citizen political participation, on the responsiveness of local institutions, on their own capacity to act as political leaders, and also on the efficient use of resources. A little bit of caution in giving up jobs is wise, as it may be easier to give a job away than to get it back.

THE MULTIPLE APPROACH

In addition to selling and giving away services, the local official can use indirect taxes as one way to avoid the public. Once a sales tax is part of the tax structure, then excise taxes for other commodities and services, under the guise of a sales tax extension, are relatively easy to establish. Consumption of utilities, entertainment, hotel rooms, drinking, and smoking can all be taxed. The link to the sales tax can provide for almost automatic increases, as the sales tax increases. Such taxes may hit the poor harder than the rich. Whether or not they are regressive can be relatively unimportant as long as the taxes are nominal and do not hurt. In Oakland, for example, a family with $6,000 annual gross income would pay less than one-half of one percent of their income for both the city's sewer service charge and utilities consumption tax. When properly designed, such small excises are not felt, the tax public remains small, and regardless of the original justification of the tax, it can be used for general expenditure purposes.

Devices such as user charges and indirect taxes keep the tax structure complex and fragmented, which may in turn contribute to the illusion of painless taxation. One problem with the public-avoidance approach featuring many, individually painless taxes is that the taxpayer may soon begin to feel a cumulative pain. Lyle Fitch's description (1964: 117) of the local scene indicates that we may be approaching that pain threshold: "Local governments, within the generally narrow confines of state-imposed restrictions, have shown considerable ingenuity in tapping pools of potential revenue, however small; few things on land, sea, or in the air, from pleasures and palaces to loaves and fishes, escape taxation somewhere."

Another problem with the painless approach is that it encourages small tax publics. The few citizens who pay attention to taxes may gain favorable rate schedules or measures of tax liability. They pay, but relative to the rest of the public, they may pay less. The public usually does not realize that there is tax discrimination and keeps paying without complaint. Considering the imperatives of the revenue situation, it may be that such minor discrimination is a small cost to pay for overcoming resistance.

Probably more critical is that the approach seldom provides enough resources. In order to avoid more widespread dissatisfaction caused by the cumulative effects of many, relatively painless revenue sources, the effective official must plan an upper bound on the number of such sources. The approach maintains; it does not provide the resources to expand or cope with today's significant social and economic problems. In other words, public officials can use the approach to meet the next round of salary demands. Yet many of the public's demands will not be met, as there will be nothing left for lighting, curbs, library service, and recreational facilities. Therefore, for many of our cities, officials will have to take a multiple approach. The official will try to get outside funds, he will avoid the public and work with the small tax publics for small tax increases, and he will seek the support of the community for large revenue increases.

The multiple approach means that officials would have enough money to start the piecemeal rebuilding of our cities. For example, in Oakland, $20 millions a year in extra funds could significantly improve conditions. Such a sum is about one-third of the City of Oakland's 1970 budget and would provide the resources to improve traditional services, to innovate in the methods of delivering services, and to expand into new service areas. Of course, officials and citizens would have to be willing to increase their taxes, but $20 millions could be obtained adopting a municipal income tax ($12 millions), raising the property tax ($5 millions), and adding new, but painless, excises ($3 millions). Moreover, if tax sharing became a reality and officials received enough money to cover the usual annual increases in operating costs, then Oakland would be in fiscal shape to start its face lifting.

What the multiple approach means operationally is that public officials will have to become risk-takers. They may have to gamble on a local income tax, or bonds, or even raising the property tax rate. Unfortunately, high tax rates are often taken as an indicator of an inefficient administration and officials are embarrassed when their rates are the highest in the state. I am certainly sympathetic with the psychological hurdle that officials must surmount if rates are to be raised. But there is also a defeatism in waiting for things to get worse before they get better. The property tax rate can be raised without a crisis. If local officials doubt this, they have only to look back at experience in the last thirty years. The political limit on the property tax is primarily a psychological limit that can be overcome if local officials are willing to take action. Similarly, bonds can be passed. Citizens who remember the Depression and urge pay-as-you-go financing do not appreciate its limitations. Certainly pay-as-you-go financing can reduce the risk of reckless spending, save interest, and perhaps attract industry (Deran, 1961: 30-33). But at the same time, the postponement of capital improvements can pressure the tax rate by incurring increased maintenance and operating costs. Poor cities may find themselves in a vicious circle in which there is not enough slack in the budget to finance capital improvements, so they are deferred, maintenance goes up, and budgetary slack is further reduced. In such a situation, officials of cities with excellent bond ratings should certainly make the attempt to build coalitions to pass bonds.

CONCLUSION

Throughout my discussion, I have argued from two premises: (1) revenue maximization is a criterion of primary importance, and (2) revenue maximization is accomplished by local action. Obviously other criteria such as efficiency and equity should also be considered, and certainly there may indeed be viable alternatives to local action. It seems hardly any advice at all to suggest that what appears as a fiscal crisis is in reality a crisis of political leadership. Some officials would probably prefer a suggestion for a new tax resource which could easily gain public acceptance and bring in a great deal of revenue.

Although I have made a few suggestions, none of them is new. Most of the real options available to local officials have already been discussed for years. As long as we continue to define fiscal reality solely in economic and resource terms, innovative suggestions will not appear.

What do public officials need to know that social science research might be able to tell them? Specifically, we need to understand the conditions for citizen support of taxes. When will elites and the general public support revenue issues? If the conditions for tax support are haphazard and subject to the vagaries of the citizen's upset stomach, should referendums be eliminated? Do some citizens have a general propensity for supporting tax increases, or is such support grounded in specific policy issues? To answer these questions, we need quantitative measures of tax support which scale the scope and depth of citizen support. We also need to understand how the structure of the political system affects the kinds of taxes levied on its citizens. Do cities without a strong political organization rely more on indirect rather than direct forms of financing? Cities, lacking strong political structures, may not be able to pass an income tax ordinance while cities with strong mayors may be in a better position to do so. Or is the choice of tax tactics a function of political leadership rather than political structure? Aggressive leaders may prefer direct ways of coping with the public while more passive personalities may prefer indirect ways. At least, we could investigate whether officials are turned out of office after they have increased the tax burden. If the results were negative, then officials might not be so afraid of raising the property tax. In short, it is time for social scientists to subject myths surrounding municipal taxation policies to some empirical tests.

Politics must become a more significant part of the picture of fiscal reality. Currently, the politics of local revenue is mostly nonpolitics. It requires the replacement of acquiescence with concern and political skill by both leaders and citizens. As Harlan Cleveland (1961: 126) so aptly put it: "Local governments are not broke; they just think they are. Our metropolitan cities and suburbs are bankrupt, all right, but not in resources. They are bankrupt, most of them, in imagination, organization, leadership and will."

REFERENCES

ADAMS, R. F. (1967) "On the variation in the consumption of public services." In H. E. Brazer (ed.) Essays in State and Local Finance. Ann Arbor: Institute of Public Admistration, University of Michigan.

Advisory Commission on Intergovernmental Relations (1968) Sources of Increased State Tax Collections: Economic Growth vs. Political Choice. M-41. Washington, D.C.: (October).

ALDERFER, H. F. (1955) "Is authority financing the answer?" in Financing Metropolitan Government. Princeton: Tax Institute, Inc.

BAUER, R. A., I. de S. POOL and L. A. DEXTER (1968) American Business and Public Policy. New York: Atherton Press.

BERLE, A. A., Jr. (1960) "Reflections on financing governmental functions of the metropolis." Proceedings of the Academy of Political Science XXVII. New York: Columbia University (May): 66-79.

BIRDSALL, W. C. (1965) "A study of the demand for public goods." In R. A. Musgrave (ed.) Essays in Fiscal Federalism. Washington, D.C.: Brookings Institution.

BOLLENS, J. C. et al. (1961) Exploring the Metropolitan Community. Berkeley: University of California Press.

BOSKOFF, A. and H. ZEIGLER (1964) Voting Patterns in a Local Election. Philadelphia: J. B. Lippincott.

BRAZER, H. E. (1967) "The federal government and state-local finances." National Tax Journal XX (June): 155-164.

––– (1964) "Some fiscal implications of metropolitanism." In B. Chinitz (ed.) City and Suburb: The Economics of Metropolitan Growth. Englewood Cliffs, N.J.: Prentice-Hall.

––– (1959) City Expenditures in the United States. Occasional Paper 66. New York: National Bureau of Economic Research.

BREAK, G. F. (1967) Intergovernmental Fiscal Relations in the United States. Washington, D.C.: Brookings Institution.

BROWNLEE, O. H. (1961) "User prices vs. taxes." In Public Finances: Needs, Sources, and Utilization. National Bureau of Economic Research. Princeton: Princeton University Press.

BUCHANAN, J. M. (1967) Public Finance in Democratic Process: Fiscal Institutions and Individual Choice. Chapel Hill: University of North Carolina Press.

CAMPBELL, A. K. and S. SACKS (1967) Metropolitan America: Fiscal Patterns and Governmental Systems. New York: Free Press.

CARTER, R. F. and W. G. SAVARD (1961) Influence of Voter Turnout on School Bond and Tax Elections. U.S. Department of Health, Education, and Welfare. Cooperative Research Monograph Number 5. Washington, D.C.: Government Printing Office.

CLEVELAND, H. (1961) "Are the cities broke?" National Civic Review L (March).

Committee for Economic Development (1966) Modernizing Local Government: To Secure a Balanced Federalism. New York. July.

DAVID, E. L. (1967) "Public preferences and state-local taxes." In H. E. Brazer (ed.) Essays in State and Local Finance. Ann Arbor: Institute of Public Administration, University of Michigan.

DAVIS, O. A. and G. H. HAINES, Jr. (1966) "A political approach to a theory of public expenditure: the case of municipalities." National Tax Journal XIX (September): 259-275.

DERAN, E. Y. (1961) Financing Capital Improvements: The "Pay-As-You-Go" Approach. Berkeley: Bureau of Public Administration, University of California.

DOWNS, A. (1960) "Why the government budget is too small in a democracy." World Politics XII (July).

FITCH, L. C. (1964) "Metropolitan financial problems." In B. Chinitz (ed.) City and Suburb: The Economics of Metropolitan Growth. Englewood Cliffs, N.J.: Prentice-Hall.

GALBRAITH, J. K. (1958) The Affluent Society. New York: New American Library.

GENSEMER, B. L., J. A. LEAN and W. B. NEENAN (1965) "Awareness of marginal income tax rates among high-income taxpayers." National Tax Journal XVIII (September): 258-267.

HELLER, W. W. (1968) "A sympathetic reappraisal of revenue sharing." In H. F. Perloff and R. P. Nathan (eds.) Revenue Sharing and the City. Baltimore: Johns Hopkins Press for Resources for the Future, Inc.

HORTON, J. E. and W. E. THOMPSON (1962) "Powerlessness and political negativism: a study of defeated local referendums." American Journal of Sociology LXVII (March): 485-493.

JANOWITZ, M., D. WRIGHT and W. DELANEY (1958) Public Administration and the Public—Perspectives Toward Government in a Metropolitan Community. Ann Arbor: Institute of Public Administration, University of Michigan.

KEY, V. O., Jr. (1961) Public Opinion and American Democracy. New York: Alfred A. Knopf.

League of California Cities (1966) Statement of Principles and Alternatives Regarding Tax Reform. October.

LINEBERRY, R. L. and E. P. FOWLER (1967) "Reformism and public policies in American cities." American Political Science Review LXI (September).

MARGOLIS, J. (1961) "Metropolitan finance problems: territories, functions, and growth." In Public Finances: Needs, Sources, and Utilization. National Bureau of Economic Research. Princeton: Princeton University Press.

MAXWELL, J. A. (1965) Financing State and Local Governments. Washington, D.C.: Brookings Institution.

MUELLER, E. (1963) "Public attitudes toward fiscal programs." Quarterly Journal of Economics LXXVII (May).

Municipal Finance Officers Association (1967) "National conference on local government fiscal policy." Municipal Finance XXXIX (February): 93-96.

MUSGRAVE, R. A. (1959) The Theory of Public Finance: A Study in Public Economy. New York: McGraw-Hill.

NETZER, D. (1968) "Federal, state, and local finance in a metropolitan context." In H. S. Perloff and L. Wingo, Jr. (eds.) Issues in Urban Economics. Baltimore: Johns Hopkins Press.

Newsweek (May 26, 1969): 68.

PEACOCK, A. T. and J. WISEMAN (1961) The Growth of Public Expenditure in the United Kingdom. Princeton: Princeton University Press.

PECHMAN, J. A. (1965) Financing State and Local Government. Reprint 103. Washington, D.C.: Brookings Institution.

Research Advisory Committee (1966) Twin Cities Metropolitan Tax Study: Recommendations of the Research Advisory Committee. Minneapolis. December.

RUGGLES, R. (1968) "The federal government and federalism." In H. S. Perloff and R. P. Nathan (eds.) Revenue Sharing and the City. Baltimore: Johns Hopkins Press for Resources for the Future, Inc.

SCHEFFER, W. F. (1962) "Problems in municipal finance." Western Political Quarterly XV (September).

Southern California Research Council (n.d.) Taxation by Local Government. Report 13: Los Angeles County: A Case Study.

STOCKFISCH, J. A. (1968) "The outlook for fees and service charges as a source of revenue for state and local governments." In 1967 Proceedings of the Sixtieth Annual Conference on Taxation. Columbus: National Tax Association.

——— (1960) "Fees and service charges as a source of city revenue: a case study of Los Angeles." National Tax Journal XIII (June).

STONE, C. N. (1965) "Local referendums: an alternative to the alienated-voter model." Public Opinion Quarterly XXIX, 2: 213-222.

Tax Foundation, Inc. (1966) Fiscal Outlook for State and Local Government to 1975. Research Publication Number 6. New York: Tax Foundation, Inc.

Tax Institute, Inc. (1955) Financing Metropolitan Government. Princeton.

TEMPLETON, F. (1966) "Alienation and political participation." Public Opinion Quarterly XXX, 2: 249-261.

THOMPSON, W. (1968) "The city as a distorted price system." Psychology Today II (August).

——— (1965) A Preface to Urban Economics. Baltimore: Johns Hopkins Press for Resources for the Future, Inc.

VANDERMEULEN, A. J. (1964) "Reform of a state fee structure: principles, pitfalls, and proposals for increasing revenue." National Tax Journal XVII (December).

VICKERY, W. W. (1963) "General and specific financing of urban services." In H. G. Schaller (ed.) Public Expenditure Decisions in the Urban Community. Washington, D.C.: Resources for the Future, Inc.

Wall Street Journal (May 21, 1969): 1.

WILSON, J. Q. and E. C. BANFIELD (1964) "Public-regardingness as a value premise in voting behavior." American Political Science Review LXIII (December).

WOOD, R. C. (1961) 1400 Governments: The Political Economy of the New York Metropolitan Region. Cambridge: Harvard University Press.

5

Metropolitan Disparities and Fiscal Federalism

JOHN RIEW

☐ THE FINANCIAL NEEDS of states and their local subdivisions, long stressed by fiscal economists, are now widely recognized. Major legislative actions aimed at assisting state and local governments seem likely when the federal budget is relieved of its current burden of war expenditures. Such a conjecture is strengthened by the latest move of the administration proposing a modest but specific action on revenue sharing. Intensive discussions indeed have been underway for some time concerning the type of actions to be taken (see U.S. ACIR Volume 1: 1967, Appendix B, Table B-2, 312-318). The proposals advanced thus far, however, fail to give adequate consideration to the effects they would have on fiscal structures within metropolitan areas. Serious financing problems confront most of America's major urban areas, but the problems are not

AUTHOR'S NOTE: *The author is Associate Professor of Economics at the Pennsylvania University. He is indebted for valuable comments and suggestions to Professors Harold M. Goves, Martin Bronfenbrenner, Edward C. Budd, Warren Robinson, and Ned Shilling. The cooperation and assistance of Professor Kenneth Quindry, formerly the research director of the Wisconsin Tax Department, is also gratefully acknowledged.*

uniform within these areas. Failure to understand the intraurban or intrametropolitan fiscal pattern and the effect of various sharing (or crediting) schemes on the latter clearly will leave unsettled some of the most crucial problems. This paper attempts to bring into focus the current pattern of fiscal disparities among metropolitan-area municipalities and to discuss some of its policy implications.

In the interest of an in-depth study, we have chosen Milwaukee, a specific area, as the object of analysis. One of the major metropolitan areas in the United States, it shares with others many similar. social and economic problems. The main advantage of the choice is that Wisconsin offers unusually good sources of information for our purpose. The state has reliable equalized values of property, and this greatly facilitates comparisons of tax burden. Complete expenditure data are reported annually by all municipalities and school districts. It has income tax returns available for research purposes. Our data in most cases not only enable cross-sectional analysis of the disparities, but provide their recent trends.

In the present study, we shall define our metropolitan area synonymously with the urban area of Milwaukee as defined by the Census Bureau. The Standard Metropolitan Statistical Area, a possible alternative, contains large rural segments which are only remotely related economically with the main body of the metropolis. Thus, the Milwaukee metropolitan area under our survey consists of nineteen municipalities of Milwaukee County and ten incorporated municipalities of adjacent counties of Waukesha and Ozaukee. The twenty-nine municipalities are then grouped by function and by income class. Variation in income is more significant among residential units; hence, classification by income is limited to the residential category. Included in our survey area, thus, are the central city of Milwaukee, five balanced municipalities, two industrial municipalities, and twenty-one residential municipalities —the last being further divided into six high-income, seven medium-income, and eight low-income units (see Appendix Table A-1 for average size of population, land area, and population density for each class of municipalities). Outside the central city, the functional classification is based on the ratio of business property to total property within each municipality. Thus, industrial municipalities are those in which total equalized value of business real properties (mercantile, manufacturing, and utility) constituted more than sixty percent of total real property in 1966, while balanced municipalities are those in which this ratio was between forty and sixty percent. Classification by income of residential municipalities was made on the basis of the difference in the average 1965 adjusted gross income per tax return for each municipality (the tax return for the state income tax—see the footnote to Table 5).

METROPOLITAN FISCAL DISPARITIES

TAXES

Local tax revenue in the United States consists mainly of property taxes. In 1966, the total amount collected in property taxes was $23.8 billion, 87.1 percent of the total local tax revenue. The weight of the property tax is even greater in Wisconsin. In that state, the *general property tax* is practically the only tax levied by the local government. Thus, the tax rate typically has been high in most of the municipalities.

In property tax rate, Table 1 shows the central city stands distinctly above all other groups in municipalities. The full value rate for the central city in 1966 was 40.7 mills, far above the suburban average of 28.4 mills. It reveals also that the spread between the city and each of the suburban groups has increased during the ten years, 1957-1966. During this period, the central city's tax rate rose by forty-eight percent as compared with the suburbs' thirty-one percent. Data not presented here show that the city-suburb disparities grew even larger during the more recent period, 1960-1966; the six-year increase of twenty-eight percent for the central city is about twice the rate of increase for most of the suburban groups.

The property tax in Wisconsin is a composite of the local, school, and county levies (the state levy is negligible and disregarded here). In the combined local and school levy for 1966, the central city has the highest rate of 30.3 mills (see Table 2). This compares with others which range from 17.7 mills (industrial units) to 22.0 mills (low-income residential units). The disadvantage of the city involves mainly the low local levy. Its 13.8-mill rate towers over all suburban groups, among whom the levy varies within the low range of 1.7 to 4.2 mills. The relatively high school levy, on the other

TABLE 1

MEAN FULL-VALUE PROPERTY TAX RATES AND
THEIR INCREASE AMONG TYPES OF MUNICIPALITIES
1957-1966

	Full-Value Tax Rate 1957	1966	Increase[a] 1957-1966
	(in mills)		(percentage)
Central City	27.5	40.7	48.0
Balanced	22.1	29.5	29.8
Industrial	20.3	24.3	22.6
Residential			
High-income	21.9	29.4	34.6
Medium-income	22.2	28.4	29.4
Low-income	20.9	27.9	33.9
Urban Area Suburb	21.6	28.4	31.3
(median)	21.5	26.1	31.1

SOURCES: *City Taxes, Village Taxes,* and *County Clerk's Abstract of Assessments and Taxes,* State Department of Revenue, Madison, Wisconsin.

[a] The figures shown are the mean of the percentage increases for individual municipalities in each group.

TABLE 2

MEAN FULL-VALUE PROPERTY TAX RATE BY PURPOSE
AMONG TYPES OF MUNICIPALITIES 1966

	Local	School	Local and School
	(in mills)		
Central City	13.8	16.5	30.3
Balanced	4.2	15.9	20.1
Industrial	1.7	16.0	17.7
Residential			
High-income	2.2	18.3	20.5
Medium-income	1.9	19.0	20.9
Low-income	3.2	18.8	22.0
Urban Area Suburb	2.7	18.0	20.7
(median)	2.1	18.8	20.3[a]

SOURCES: Same as Table 1

[a] This is the median of the *combined rates* for all of our urban-area municipalities and does not necessarily coincide with the sum of the medians of local rates and school rates.

hand, varies from 15.9 to 19.0 mills. This smaller rate variation and the dominance of school expenditures in the local budget make the school levy an equalizing element in the overall property taxation. The relative picture in 1957, our data show, was largely similar. But the city-suburb disparity in the combined rates has increased considerably during the subsequent ten years, attributable mainly to the increased spread in the local levy.

In county levy, Milwaukee County has always been high. There, the levy in 1966 was 10.2 mills; in contrast, the adjacent counties of Waukesha and Ozaukee levied lower rates of 3.7 and 3.3 mills. Milwaukee County spends heavily for *charities and corrections, health, recreation,* and *highways.* If larger outlays in these services are incurred as a necessary part of the metropolitan economic life, responsibility for financing these functions should not be, it seems, placed solely upon one segment (though the major one) of the area. Tax distribution of the county levy, as it is, involves an inequity. While low-income cities of Milwaukee County pay 10.2 mills, some well-to-do municipalities outside the county, but as integral a part of the metropolitan area as the former, pay their own counties' low rates of less than four mills. It may be argued that some of the county services, charities, for instance, are exclusively for the benefit of the county residents. On the other hand, such items as highways (especially limited access highways) typically offer greater benefits to those in the outlying urban communities. In such a case, the latter are genuine beneficiaries who do not share in the cost.

The coefficient of variation which measures the overall variation in tax rate among all of our urban-area municipalities shows a slight increase, from 13.8 in 1960 to 16.2 in 1966. (The coefficient of variation—standard deviation as a percentage of the mean—is a more appropriate measure than standard deviation for our purposes here, since, as shown in Table 3, means have been changing.) What deserves great attention from the foregoing exposition is the fact that the central city faces a deepening fiscal strain and has been forced to increase its relative tax effort in recent years (see Table 3). That this is not unique to Milwaukee but rather typical of the major urban areas of the United States is revealed in the latest Advisory Commission on Intergovernmental Revenues (ACIR) study of the thirty-seven largest metropolitan areas (U.S. ACIR Volume 2, 1967: 4-5 and Appendix D).

TABLE 3

FULL-VALUE PROPERTY TAX RATES
CENTRAL CITY, URBAN-AREA SUBURBS, 1957-1966

	Central City	Suburbs[a]	CC as Percentage of Suburbs
	(in mills)		
1957	27.5	21.6	127.3
1960	31.9	25.2	126.7
1963	36.9	27.5	134.2
1966	40.7	28.4	143.3

SOURCES: Same as Table 1.

[a] The mean of the full-value rates for individual municipalities.

TABLE 4

MEAN PER CAPITA EQUALIZED PROPERTY VALUES[a]
AMONG TYPES OF MUNICIPALITIES, 1966

	Per Capita Property Values	
	(in dollars)	*(index)*
Central City	5,454	100.0
Balanced	8,982	164.7
Industrial	25,239	462.8
Residential		
High-income	10,937	200.5
Medium-income	7,180	131.6
Low-income	5,799	106.3
Urban Area Suburb	9,219	169.0
(median)	7,272	133.3

SOURCE: Supervisor of Assessments, *Statistical Report of Property Values, 1966* to State Department of Revenue, Madison, Wisconsin.

[a] The 1966 population data for municipalities (needed for per capita figures) were interpolated from the 1963 and 1970 estimates by the South Western Wisconsin Planning Commission. The Commission obtained the 1963 estimates from their origin-and-destination travel surveys. The 1970 estimates were based upon the development of each civil division in accordance with the staged land use pattern proposed in the adopted regional land use plan. Our figures include taxable real and personal properties. Appendix Table A-2 shows the property valuations classified by type and use.

FISCAL RESOURCES

Main tax bases (including potential ones) for local government are property, income, and sales. Because of data availability and the likelihood that a sales tax at the local level has serious problems for business migration, we shall concentrate on property and income.

As for property, the decisive advantage, Table 4 shows, goes to the industrial municipalities. Their per capita valuation of $25,237 in 1966 is more than twice that of the high-income residential units. The central city, the lowest of all ($5,454) and the low-income residential units ($5,799) both carry an amount about one-half that of the high-income residential units. In between are the balanced units and the medium-income residential units, with their per capita values of $8,982 and $7,180.

Between 1960 and 1966, all types of municipalities have experienced an increase in their per capita valuation, at least in current dollars. The rise is substantial for residential classes. The high-, medium-, and low-income residential units respectively gained sixteen, fifteen, and thirteen percent, while the balanced and industrial units gained less rapidly, nine and seven percent. The central city showed a token gain of 3.8 percent. Considering the price rise, the city could not even hold its own in per capita valuation during these years of the sixties. Weighing heavily in this phenomenon are sluggish demand in the real estate market and continuous absorption of private properties into public uses in the central city. Building permits in the city, for instance, declined steadily from 1,283 units (1,116 single-family homes and 292 duplexes) in 1960 to 610 units (466 singles and 144 duplexes) in 1966. As for erosion of taxable properties, the property values taken off the tax roll *each year* during the 1960-1966 period averaged approximately $4.9 million in assessed valuation (assessment ratio varied from 52.5 to 54.6 percent during the period), of which nearly sixty percent represented the use for expressways. (Sources: Annual Report of the Department of Building Inspection and Safety Engineering, the City of Milwaukee and the records available at the Tax Commissioner's Office, the City of Milwaukee.)

Thus, while the central city in 1960 exceeded the per capita property values in low-income residential municipalities, the order is now reversed. In 1960, furthermore, the per capita valuation of the central city was fifty-five and eighty-two percent of the per capita valuation for the high- and medium-income residential municipalities; figures are now down to fifty and seventy-four percent respectively. Variation in per capita valuation among all urban area municipalities remained stable in recent years (the coefficient of variation being 50.2 for 1960 and 49.7 for 1966).

When the local government relies mainly on property as its tax base, income as a fiscal resource may seem irrelevant. But it may be looked upon as an important potential tax base. Naturally, the most privileged in income (see Table 5) are the high-income residential municipalities. Their mean income (noted on tax returns) of $12,273 is over two and one-half times that of the central city and more than twice that of most other groups. The property-rich industrial municipality types now reveal themselves as income-poor. The middle-income residential municipalities, though not comparable to the high-income ones, stand distinctly above the others (Appendix Table A-3 shows ranges and medians of the adjusted gross income for each class of municipalities).

TABLE 5

MEAN ADJUSTED GROSS INCOME PER TAX RETURN AND PER CAPITA[a] AMONG TYPES OF MUNICIPALITIES, 1965

	Income Per Tax Return		Income Per Capita	
	(dollars)	(index)	(dollars)	(index)
Central City	4,697	100.0	2,387	100.0
Balanced	5,532	117.8	2,446	102.5
Industrial	5,112	108.8	2,046	85.7
Residential				
High-income	12,273	261.3	7,180	300.8
Medium-income	6,642	141.4	3,146	131.8
Low-income	5,436	115.7	2,400	104.3
Urban Area Suburb	7,197	153.2	3,606	151.0
(median)	6.049	128.8	2,681	112.3

SOURCE: Computed from the figures obtained in IBM files, STate Department of Revenue, Madison, Wisconsin.

[a] This is adjusted gross income (AGI) for the state income tax and differs little from the federal adjusted gross income. In 1966, the AGI for state income tax purposes was $7,849.5 million and varied only .35 percent from the federal AGI of $7,877.0 million.

Compared in *per capita* terms, the income differential between the central city and high-income residential municipalities is larger, a reflection of a smaller family size for such municipalities. When we take income as a measure of family welfare or affluence, we would want to consider the income of the family as well as its needs. Thus, if basic needs are determined largely by the size of the family, income per capita rather than income per tax return or per family may be preferable (see David: 1959, 393-399). On the other hand, the reduced spreads in per capita income between the central city and the balanced and low-income residential municipalities suggest that the family size is generally larger for the latter groups. (Since the number of persons gainfully employed per family varies among municipalities, the more accurate account can be made by considering this factor also. Thus, the lesser spread in per capita income between the central city and the low-income residential units, for instance, may be due to larger average family size for the latter, or their lower ratio of gainfully employed per family, or both.)

Disparities in income between the city and the suburban groups, with exception of the industrial municipalities, have all increased, absolutely and relatively between 1959 and 1965. The coefficient of variation in per capita income among all urban area municipalities, moreover, rose from 62.5 to 69.1 during this period. Burkhead (1961) some years ago hypothesized a condition of increasing fiscal "homogeneity" with respect to such characteristics as tax effort, ability, and expenditures for governmental units within the major metropolitan areas of the United States and believed that his empirical study of the Cleveland area supported his hypothesis. This writer argued then that such a trend may be only transitory and that what happens after a metropolis (delimited at a given time) becomes more urbanized is uncertain (see Burkhead, 1961: 337-348; Riew, 1962: 218-220).

EXPENDITURES

The expenditures are treated separately for general (nonschool) and school purposes. Capital outlays, approximately one-fourth of total expenditures in recent years, are not included in the present analysis.

In general municipal services, the central city spends substantially more per capita than the large majority of the suburbs, ranking eighth among the twenty-nine municipalities. In terms of specific functions, the central city greatly exceeds the suburbs in *police, health and sanitation,* and *fire* (see Appendix Table A-4). In fact, these are the three largest expenditure items for the central city; they respectively claimed thirty-one, twenty, and sixteen percent of the total operating budget. The suburbs, on the other hand, exceed the central city in highway and general government;

TABLE 6

MEAN PER CAPITA OPERATING EXPENDITURES[a]
AMONG TYPES OF MUNICIPALITIES, 1966

	Per Capita Operating Expenditures	
	(dollar)	*(index)*
Central City	*81.3*	*100.0*
Balanced	62.5	77.0
Industrial	142.9	175.9
Residential		
High-income	118.5	145.9
Medium-income	54.3	66.8
Low-income	47.9	58.9
Urban Area Suburb	*72.5*	*89.2*
(median)	63.5	78.1

SOURCE: *Annual Report of Municipal Clerk, 1966* to State Bureau of Municipal Audit, Madison, Wisconsin.

[a] For breakdowns of expenditures by function, see Appendix Table A-4.

for the suburbs, highway, police, and general government constituted the three most important items, claiming twenty-eight, twenty-one, and nineteen percent of their operating budgets respectively.

We find a similar picture when expenditures are compared in terms of property valuation. The largest differences between the central city and the suburbs, our data indicate, are again in police, health and sanitation, and fire. The latter seem to be the areas of service in which high densities of population and property, greater concentration of business and perhaps significantly the nonresident commuters, all impose additional burden on the central city (see Appendix Table A-1). An earlier study by Brazer (1959) reveals that the density of population and the ratio of the core city's population to that of the metropolitan area are associated to a substantial extent with the level of per capita expenditures of the city. (For similar studies, see Hawley, 1951a; Margolis, 1961.)

In per capita terms, the largest spenders are the industrial and high-income residential units. They greatly exceed the central city, especially in highway, police, and general government. Margolis' (1957) earlier study of the San Francisco Bay area showed somewhat different results in that the high-value "dormitory" municipalities spent substantially less per capita than the central city, but he divided the dormitory (residential) communities between high and low values whereas this study divided them into three categories, high-, medium-, and low-income groups. That the central city spends less per capita in general government may be explained largely by economies of scale. The higher per capita figures on highway and police for the industrial and high-income residential groups conform to the results of Hirsch's study of the St. Louis area. His analysis indicates that variations in per capita expenditures on both highway and police are highly correlated with property valuations (see Hirsch, 1961). Expenditure data involving the Cleveland area point to a generally similar pattern (Sacks and Helmuth, 1961: 68-79).

Since 1960, the level of expenditures per capita has increased for all classes of municipalities, and most significantly for residential municipalities. Differences in per capita expenditures between the central city and the more well-to-do municipalities have increased, the latter showing higher rates of increase. The increased differences are also observed between the high-income and the low-income residential groups.

Analysis of school expenditures involves the problem that a school district and a municipality do not necessarily share common boundaries. The report of a school district pertains to the district with no breakdown by municipality. Thus, where a joint school district is involved, expenditures and enrollment were apportioned among municipalities according to their relative shares of population aged one through eighteen in the school district.

The central city, although leading in general expenditures, falls greatly behind the suburbs in school expenditures. With its per pupil operating expenditures of $508 (Table 7), the central city of 1965-1966 ranked twenty-eighth among the twenty-nine municipalities in the urban area of Milwaukee. The high-income residential suburbs, with their figure of $726 exceeded all others by significant margins. The balanced, industrial, and medium-income residential suburbs spent comparable amounts in the $630-640 range while the low-income residential suburbs, trailing others, still exceeded the central city.

TABLE 7

MEAN PER PUPIL AND PER CAPITA SCHOOL OPERATING EXPENDITURES AMONG TYPES OF MUNICIPALITIES, 1965-1966

	Per Pupil Expenditures		Per Capita Expenditures	
	(dollars)		*(dollars)*	
Central City	*508*	*100.0*	*70*	*100.0*
Balanced	636	125.2	103	147.6
Industrial	630	124.0	104	149.0
Residential				
High-income	762	150.0	149	212.9
Medium-income	642	126.4	121	173.3
Low-income	578	113.8	115	164.7
Urban Area Suburb	*648*	*127.6*	*121*	*173.3*
(median)	631	124.2	118	169.0

SOURCE: *Annual Report of School District, 1965-1966* to State Superintendent of Public Instruction, Part I, II, and IV, Madison, Wisconsin.

To be noted also in Table 7 is the great difference in the expenditure variation between per pupil and per capita terms. The suburbs on the average exceeded the central city by twenty-eight percent in *per pupil* expenditures, but in *per capita* expenditures, the suburb exceeded the city by seventy-three percent.

The main explanation for this lies in demographic differences. The central city has by far the smaller share of school age population which in turn explains its lower ratio of pupil enrollment to total population. In 1965-1966, the ratio of the population aged one through eighteen to the total population was thirty-three percent for the central city in contrast to forty-two percent for the suburbs and the pupil-population ratios in that year were 13.7 and 19.1 percent for the city and the suburbs respectively (see Appendix Table A-5). The ratio of pupil enrollment to population may also be influenced by other factors, such as parochial enrollment and student dropouts. To separate out these aspects, the ratios of pupil enrollment to school age population (more precisely the population aged one through eighteen) are compared among municipality groups. Although the central city and the balanced communities appear to have somewhat larger shares of parochial enrollment and possibly of dropouts also, differences in these characteristics, as we may note in Appendix Table A-5, are not significant in affecting this pupil-population ratio.

In 1959-1960, the central city was the lowest in school expenditures, the high-income residential by far the highest, and the low-income residential relatively low but still higher than the central city. The coefficient of variation in per pupil expenditures among all individual municipalities, however, decreased from 19.6 in 1959-1960 to 15.1 in 1965-1966. The spread in per pupil expenditures between the high-income residential suburbs and other less affluent suburbs has decreased relatively. This trend toward some equalization in school expenditures may be attributable mainly to the demonstration effect among the urban school districts; the increase in school levy, our data show, has been greater for the less affluent communities in recent years.

In general expenditures, the central city spends more per capita than the suburbs—by about eleven percent. On the other hand, the city spends less in education and this reverse city-suburb difference is much greater than the difference in general expenditures; the suburbs on the average exceed the city by about twenty-eight percent in per pupil operating expenditures. A very similar finding concerning the city-suburb comparisons of general and school expenditures may be noted also in a recent study of the Philadelphia area (Epps, 1968: 15-18). The U.S. Advisory Committee on Intergovernmental Relations (1967: 64-75) study, cited earlier, furthermore suggests that the picture depicted above applies, by and large, to most of the nation's major urban areas.

At this point, a comment on the meaning of expenditure variation is in order. The meaning of tax disparities is clear enough. A lower tax rate represents a lower burden and, other things being equal, is preferable. Rather obvious also is the meaning of resource disparities. A high property valuation or income, say per capita, would indicate a greater fiscal capability, although in the case of the former, different types of property may entail different costs in public services. Interpretation of expenditure disparities, however, is confounded by differences in environmental conditions. Here, the latter become an important variable in the cost function.

This is especially so with general expenditures. If a community X spends twice as much as community Y in police or fire, for instance, it does not necessarily mean that the former provides its residents twice as much protection from crimes or fire hazards as does the latter. The environmental differences also affect the cost of operating public schools, but with the latter, the environmental factor probably is relatively insignificant. More significant environmental differences, as far as public schools are concerned, would be population density and the associated differences in the transportation cost. A recent study, however, reveals that differences in the average transportation cost between school districts covering a large area and those covering a small area are considerably less than commonly believed, due largely to the fact that the fixed costs—depreciation, driver salary, insurance, garage rental, etc., comprise the major part of the total cost (Riew, 1966: 286). As for education, the level of expenditures, we may say, can be regarded as a fairly reasonable indicator of the quality or standard of education. That the city is long in general expenditures while short in school outlays thus must be noted in the light of these considerations.

SUMMARY AND IMPLICATIONS

It goes without saying that no single metropolitan area represents all others. There are, however, some important problems today which confront most major metropolitan areas. Thus, certain findings and inferences drawn from a study of one area may be useful in the solution of the problems facing others.

The present study reveals that there exist significant fiscal disparities among the municipalities of the Milwaukee metropolitan area. Taxes, fiscal resources, and expenditures vary greatly between the central city and the suburbs, as well as among the suburban municipalities.

The *tax rate* disparity between the central city and the suburbs has increased significantly during the past several years; among suburban municipalities the variation in tax rate has shown mild increase. In *property valuation* also, the disparity between the city and the suburbs increased somewhat, while among suburban municipalities the variation in per capita property valuation remained generally stable. In per capita *income,* disparities have increased between the central city and the suburb as well as among the suburbs. Although some lessening of disparities in school *expenditures* is noted, no similar evidence is indicated in nonschool expenditures.

While the present analysis of fiscal disparities involves comparisons among all groups of our urban area municipalities, it clearly reveals that the more important disparities exist between the central city and the suburbs. Among groups of municipalities, the central city ranks the lowest in per capita property valuation and family income while ranking by far the highest in property tax rate. In public education, the central city spends the lowest amount per pupil, whereas in general expenditures per capita, it ranks high only next to the industrial and high-income residential suburbs. All aspects considered, the situation with regard to the central city financing is serious indeed, and the prospects are even grimmer. (Simple extrapolations of revenue and expenditures based on seven years prior to 1967 suggest that the city's budget in 1975, under the present tax structure, would require an increase in the full-value property tax rate by 13.1 mills over the 1966 rate of 40.7 mills. By comparison, the city of Wauwotosa, a medium-income suburb, whose 1966 full-value rate was 26.3 mills, is expected to face a smaller increase of 8.1 mills. Provided that the state fund for the recently enacted property tax credits rises continually to maintain the present level of relief [relative to local taxes], the effective rise in the rate with the 13.1-mill increase will be 10.3 mills. Even so, the central city's tax rate would become much higher and its relative tax disadvantage would increase considerably.)

The property tax in the central city is overworked; the statistics bear out that the present rate is extremely high, absolutely and relatively. This involves the question of equity. The concern here stems largely from the notion that municipalities which make up an economically integrated community should share the burden of financing public services of common need. If the central city and the activities therein are an integral part (more likely the major part) of the metropolitan economic life and it is within this context that the suburban municipalities function and exist, the central city's tax rate which exceeds the suburbs' by great margins raises a question of fiscal inequity. Related to the issue, Harold Groves (1967) appropriately states: "Differences in economic capacity among cities are as common and inevitable perhaps as those among people. We bear with these differences in large degree because we regard them as a necessary feature of a free society. But when what is in some sense an organic unity is artificially divided to create rich and poor units of government, the legitimacy of the product invites skepticism." Tiebout's (1956) defense (or advocacy) of a fractionalized metropolis in the name of the maximum welfare of consumer-voters (the consumer-voters being offered a wide range of choice among municipalities with different sets of revenue-expenditure patterns) fails to tackle this basic issue of equity involving the property taxation at the local level which asserts an extraterritoriality. A high property tax in the central city, many agree, is a serious hindrance to housing development and other construction activities in the city, and is an important factor contributing to governmental splintering in a metropolitan area. (For further discussion on the issues of equity, efficiency, and other effects of the property tax, see Netzer, 1966 and elsewhere in this volume; Brazer, 1962).

Increase in municipal charges and fees may be explored but the possibilities here are limited. While greater efforts may be exerted for active assistance from the state and the federal government, the city must seek its own way out. The more promising long-run solution seems to lie in a tax reform. The use of a nonproperty tax, especially a municipal income tax (municipal income tax, unlike municipal sales taxes, would not pose serious problems of business migration from the central city), coupled with generous tax credits by higher levels of government, is particularly desirable.

The tax credit heretofore has been discussed largely in the context of the federal-state coordination.[1] A federal tax credit of a local (municipal) income tax, *levied at the source of earnings*, the author believes, deserves serious consideration. Under such a scheme, some portion of the taxes paid to local government would be deducted from, hence, serve as a credit for, a man's federal tax bill. With increased mobility and advanced communication, a local or a metropolitan economy today is becoming more and more an integral part of the national economy. Thus, on economic grounds, such a fiscal arrangement may be justified. Since a large part of the problem for states involves their aid to local governments (transfers from state to local governments, in the aggregate, accounted for thirty-six percent of state general expenditures, according to the U.S. Bureau of the Census in 1967), such a program would also significantly help to solve the fiscal problems of the states.

A federal credit of municipal income taxes would thus lessen the need for the credit of state income taxes. Although one does not have to replace the other, if a choice must be made between the two or if both are to be used, the priority or greater weight, it seems, must be given to the direct federal credit of municipal taxes. Under the latter program, central cities with their pressing financial problems do not have to subject themselves to the mercy of state legislatures or incure the bureaucratic costs of having another unit of government involved in the transaction.

A comprehensive study by this writer of the system of state transfers in Wisconsin (a state proud of its progressive tradition with pioneering roles in social security, state income tax, etc.) reveals that the state-to-local transfers as a whole greatly aggravate the situation with regard to the city-suburb disparities as well as overall disparities among suburban municipalities of Milwaukee (U.S. ACIR, 1967: 304-315). This is not necessarily indicative but certainly suggestive of the difficulties associated with the Heller-Pechman Plan (see Heller, 1966; Pechman, 1965). To circumvent this problem, it was suggested more recently that the federal government could "broadly restrict use of the funds to education, health, welfare, and community development programs" (Heller, 1968; see also Heller and Pechman, 1967). One wonders how effective a measure such restrictions could be as the states shift their own funds from the areas listed above to other uses. ACIR's proposition for federal credit of state income tax, which Heller considers the next best to his own plan, is subject to a similar shortcoming involving possible discrepancies between the state' discretion and local needs.

With the federal credit of municipal income tax, levied at the source of earnings, the central city has much to gain. Great concentration of business properties (see Appendix Table A-2) and conglomeration of public (federal, state, and local) agencies and various nonprofit establishments would assure the city of a very promising tax base. The balanced and industrial municipalities would also gain to varying degrees. Such a program, on the other hand, would mean a definite disadvantage for residential municipalities; which comprise by far the great majority of the urban area municipalities—twenty-one out of twenty-nine in the case of Milwaukee.

As of mid-1967, more than 170 cities with 10,000 or more inhabitants, plus an uncountable number of smaller taxing entities, derived revenue from taxes levied on some form of income. Nearly three-fifths of the 170 cities levied a rate of one percent and only a few imposed a rate as high as two percent. For instance, Philadelphia, Newport (Kentucky), and Gadsden (Alabama) levy flat rates of two percent while New York and Baltimore use graduated schedule rising from 0.4 to two percent and one to 2.5 percent respectively (since mid-1967, the Baltimore tax is levied only on residents in the form of a fifty percent surtax on the graduated income tax of Maryland). (See Tax Foundation, Inc. *City Income Taxes,* New York, 1967, pp. 11-13 and Appendix A. p. 41.) Washington, D.C.'s graduated net income tax which ranges from two to six percent is more closely akin to a state tax than to the municipal income taxes. A fifty percent federal tax credit would enable the municipalities levying a two-percent income tax to raise the rate to four percent and those levying a one percent tax to two percent. Such a program would significantly improve the local fiscal balance. The Philadelphia income tax, for instance, with its two percent rate, yielded $119 million, forty-seven percent of the total city tax revenue for 1966-1967 (U.S. ACIR, 1968: 95-97). With the rate doubled, the revenue from the income tax would have doubled without increasing the total (federal and municipal) income taxes for the taxpayer and would have made up sixty-four percent of a larger total tax revenue for the city.

How generous the tax credit should be would depend in part on the financial ability of the federal government. But if we assume that the two percent rate is the realistic maximum for the inevitably competitive municipal income tax, a federal tax credit of forty to fifty percent would not be overly burdensome to the federal treasury. Ignoring the question of how soon a revenue sharing plan of any kind can become a reality, the program cited above, to say the least, would be less costly than the Heller proposal of an annual allotment of two percent of the federal income tax base. Even if all municipalities levied a four percent income tax (in which case the effective rate is two percent with a fifty percent credit), the total credits allowed would be less than two percent of the federal individual income tax base because: (1) municipal income tax bases are inevitably less extensive in coverage and (2) the creditable taxes levied at the source of earning, allow no doublecounting.

The proposed plan admittedly provides no equalization among states and, for that matter, not even within states. That this is necessarily a demerit, however, is to be debated. In this connection, one must resolve the question concerning the desirability of letting the more productive areas be hampered by their obligation to promote the social overhead capital in the less productive areas when it is yet to be learned whether the resource shifts involved bear positive returns in the relative sense. A. D. Scott (1950) earlier wrote of interstate transfers as providing "amenities to poor people in resource-poor states, a situation which may be undesirable in the long-run." He states (1950: 416-422), "The maximum income for the whole country and the highest average personal income . . . can be achieved only when resources and labor are combined in such a way that the marginal product of a similar unit of labor is the same in all places. An increase will result when labor is transferred from places where its marginal product is low to places where it is high." In rebuttal, Buchanan (1952) stresses a point that if labor is less productive in a given region, it is the result of a lack of entrepreneurial talent and capital, both public and private. Equalizing transfers, by increasing public capital and attracting private capital in poor regions, would tend to increase the efficiency and the income of labor in these regions. In his words, "attention has been concentrated on the mobility of one of the two basic productive factors, labor. A relative surplus of labor implies a relative shortage of capital resources, and the movement of capital is as important as that of labor in attaining long-run resource equilibrium." This reasoning leads to a basic question as to why in a competitive economy, certain regions, in spite of their relative labor surplus, fail to attract or to sustain the capital and whether equalizing transfers would not perpetuate inefficiencies rather than help attain the resource equilibrium. Recent literature on revenue sharing largely ignores the issue of resource allocation. Noteworthy among exceptions is George Break's treatment of the matter in his Intergovernmental Fiscal Relations in the United States (Brookings Institution, 1967: 144-148). Although parts of his discussions seem unconvincing, he does delve into various aspects related to the issue. In a dynamic economy characterized by rapid advances in technology, emerging of new resources and of new products, fast changing preferences, and so forth, an appropriate policy may be the one which emphasizes more improvement in the mobility of productive factors.

APPENDIX

TABLE A-1

MEAN POPULATION AND LAND AREA
POPULATION AND PROPERTY VALUATION PER SQUARE MILE
AMONG TYPES OF MUNICIPALITIES, 1966

Municipalities By Type	Number of Municipalities	Population	Land Area Square Mile	Population Per Square Mile	Property Values Per Square Mile
					(dollars)
Central City	1	783,758	91.1	8,603	47,189
Balanced	5	24,849	5.1	4,212	35,655
Industrial	2	9,334	15.2	2,374	71,936
Residential					
High-income	6	9,012	3.0	4,136	31,779
Medium-income	7	19,084	14.7	1,989	15,628
Low-income	8	19,188	16.6	2,281	14,081
Urban Area Suburb	28	17,288	11.0	2,956	26,245
(median)		12,599	5.5	2,444	18,738

SOURCE: For population estimates and property valuations, see the footnote and sources under Table 4, page 142. The square mileage figures for municipalities with 5,000 or more inhabitants were obtained from *Municipal Year Book*, 1961, Geographical Division, Bureau of the Census, pp. 85-140. For smaller municipalities, a polar planimeter was used to measure square mileage from the county maps available at the Wisconsin State Highway Commission.

TABLE A-2

MEAN PER CAPITA EQUALIZED PROPERTY BY TYPE AND USE AMONG TYPES OF MUNICIPALITIES, 1966

	Personal Property [a]	Real Property Mercantile	Real Property Manufacture	Real Property Residential	Total [b]
		(in dollars)			
Central City	1,002	1,074	896	2,474	5,454
Balanced	1,967	1,135	2,091	3,777	8,982
Industrial	7,741	1,116	13,110	3,048	25,239
Residential					
High-income	127	752	238	9,814	10,937
Medium-income	589	761	446	5,232	7,180
Low-income	725	531	596	3,745	5,799
Urban Area Suburb	1,284	787	984	5,373	9,212
(median)	607	666	180	4,558	7,272

SOURCE: Supervisor of Assessments, *Statistical Report of Property Values, 1966,* to State Department of Revenue, Madison, Wisconsin.

[a] Taxable personal properties consist mostly of business properties such as machinery, equipment, inventory, etc.

[b] Figures in this column include farm properties which still comprise a minor part of the taxable properties in a number of our urban-area municipalities.

[c] The mean figures for the urban-area suburb are significantly affected by the exceptionally high figures for the two industrial municipalities. Thus, the more meaningful averages for the suburb would be the median values. It may be noted again that the mean for the suburb is the unweighted average of the values for individual municipalities.

TABLE A-3

RANGES, MEDIANS OF ADJUSTED GROSS INCOME PER RETURN [a] AMONG TYPES OF MUNICIPALITIES, 1965

	Range	Median	
	(dollars)	*(dollars)*	*(index)*
Central City	4,697	4,697	100.0
Balanced	4,909 - 7,340	5,198	110.7
Industrial	4,803 - 5,421	5,112	108.8
Residential			
High-income	7,827 - 19,892	11,505	244.9
Medium-income	6,213 - 7,149	6,527	139.0
Low-income	4,859 - 5,885	5,438	115.8
Urban Area Suburb	4,803 - 19,892	6,049	128.8

SOURCE: Computed from data in the IBM files, State Department of Revenue, Madison, Wisconsin.

[a] This is the adjusted gross income for state income tax purposes and differs little from the federal adjusted gross income (see the footnote to Table 5).

TABLE A-4

MEAN PER CAPITA OPERATING EXPENDITURES BY FUNCTION AMONG TYPES OF MUNICIPALITIES, 1966

	General Government	Police	Fire	Health and Sanitation	Highway	Recreation	Total [a]	Total as % of CC
				(in dollars)				
Central City	11.9	24.8	13.1	16.4	11.9	3.2	81.3	100.0
Balanced	11.2	13.2	9.9	10.9	15.2	2.1	62.5	77.0
Industrial	25.0	29.1	38.7	21.5	25.7	3.0	142.9	175.9
Residential								
High-income	22.6	32.9	9.3	19.4	32.1	2.2	118.5	145.9
Medium-income	10.2	10.4	6.6	6.6	18.9	1.6	55.8	67.8
Low-income	9.8	9.2	6.0	4.6	15.0	3.2	47.9	58.9
Urban Area Suburb	13.6	16.7	9.5	10.6	20.0	2.4	72.5	89.2
(median)	11.4	13.4	5.9	9.1	15.5	2.1	63.5	71.9

SOURCE: *Annual Report of Municipal Clerk, 1966* to State Bureau of Municipal Audit, Madison, Wisconsin.

[a] The total is the sum of the individual items presented and does not include miscellaneous expenses.

TABLE A-5

MEAN RATIO OF POPULATION 1-18 AND ADA TO TOTAL POPULATION AMONG TYPES OF MUNICIPALITIES, 1965-1966

	Population 1-18 Total Population	Average Daily Attendance Total Population	Average Daily Attendance Population 1-18
	(percentage)		
Central City	32.5	13.7	42.2
Balanced	38.2	15.9	42.0
Industrial	34.3	15.8	44.7
Residential			
High-income	42.4	18.7	44.4
Medium-income	42.3	19.2	45.6
Low-income	44.9	18.5	44.4
Urban Area Suburb	41.8	19.1	44.3
(median)	40.9	19.1	44.9

SOURCE: Computed from data in *Annual Report of School District, 1965-1966,* to State Superintendent of Public Instruction, Part I, II, and IV, Madison, Wisconsin.

NOTE

1. See Nathan and Hollander (1959), Heller (1959), Maxwell (1962), Bridges (1965), Committee for Economic Development (1965). Advisory Commission on Intergovernmental Relations recommended a credit of state and local income taxes, but as the title suggests, the discussion was directed mainly to federal-state relationships. See U.S. Advisory Commission on Intergovernmental Relations (1965: 18-19).

REFERENCES

BRAZER, H. (1962) "Some fiscal implications of metropolitanism." Pp. 71-78 in G. S. Birkhead (ed.) Metropolitan Issues: Social, Governmental, Fiscal. Syracuse: Syracuse University Press.

——— (1959) City Expenditures in the United States. National Bureau of Economic Research, Occasional Paper 66.

BRIDGES, B., Jr. (1965) "Allowance for state and local nonbusiness taxes." Pp. 203-214 in R. S. Musgrave (ed.) Essays on Fiscal Federalism. Washington, D.C.: Brookings Institution.

BUCHANAN, J. A. (1952) "Federal grants and resources allocation." Journal of Political Economy: 208-217.

BURKHEAD, J. (1961) "Uniformity in governmental expenditures and resources in a metropolitan area—Cuyahoga County." National Tax Journal (December).

DAVID, M. (1959) "Welfare, income, and budget needs." Review of Economics and Statistics (November).

EPPS, R. W. (1968) The Metropolitan Money Gap. Business Review (June) Federal Reserve Bank of Philadelphia.

GROVES, H. (1967) "Innovation in tax-sharing—the Wisconsin experience." In Revenue Sharing and Its Alternatives: What Future for Fiscal Federalism? Joint Economic Committee, Ninetieth Congress, First Session, Volume I.

HAWLEY, A. H. (1951) "Metropolitan population and municipal government expenditures in central cities." Journal of Social Issues, 7: 100-108.

HELLER, W. W. (1968) "A sympathetic reappraisal of revenue sharing." In H. S. Perloff and R. P. Nathan (eds.) Revenue Sharing and the City. Baltimore: Johns Hopkins Press.

——— (1966) New Dimensions of Political Economy. Cambridge: Harvard University Press.

——— (1959) "Deductions and credits for state income taxes." Pp. 419-433 in Tax Revision Compendium, submitted to the House Ways and Means Committee, Eighty-sixth Congress, First Session, Volume I.

——— and J. A. PECHMAN (1967) "Questions and answers on revenue sharing." Pp. 111-117 in Revenue Sharing and Its Alternatives: What Future for Fiscal Federalism? Joint Economic Committee, Ninetieth Congress, First Session, Volume II.

HIRSCH, W. Z. (1961) "Factors affecting local governmental expenditure levels." Pp. 330-347 in J. C. Bohlens (ed.) Exploring the Metropolitan Community. Berkeley: University of California Press.

MAXWELL, J. A. (1962) Tax Credits and Intergovernmental Fiscal Relations. Washington, D.C.: Brookings Institution.

MARGOLIS, J. (1961) Public Finances: Needs, Sources and Utilization. Princeton: Princeton University Press.

––– (1957) "Munici-fiscal structure in a metropolitan region." Journal of Political Economy.

NATHAN, R. N. and E. D. HOLLANDER (1959) "The role of individual income taxes in federal-state-local fiscal relationships." Pp. 219-227 in Tax Revision Compendium, submitted to the House Ways and Means Committee, Eighty-sixth Congress, First Session, Volume I.

NETZER, D. (1966) Economics of Property Tax. Washington, D.C.: Brookings Institution.

PECHMAN, J. A. (1965) "Financing state and local governments." In American Bankers Association, Proceedings of a Symposium on Federal Taxation.

RIEW, J. (1966) "Economies of scale in high school operation." Review of Economics and Statistics (August).

––– (1962) National Tax Journal (June).

SACKS, S. and W. F. HELLMUTH, Jr. (1961) Financing Government in a Metropolitan Area. New York: Free Press.

SCOTT, A. D. (1950) "A note on grants to federal countries." Economica (November).

TIEBOUT, C. M. (1956) "A pure theory of local expenditures." Journal of Political Economy (October): 416-424.

U.S. Advisory Commission on Intergovernmental Relations [ACIR] (1968) State and Local Finances—Significant Features, 1966-1969. Washington, D.C.

––– (1965) Fiscal Balance in the American Federal System. Volumes 1 and 2 (October). Washington, D.C.

––– (1965) Federal-State Coordination of Personal Income Taxes. Washington, D.C.

U.S. Committee for Economic Development (1967) "A fiscal program for a balanced federalism." Pp. 1105-1107 in Revenue Sharing and Its Alternatives: What Future for Fiscal Federalism? Joint Economic Committee, Ninetieth Congress, First Session, Volume II.

6

Reflections on the Case for the Heller Plan

LYLE C. FITCH

☐ MY COMMENTS are addressed to what might be called the pure form of the "Heller Plan"—a per capita block grant to states, with no conditions or restrictions. Heller, recognizing some of the weaknesses of the "pure form," has suggested some modifications. My own inclination is to go further and in a different direction. I suggest that any general-purpose federal grant should be used, not solely as a revenue-sharing measure, but also as a means of stimulating improvement in state and local government organization and planning.

Persuasive as I find Evangelist Heller, I am not quite ready to accept the pure gospel which calls for distributing the national dividend in the form of a per capita grant to states, with no conditions or restrictions.

The essence of the issue, it seems to me, is in the public decision-making processes of allocating resources and whether the state and local governments, as now constituted, are the best decision makers we can command to dispose of the "national

EDITOR'S NOTE: *Reprinted from* Issues in Urban Economics, *edited by Harvey S. Perloff and Lowdon Wingo, Jr. Published for Resources for the Future, Inc. by the Johns Hopkins University Press. Copyright © 1968. Reprinted by permission of the author and the publisher.*

dividend." We have many alternative decision makers to chose from, including the federal government, whose decisions might call for direct expenditures or grants for specific purposes; taxpayers, whose decisions would be activated by tax reduction; the poor, assisted by some form of minimal income-guarantee program.

Certainly, state and local governments, beset by pressures of population growth and rising demands for services, need the money. Further, as HEW Secretary John Gardner has said: "The problems of urban society are of such magnitude and are so diverse that it is increasingly difficult to deal with them by remote direction from Washington. What is needed is the development of greater local responsibility and people qualified to assume this responsibility."

Let us stipulate that the principal future concerns of domestic government will be posed by the measure of urban expansion—the United States urban population will increase by seventy-five to one hundred percent in the last third of the twentieth century. One implication is that we shall have to do more building—of housing, urban infrastructure, and capital plant—between now and the end of the century than we have done in the past 350 years. And urban problems—the crawl of sprawl, traffic congestion, pollution, the shortage of recreation, the lagging response to housing needs—will increase in geometric proportion to population growth.

The crucial question is, who will determine the course of future urban expansion and set standards therefor? Thus far, most of the urban population growth has been accommodated by short-run market forces. The result is the specter of endlessly sprawling urban development reaching over hill and field, careless of efficiency, indifferent to beauty, heedless of any human values save the immediate need for a roof and four walls; holding out no aspirations or incentives for innovation in design, technology, or organization; paying little attention to efficient and harmonious spatial relationships between residence, employment, recreation, and activity centers.

But can adequate leadership be expected from the fifty states which would be the primary recipients of the federal largesse under the simple per capita grant plan and, beyond them, from the 90,000-odd local governments?

We should first consider Martin's (1965: Chapter 6) observation that "only a few states have ever assumed significant program responsibilities in these areas. . . . The states have been slow to take meaningful hold of urban problems in the past, and there is little sign of any real intention to do so now." The Advisory Commission on Intergovernmental Relations, in a report of March 1967, says much of the same thing, commenting that only a handful of states have moved to meet the problems of their urban areas and that state governments are on the verge of losing control over the mounting problems of central-city deterioration and the rapid growth of metropolitan areas. Martin (1965: 169) observes further that while the states are critical "of the growing practice of direct dealing between Washington and the cities, which they regard as both a perversion of the federal system and a pointed threat to state sovereignty," the states have displayed little interest in taking action. "It is not that the states wish to play an active role, but that they wish to be thought of as wishing to."

But the questionableness of the states as instruments of fiscal salvation does not stem alone from their historic disinterest in urban affairs. Wilson (1966) and others (Bender, 1966; Tydings, 1966; Sherill, 1965; Simon, 1965; 1964; Allen, 1949; Armbristor, 1966) have been pointing out that it is at the state level that corruption is still most notorious and widespread.

> More boodle is flying around with no one watching in state capitals than in city halls; and state governments continue to be badly decentralized, with formal authority divided among a host of semi-autonomous boards, commissions, and departments. The states have rarely been subjected to the kinds of reform which over the years have gradually centralized formal authority in the hands of a professional manager or a single strong mayor" [Wilson, 1966: 35].

Economists are, of course, by training more sensitive to violations of economic law than to such nasty transgressions against economic order as corruption—the invisible hand, whatever its other disabilities, is never caught robbing the till. But, surely the propensity toward fiscal propriety, or the lack of it, should be a consideration in selecting a chosen instrument to dispose of the national dividend. To be sure, the degree of corruption is not such as to incapacitate most state governments (the highways do get built) and there are some—a few—which display the inventive and innovational qualities that Heller extols. But, one darkly suspects that, on balance, ingenuity and innovation may be used more to get hold of money than to use it for promoting the general welfare.

Finally, there is the question of whether many of the states have the geographic jurisdiction, resources, and general sense of unity and of common interest to deal adequately with the problems of their growing cities and metropolitan areas, let alone the looming problems of megalopolis (see Bahrenburg, 1964).

On turning to local governments, we find fragmentation, political weakness, and antiquated organization which greatly impair them for useful service as decision-making instruments to meet the problems of this high-flying age. Urban governments, particularly in the older sections of the country, were devised for the conditions of the nineteenth century. Continued population growth, along with the advent of the motor vehicle and other modern communications devices, have made obsolete old urban government boundary lines and have created a complex of problems with which present urban governments were never designed to cope.

The Committee for Economic Development's (1966: 8) recent policy statement on Modernizing Local Government says, "American institutions of local government are under severe and increasing strain. . . . Adaptation to change has been so slow, so limited and so reluctant that the future role—even the continued viability—of these institutions is now in grave doubt." CED's credentials for responsibility and respectability have seldom been questioned, and its deliberations on local government were assisted by the usual pride of academic lions. Therefore, we have to pay some attention to indictments such as the following:

Very few local units are large enough—in population, area, or taxable resources—to apply modern methods in solving current and future problems. Even the largest cities find their major problems insoluble because of limits on their geographic areas, their taxable resources, and their legal powers.

Overlapping layers of local governments abound—municipalities, townships, school districts, special districts—which in certain areas may number ten or more. They may all have power to tax the same land, but frequently no one has the power to deal with specific urban problems.

Public control of the local governments is ineffective or sporadic, and public interest in local politics is lagging. Contributing factors are the confusion resulting from the many-layered system, profusion of elective offices without policy significance, and increasing mobility of the population.

Most units are characterized by notably weak policy-making and antiquated administrative machinery. Organizational concepts considered axiomatic in American business firms are unrecognized or disregarded in most local governments.

Personnel are notoriously weak. Low prestige of municipal serv-ice, low pay scales, and lack of knowledge and appreciation of professional qualifications all handicap the administrative process (Committee for Economic Development, 1966: 11 ff.).

The subject of metropolitan-scale government warrants a further comment. It would appear that some have given up on the possibil-ity of getting effective metropolitan-scale government at all, and it is further asserted that smaller governments have a greater capacity to meet consumer preferences, since it is easier for people to locate in jurisdictions offering service levels which best meet their partic-ular needs (the Tiebout effect). A supporting view is that collective decisions reached by small aggregations of individuals are likely to be more nearly optimal than decisions of larger aggregations. The view is one of the urban governmental process as a two-person game between taxpayers-consumers and governments-suppliers with the taxpayers able to say, "Don't cheat. I know what I dealt you!"

The preference for smaller governments runs into a fact of life: that smaller governments tend to be less efficient and to command less constituent interest than larger governments; local governments in general get less attention from constituents than state govern-ments get less than the national government. And it appears that the smaller the government, the stronger the veto power of the negative elements therein.

The plea for "grass roots" government is likely to be, in fact, a plea for less government. To cite Martin (1965: 190) again:

"Bring government back home" and "return government to the peo-ple" are battle cries equally familiar in the cause of local autonomy and in that of states' rights. . . . In point of fact it is quite clear that the goal sought often is not local autonomy or states' rights as such but rather less government, and moreover government more imme-diately and more directly subject to control.

If funds are handed out to bodies without competent planning and administrative machinery, there is no particular reason to think that they will be disbursed by any system of considered priorities. They may even be used to reduce taxes, leaving public services unimproved. Nathan (forthcoming), in a recent study for the Com-mittee for Economic Development of probable responses in several low-income states, concludes that this is the most likely outcome in such states. Well, Heller has said, at least we will have substituted superior and cheaper-to-administer taxes for existing inferior state and local taxes, and this would be a gain. But I submit that it is not much of a gain, considering the fact that the social costs of deficient education and other social services have to be picked up somewhere else, as when the undereducated move into urban centers in wealthier states.

Given such prospects, and such testimonials of inadequacy, incompetence, and propensity to peculation, are we nonetheless in the position of those who have to patronize the crooked roulette wheel because it is the only one in town?

Surely if federal tax machinery provides the wherewithal for a "national dividend" it would be profligate to use federal funds simply to bolster up existing institutions. If we are going to depend, as I think we must, on the decision-making and innovational capacities of state and local governments, let us seek to improve those capacities. Let us fashion federal grants to achieve this end.

Tying federal grants to minimum standards of administration or planning is nothing new. Thus:

- The Bureau of Public Roads has required federal-aid highways to meet design and construction standards. Pursuant to a 1962 amendment to the federal-aid highway law, the Secretary of Commerce after July 1, 1965, will not approve any programs for projects in an urban area of more than 50,000 population unless he finds that such projects are based on a continuing comprehensive planning process carried on cooperatively by state and local communities.

- The Hill-Burton Act of 1946 requires preparation of general state plans for hospital development as a condition of federal grants for hospital construction.

- The Housing Act of 1954 introduced a requirement for comprehensive community planning and the submission of community "workable plans" as a condition of urban renewal grants. While no one would claim that this stipulation has wrought wonders, while many of the "workable plans" have been rudimentary and the provisions of many have not been complied with, the requirements have undoubtedly made urban governments more aware of the elements of urban renewal and of the virtues of planning and have touched off a boom in the demand for planners.

A number of recent congressional acts have undertaken to encourage metropolitan areawide planning and organization.

- Section 702 of the Housing Law provides planning grants to metropolitan areas. This has led to the establishment of more than one hundred metropolitan planning commissions.

- Legislation providing grants for water pollution control increases the amount of the federal contribution from thirty to fifty percent of construction costs for a metrowide plan, as opposed to a purely local-jurisdiction project.

- The Housing Act of 1966 requires that after June 30, 1967, to the extent possible, any local government seeking federal aid for development funds will have to be reviewed by an agency which is designated as a metropolitan or regional planning agency and, further, that the designating body to which the planning agency is responsible be composed of elected chief officials of the region. Under the impetus of this legislation several score of these metro-planning agencies have been organized.

AGENDA FOR IMPROVED STATE AND LOCAL GOVERNMENT

Most state governments need substantial reorganization and reform of personnel and budgeting practices if they are effectively to discharge their decision-making and administrative responsibilities inherent in the federal system. Most generally needed items include:

- abolition of quasi-independent administrative boards and commissions (frequently they have earmarked funds), insulated from any responsibility for state welfare as a whole;

- concomitantly, centering administrative responsibility in the office of the governor, and equipping the office with planning, budgeting, and administrative expertise;

- limitation of legislative responsibility to matters of broad policy and budget approval; abolition of legislative budgets and exercise of administrative powers by legislatures or by individual legislators;

- comprehensive merit personnel systems;

— comprehensive budgets covering all funds and expenditure categories, preferably based on program budget concepts (these have been notably unsuccessful thus far for reasons having little to do with their intrinsic merits);

— strict conflict-of-interest laws;

— constitutional provisions affording maximum latitude to local governments, which meet reasonable standards of adequacy.

For an agenda of local government reform, I refer to the CED (1966: 16 ff.) prescription. These include:

— reduction in number of local governments by at least eighty percent, and severe curtailment of overlapping layers of local government (townships and most types of special districts are obvious targets for elimination);

— limitation of popular election to members of legislative bodies and the chief executive in the "strong mayor" type of municipal government;

— a single strong executive—elected mayor or city manager;

— modern personnel systems;

— use of county, or combinations of county, jurisdictions to attack metropolitan problems;

— use of federal—and state—grants-in-aid to encourage local government administrative reforms, particularly reforms having to do with consolidation and organization to meet metropolitan problems.

SO WHAT CAN WE DO WITH
FEDERAL GRANTS

Having agitated for use of federal grants as incentives to improve state and local government decision-making and administrative capacities, I must confess to being short of suggestions for ways to accomplish this objective. It would be simple if the federal government could say to the states and localities: "Reorganize; eliminate corruption; introduce PPBS (programming-planning-and-budgeting systems); hire good personnel" as a condition for getting per capita grants. But the matter in fact is not so simple. Consider the job of writing the criteria of acceptable standards. Then consider the job of evaluating the states and localities to determine if they have met the criteria (including allowing for inevitable time lags). Finally, consider the fights with state congressional delegations in cases where it is necessary to determine that a state and its local governments have not met required standards.

Analogous, if less comprehensive, stipulations have been made before. True, they have evoked yowls of protest from offended politicians, as did the requirement for placing unemployment compensation systems under civil service; however, protesters against improved administration are seldom on the side of the angels. I, therefore, think that we can tie federal grants to needed innovations in governmental organization and administration despite the difficulties cited above. In the absence of such incentives, roadblocks to improvement are, generally speaking, virtually impassable.

I shall toss out a few suggestions as starting points for further discussion.

- Let the Congress declare, in a preamble to legislation establishing a national dividend in whatever form, that one of the prime objectives of the act is to increase the planning, decision-making and administrative capacities, as well as the financial capacities, of the state and local governments.

- As a condition for receiving grants, the states should prepare plans for administrative improvement of state and local government over, say, four-year periods; and thereafter file annual reports of progress in implementing such plans. Plans should address themselves to, among other things, recasting local governments to meet present and emerging problems of urbanism.

- State governments should present plans for using the federal "dividend" payments, such plans to be drawn up in consultation with their major local governments. In California, the League of California Cities has requested the state government to work with the League in preparing plans for the disposition of any federal government unconditional grant. However, I would be skeptical of imposing any rigid requirement of this nature which might end up as an instrument for perpetuating local government structures, forms or practices which need reforming.

- Each state and local government, to qualify for grants, should institute personnel systems based on merit and professional competence. (I have avoided using the term "civil service" in recognition of the fact that many present civil service systems have become rigid, unresponsive, and uncontrollable administrative straitjackets; notwithstanding, the answer still seems to lie in the direction of improving career-service systems rather than abolishing them.)

• The notion of tax "effort" by the states and localities should be built into any formula for federal rebates to the states, providing an inducement for equalization of state and local tax rates relative to ability to pay.

THE REUSS BILL

A bill (H. R. 1166) introduced in the House of Representatives by Congressman Henry S. Reuss of Wisconsin, on January 10, 1967, seeks the same objectives and for the same reasons set forth in the above discussion. It also is more specific on administrative techniques. Under the Reuss Bill, block grants would be made conditional upon the submission by the states of acceptable programs of government modernization; the review and evaluation bodies would be regional coordinating committees and the Advisory Commission on Intergovernmental Relations, which would certify as eligible programs reflecting "sufficient State creative initiative so as to qualify that State for Federal block grants." Among the items to be considered in drawing up such plans are the following:

— arrangements for dealing with interstate regional, including metropolitan, problems;

— strengthening and modernizing state governments;

— strengthening and modernizing local rural, urban, and metropolitan governments;

— proposed uses of federal block grants, including provisions for passing on at least fifty percent to local governments.

My principal reservation is predicated on a question of whether the ACIR is the most suitable body to make final determinations on acceptable programs. My fear is that this would involve the

ACIR in a good many controversies with large amounts of money at stake, and that this might dilute the high quality of the work the ACIR has been doing; it might drastically change the nature of the Advisory Commission on Intergovernmental Relations or even destroy it as an organization. These fears may be exaggerated but such possibilities should be carefully considered, along with alternative administrative arrangements.

POSTSCRIPT

As we debate the problems of metropolitan areas the urban torrent is overrunning the boundaries of these areas. Jean Gottman, some years ago, identified a megalopolis stretching from Boston to Norfolk. Wattenberg and Scammon (1965: 87) point out that it is

> already apparent that the eastern megalopolis will not stand alone for much longer. We can now see the beginnings of other megalopoli in different areas. One is clearly forming along the southwestern end of Lake Michigan. . . . Another, smaller megalopolis is growing from Miami north to Fort Lauderdale and beyond—still another megalopolis may some day connect Los Angeles and San Francisco-Oakland (with Fresno in the middle) . . . and a fifth may eventually link Detroit and Cleveland, with Toledo in the middle, or link Cleveland and Pittsburgh with Youngstown in between.

As Bertram Gross has observed, these complexes will be unprecedented aggregations not only of population but also of wealth, knowlege, and power. But this is not to argue for a quantum leap from present forms into megalopolis. It is to remind us that time is running out, and that unless we act to improve planning and decision-making mechanisms, the prospect ahead is a continuation and probably an exacerbation of mindless, formless, inefficient, and unbeautiful urban sprawl, inferior government services, and growing frustration with the widening gap between the performance and the potential of the affluent society.

REFERENCES

Advisory Commission on Intergovernmental Relations (1967). Eighth Annual Report.

ALLEN, R. S. [ed.] (1949) Our Sovereign State. New York: Vanguard Press.

ARMBRISTOR, T. (1966) "The octopus in the state house." Saturday Evening Post (February 12).

BAHRENBURG, B. (1964) "New Jersey's search for an identity." Harper's Magazine (April): 87-90.

BENDER, W. J. (1966) "The corruption problem." National Civic Review (March).

Committee for Economic Development (1966) Modernizing Local Government.

MARTIN, R. C. (1965) The Cities and the Federal System. New York: Atherton Press.

NATHAN, R. P. (forthcoming) Paper dealing with the potential impact of general aid in four selected states. Paper prepared for the Committee for Economic Development.

SHERILL, R. (1965) "Florida's pork chop legislature." Harper's Magazine (November): 82-97.

SIMON, P. (1965) "Cleaning up the Illinois legislature: a follow-up report." Harper's Magazine (September): 125.

——— (1964) "A study in corruption." Harper's Magazine (September).

TYDINGS, J. D. (1966) "The last chance for the states." Harper's Magazine (March): 71-79.

WATTENBERG, B. J. [in collaboration with R. M. SCAMMON] (1965) The U.S.A.: An Unexpected Family Portrait of 194,067,296 Americans Drawn from the Census. Garden City: Doubleday.

WILSON, J. Q. (1966) "Corruption: the shame of the states." Public Interest (Winter).

7

Federal Grants and the Reform of State and Local Government

TERRANCE SANDALOW

☐ INCREASINGLY in recent years, discussion of the appropriate division of responsibilities between the nation and the states has shifted from ideological to pragmatic grounds. One consequence of that shift has been to bring into sharper focus the dilemma which confronts the growing number of those who believe that the existing structure of government is inadequate to the pressing tasks which face the nation, especially those tasks which center upon the metropolitan areas now inhabited by two-thirds of the nation's population. A rapidly developing consensus among persons of widely divergent political perspectives accepts the view summed up in Walter Heller's (1966: 121) statement that "there is neither the administrative capacity nor the problem-solving wisdom at the top" to deal adequately with all the nation's ills. Neither Detroit, nor Minneapolis, nor the metropolitan areas of which they are a part, can successfully be run from Washington. Any attempt to do so, as even the restrained moves in that direction during recent years have made us painfully aware, is likely to be insufficiently sensitive to the variant needs of different localities, to produce debilitating bureaucratic red tape, and to fail in coordinating programs whose ultimate success can be meaningfully determined only by the extent of their contribution to a larger whole.

LOCAL GOVERNMENT INABILITY TO DEAL WITH "URBAN CRISIS"

At the same time, state and local governments lack the strength to cope adequately with the several facets of what has come to be called the "urban crisis." In part, their weakness is attributable to inadequate fiscal resources. But were that the only source of their deficiencies, the problem of allocating governmental responsibilities would be relatively easy: the fiscal resources of the federal government might simply be made available to state and local governments, presumably without "strings," as contemplated by the spate of recent revenue-sharing and tax credit proposals. The dilemma lies in the conviction of many observers that even with additional financial resources, state and local governments would not have the capability to deal adequately with the "urban crisis." The reasons which underlie these convictions are depressingly familiar: archaic legal and fiscal structures, a shortage of competent personnel, and governmental organization which has not been shaped to deal with current technological, economic, and social realities. The resulting inability of state and local governments to develop and administer programs for meeting the problems by which they are beset inevitably generates pressure upon Washington to do so, though the federal government's capacity to contribute to the solution of these problems is itself limited.

It may be useful to recognize that this apparent dilemma is not confined to the United States. In England, local government in the London area has recently been significantly reshaped and a Royal Commission has proposed a sweeping reorganization of local government in the rest of the country. Although it is not yet clear whether Parliament will enact the Commission's proposals, it seems virtually certain that major reforms will occur. Sweden has already enacted legislation drastically reducing the number of local governments throughout the country and restructuring local government in the Stockholm region, the nation's largest metropolitan area. In both countries, though their circumstances differ markedly in some respects from those in the United States, the reasons given for these actions are strikingly similar to those expressed by students of local government in the United States in calling for strengthened state and local governments. Both national and local officials recurrently stress the inability of the central government to manage local affairs and the inevitable drift toward national management unless local government can be made more effective. In consequence, the central governments of both countries seem committed to programs of strengthening local governments, thereby increasing their ability to respond to domestic needs and simultaneously reducing pressure upon the national government to do so.

The Swedish and English approaches rest upon the undoubted authority of the governments of those countries to determine the organization and powers of government below the national level. Centrally directed reorganization of local government involves political difficulties, of course, but the authority and responsibility of the national government for determining the structure of local government in those countries is clear. A similar approach in the United States would confront initially the added complications which flow from our federal system. State and local governmental organization has not, in this country, traditionally been viewed as an appropriate subject of federal concern. Issues with respect to the division of powers among the branches of state government, the distribution of functions between state and local governments, and the allocation of responsibilities among local governments have, in general, been thought of as within the province of the states, subject only to minimal restrictions upon state power imposed by the federal constitution. A state may not, thus, establish municipal boundaries along racial lines, nor may it weigh the vote of one citizen more heavily than that of another in the election of a governor, or apportion its legislature on other than a population basis. In general, nonetheless, the states retain very substantial freedom to determine the structure of government within their boundaries.

The difficulty with this arrangement, as one contemplates the need to strengthen state and local governments as a means of increasing their capacity to cope with urban problems, is that it requires a bootstrap effort. Skepticism concerning the wisdom of reliance upon such an effort is by no means unjustified. Precisely the same factors which contribute to state and local incapacity to deal effectively with current urban problems impede effective reform of governmental structure. The success of any attempt to strengthen state and local governments is, therefore, likely to depend upon the development of a federal policy directed toward that end, for if the states lack the will and the means to engage in such an effort, those ingredients must be sought elsewhere and at this point in history that is most likely to be the federal government.

THE DEVELOPMENT OF A FEDERAL POLICY

The development of a federal policy requires that some attention be directed to the tools available to the federal government if a policy is to be implemented. Unlike the situation in Sweden and England, our central government—the federal government—lacks clear authority to enact legislation reorganizing state and local governments or requiring the assumption of specified responsibilities by them. In contrast with state governments, which have all powers not denied to them by their own or the federal constitution, the United States, under traditional and still prevalent constitutional theory, is a government of enumerated powers, i.e., it may exercise only those powers conferred by the Constitution. Among the powers granted is that of spending money for the "general welfare." During the nineteenth century, and into the first third of the twentieth, the extent of that power was a subject of vigorous debate. At the present time, however, there is broad acceptance of the view stated by Alexander Hamilton (1885: 372) as early as 1791:

> The only qualification of the generality of the phrase in question which seems to be admissible is this: That the object to which an appropriation of money is to be made be general and not local; its operation extending in fact or by possibility throughout the Union and not being confined to a particular spot.

In consequence, although the issue is not entirely free from doubt, congressional authority to spend for the purpose of strengthening state and local governments, say, by supporting improved personnel systems, seems reasonably clear.

The power of Congress to legislate other than by the appropriation of money is generally understood to be more limited than its power to spend, however, even after the expansive interpretation of its powers in the years since the New Deal. Congressional authority to employ coercive measures, in other words, is more restricted than its power to induce action by purchasing cooperation. We need not pause to enumerate the areas in which Congress may regulate (i.e., legislate coercively). It is sufficient to note that the Constitution does not expressly confer upon the Congress the power to determine state or local governmental structure or other matters of internal management, though the powers which are conferred (such as the power "to regulate commerce . . . among the several states") may to some extent sanction legislation directed to that end or which has that effect. Most lawyers, for example, would doubt the power of Congress to enact legislation directing state or local governments to adopt merit systems for their employees, though arguably Congress might do so with respect to state activities which affect interstate commerce.

Programmatic Values

In any event, constitutional tradition and political reality, if not necessarily the letter of the Constitution, are likely to require that less direct means for implementing a federal policy at least be explored before resorting to authoritative means. The primary means by which the federal government has sought to influence state and local policies during the present century is the grant-in-aid. Historically, however, neither Congress nor the executive branch have to any substantial degree consciously employed grants-in-aid to strengthen the governing capacity of state and local governments. Federal aid programs have been directed toward stimulating or improving the quality of discrete public services in which there is a "national interest"—highways, welfare, vocational education, housing for low-income families, etc.—rather than toward enhancing the ability of state and local governments to govern generally. In part, no doubt this emphasis is attributable to ideology. State governments (and their local progeny), as suggested above, have tended to be viewed as constitutionally semiautonomous bodies whose internal organization and management are not proper concerns of the federal government. Ideology, in this instance, is supported by politics. Both the executive branch of the federal government and the Congress are organized largely along functional lines, lines which in substantial measure are congruent with interest groupings in the private sector. The political pressures within and without the federal government which generate federal grant programs tend, in consequence, to be programmatic in origin. Private interests which seek federal support and the professionals and politicians who shape programs coalesce around particular demands—more highways or improved vocational education or more public housing units—not abstractions such as economic efficiency, political strength, or responsive government.

Concern for programmatic values, of course, encompasses concern for governmental effectiveness, but since attention has normally been focused upon the former, the federal government, when it has interested itself in strengthening state and local governments, has typically been concerned only with strengthening their ability to perform a specific aided function. The "single state agency" requirement of the public assistance titles of the Social Security Act is illustrative. Each of those titles requires that the program aided thereunder—aid to the aged, aid to the blind, aid to the permanently and totally disabled, and aid to families with dependent children—be administered by a single state agency or, if administered by local governments, be supervised by such an agency. The requirements were aimed at promoting the effectiveness of federally-aided welfare programs by centralizing the administration of those programs and easing the supervisory burdens of federal administration. No attempt was made, however, to influence administration of state-local general assistance programs and improvements in the administration of those programs have tended to lag behind federally-supported programs.

The difficulty with this approach, as we have increasingly become aware during the past decade, is that governmental programs do not exist in isolation from one another. Deficiencies in one program impact upon others. Progress toward the objectives of a rational welfare policy depends not only upon the adequacy of state-local welfare programs, including those which are not federally aided, but upon the adequacy of the educational system, vocational and family guidance services, job-training programs, and a variety of other governmental services. A federal policy aimed narrowly at strengthening the administration of federally-aided categorical welfare programs is not likely to achieve even the goals of those programs unless related programs are being administered by state and local governments equal to the task. More than "administration" is involved, moreover. Since needs vary in different parts of the country, state and local governments must also have the ability to devise programs and to identify those which require priority, and they must have the political strength to adopt and carry them out.

The Need for Comprehensive Approaches

To a limited extent, recognition of the need for more comprehensive approaches has begun to appear in recent federal legislation but the strength of the tradition and political pressures behind narrowly defined programs is reflected in the ambivalence with which the federal government has thus far acted. Section 204 of the Demonstration Cities and Metropolitan Development Act of 1966, for example, requires that every application for a federal loan or grant "to assist in carrying out . . . [various projects involving physical development, e.g., airports, highways, water supply and distribution facilities, etc.] shall be submitted for review—(1) to any areawide agency which is designated to perform metropolitan or regional planning for the area within which the assistance is to be used . . . and (2) if made by a special purpose unit or local government, to the unit or units of general local government with authority to operate in the area within which the project is to be located." The applicant is required to forward the reviewing agency's or unit's comments and recommendations to the federal granting agency and to submit also a statement that such comments and recommendations have been considered by the applicant prior to submission. The comments are to "include information concerning the extent to which the project is consistent with comprehensive planning . . . for the metropolitan area or the unit of general local government . . . and the extent to which such project contributes to the fulfillment of such planning." If Congress had stopped at this point, Section 204 would have been a straightforward, if relatively weak, measure for achieving territorial and functional integration of physical development in metropolitan areas. That, at least, was the conception of the Advisory Commission on Intergovernmental Relations, where the proposal originated. Significantly, however, Congress went on to provide—as the Commission had not recommended—that the metropolitan agency's comments and recommendations shall be reviewed by the federal agency "for the sole purpose of assisting it in determining whether the application is in accordance with the provisions of federal law which govern the making of the loan or grant." The effect of this last sentence is substantially to weaken the integrative potential of Section 204 by reinforcing the narrower programmatic biases of the federal agencies administering grant programs.

The use of grants-in-aid to promote major reorganization of government at the state and local levels would, thus, mark a substantial departure from past practice. Still, Congress has taken a few steps in that direction. Section 204, though feeble, is but one of a number of congressional measures designed to strengthen development of metropolitan perspectives on urban problems. Encouragement, mainly in the form of financial support, has been

provided for metropolitan planning and for metropolitan councils
of government. Financial inducements have been provided for areawide development projects. Congress has also, in the past several
years, cautiously begun to move against special district governments. Section 204 represents one such step. A potentially more
significant one was taken in the Intergovernmental Cooperation
Act of 1968, Section 402 of which provides that "in the absence
of substantial reasons to the contrary," general purpose local governments shall be preferred to special purpose units in awarding
grants-in-aid. These somewhat hesitant steps by the Congress to
employ the existing system of grants-in-aid to influence governmental organization at the metropolitan and local levels point to
one of the possible approaches to implementing a federal policy of
strengthening state and local governments. Governmental units with
specified characteristics might be preferred or made exclusively
eligible for the award of functional grants or such grants might be
conditioned upon reorganization of state and local government and
upon the adoption of other federally specified reforms designed to
strengthen their governing capacity. Although there is a widespread
belief in the utility of federal grants as a means of influencing state
and local policies in this way, there is also a good deal of experience which suggests that there are rather severe limits to such use
of the federal spending power. Brief examination of that experience may shed some light on the results which may reasonably be
anticipated if reliance is to be placed upon use of the existing
system of functional grants to bring about a restructuring and
reform of state and local government.

The Expansion of Federal Grants

In approaching the question, it is important to bear in mind
that although functional grants would provide the leverage by
which Congress would attempt to effect reform, reforms would not
necessarily be aimed at or have the effect of strengthening state
and local capacity to perform each of the aided functions, especially in the short run. The underlying purpose, rather, would be
to strengthen state and local governments as institutions of government and though, if the purpose were accomplished, that might
ultimately lead to more effective performance of federally aided
functions, it would not necessarily do so for each of them. The
focus upon programmatic objectives implicit in the existing system

of grants-in-aid is, indeed, one of the major obstacles to development of a coherent federal policy for achieving reorganization and
reform of state and local government. Existing grant legislation
does, to some extent, as in the public assistance titles of the Social
Security Act, influence the structure of state and local government,
but the federal requirements are, except in a few instances, geared
to the needs of a specific program. That is not necessarily synony-
mous with the strengthening of state and local institutions. For
example, emphasis upon particular program objectives has in the
past, as noted by the Advisory Commission on Intergovernmental
Relations, occasionally led the federal government to promote
formation of special districts as the most expeditious means of
accomplishing a federal goal, thereby contributing to the fragmentation of government at the local level and the weakening of
general purpose governments which constitute such significant
aspects of our current difficulties.

The celebrated dispute between the state of Oregon and the
Department of Health, Education, and Welfare nearly a decade ago,
affords another illustration of the tensions that may exist between
structuring government to achieve a single programmatic goal, as
contemplated in existing grant legislation, and structuring it upon
the basis of a broader conception of the ends to be served. A
major reorganization of state government proposed by the governor
called for the establishment of six executive departments, one of
which, the Department of Social Services, would have had program
responsibilities similar to those of HEW. Separate divisions within
the department would have been responsible for administration of
programs in the following areas among others: health, mental
health, veterans' affairs, vocational rehabilitation, and public welfare. The director of the department would have been appointed
by the governor, division heads by the director. The Division of
Public Welfare would, under the reorganization plan, have served as
the "single stage agency" for administration of all federally aided
public assistance programs, but it would have been subject to
general supervision by the department. The apparent purpose of
the reorganization plan was to facilitate coordination of a group of
related services while retaining a measure of autonomy for individual programs by lodging operating responsibility for them in separate divisions. The plan was ultimately abandoned, however,
because HEW took the position that to subject the Division of
Public Welfare to supervision by the department would have been

inconsistent with the "single state agency" requirement of federal law.

A federal policy of strengthening the institutional framework of state and local government ought not, therefore, to depend upon those reforms which would be generated by efforts to enhance state and local ability to administer individual programs supported by functional grants. Such an approach would lead, at best, to pressure for organizational and other changes which facilitate administration of particular programs, but which do not necessarily serve the more general goal of strengthening state and local governments. At worst, such an approach might produce a further deterioration of the institutions of government at the state and local levels. If a federal policy is to be developed, accordingly, it must be formulated by a process other than piecemeal revision of grant-in-aid legislation. To develop a policy by the latter means would inevitably emphasize the views of those in Congress and the executive whose interests and responsibilities impose a perspective narrower than the task at hand.

Administration of Programs

The programmatic orientation of the agencies responsible for administering grants-in-aid, experience warns, is likely to create impediments to enforcement of a federal policy as well as to its development. Even if, by the exercise of imagination and a considerable leap of faith, we assume development of an appropriate federal policy with which state and local government would be required to comply as a condition of eligibility for functional grants, there is no basis for indulging the further assumption that agencies administering federal grant programs would view such conditions sympathetically. To the extent that enforcement of conditions would interfere with what the agencies view as their primary missions—highway construction, urban renewal, or welfare payments—there is, indeed, more reason to suppose that they would respond in a hostile manner which inevitably would be reflected in their enforcement efforts.

Typically, in the administration of grant programs, federal agencies enforce conditions which are related to their program objectives. Because the conditions are in general related to program objectives, moreover, the agencies tend to share common attitudes concerning them with their counterparts at the state and local levels. When that has not been true, the record of success in the enforcement of conditions is spotty at best. Thus, the failure of federal officials charged with responsibility for administration of the urban renewal program to effectively enforce statutory requirements for the relocation of displaced persons has been thought by some students of the problem to be at least partially attributable to preoccupation with slum clearance and redevelopment as *the* purposes of the program. Much the same problem exists in connection with Title VI of the Civil Rights Act of 1964, which prohibits racial discrimination in programs supported by federal funds and directs federal agencies administering such programs to enforce the prohibition. A recent study of Title VI administration concludes that

> Title VI has been almost as unpopular in some agencies as it has been in recent Congresses and the South. Each agency grants federal projects, services, land, and funds to various recipients in furtherance of its own program and the enforcement of Title VI means that the particular program will no longer be available to a noncomplying recipient. Every enforcement of Title VI which results in a denial of funds rather than in a voluntary compliance is thus in a sense working at crosspurpose with the agency's primary interest and responsibility [George Washington Law Review, 1968: 877].

Difficulties in enlisting agency interest in enforcement of conditions unrelated to narrow program objectives are a function not merely of the severity of the sanction and of the fact that the sanction works at cross-purposes with the agency's goals, but also of limited agency budgets. Effective enforcement of conditions requires manpower, not only to ascertain whether there is compliance but, more importantly, to work with state and local officials in the development of policies and programs that will bring them into compliance with federal requirements. Since agency budgets are most frequently inadequate to permit an agency to accomplish satisfactorily what it views as its primary objectives, any assignment of personnel to enforcement of conditions viewed by the agency as not directly related to those objectives represents in its view a less than optimal allocation of its resources.

Budgetary constraints may impose effective limits upon enforcement efforts even when agencies are sympathetic to the conditions they are called upon to enforce. The Office of Education has responded perhaps as warmly as any agency of the federal government to the responsibilities imposed by Title VI of the 1964 Civil Rights Act. And yet, though it has made progress in bringing about desegregation of southern schools, it has not as yet been able to achieve substantial compliance with Title VI. The reasons for this failure are many, but budgetary limitations have undoubtedly been important. During the 1966-1967 school year, the Title VI enforcement program was administered by the Equal Education Opportunities Program (EEOP) of the Office of Education. Congress appropriated slightly more than $750,000 for EEOP compliance activities during the fiscal year 1967, almost precisely one-half the amount that had been requested. EEOP's professional staff was, accordingly, limited to sixty-three persons, thirty-seven of whom were responsible for Title VI enforcement in the 1,786 southern and border state school districts under voluntary (as distinct from court-ordered) desegregation plans. (These numbers were to some extent augmented by hiring teachers and students during the summer months.) Such staff shortages, as various studies have shown, had an inevitable impact upon investigative and enforcement activities. EEOP was, for example, unable to handle administratively even those districts which had the poorest records under the guidelines which had been laid down for measuring compliance.

Related to, but not necessarily identical with, budgetary limitations, is the enormous administrative task of supervising compliance by fifty states and hundreds, perhaps thousands, of local governments. Thus, the U.S. Civil Rights Commission reports the virtual breakdown of EEOP's efforts to compile and maintain adequate statistics concerning the racial composition of schools. Similar difficulties have been encountered in connection with the relocation requirements of the urban renewal program. A recent study of the enforcement of those requirements points to the fact that

RAA [Renewal Assistance Administration of HUD] has put itself in the position of depending on the LPA [local public agency] not only to develop the specific standards implementing [the federal statutory] protection for relocatees, but also to provide the information necessary to determine in advance if those standards can be met. To compound the problem the RAA also depends on the LPA for the information necessary to evaluate LPA compliance with the standards in carrying out relocation [University of Pennsylvania Law Review, 1968: 199].

One need not assume dishonesty on the part of state and local officials, though that possibility, of course, exists, to recognize the threat to execution of national policies implicit in so great a degree of federal dependence upon the cooperation of those "regulated" by federal policy.

One further impediment to enforcement of a federal policy merits attention. Federal agencies are dependent for legislative authority and budgetary support upon a Congress elected by and, therefore, generally responsive to local constituencies. There are, in consequence, important incentives for representatives and senators to attend to the interests of state and local officials from the areas they represent and for the federal bureaucracy to respond to the representatives and senators. Significant variations exist, of course. An agency is likely to be influenced a good deal more by an important member of one of the appropriations committees or of a committee having jurisdiction over the agency's activities than it is by a relatively junior member with other committee assignments. The agency may, moreover, be subject to conflicting pressures from other members of Congress. Senators and congressmen, similarly, do not represent monolithic constituencies and their responsiveness to the interests of state and local officials may be affected by pressure exerted by other constituents. Nevertheless, it seems fair to conclude that the local orientation of the Congress is a significant braking force upon the ability of the federal government to implement national policy by enforcement of restrictive conditions upon grants-in-aid.

ESTIMATION OF IMPEDIMENTS TO FEDERAL POLICY ENFORCEMENT

An attempt to estimate the extent to which obstacles such as these would hamper effective enforcement of a federal policy inducing structural and other reforms of state and local government through the leverage of grants-in-aid would be futile without information concerning the content of such a policy, a task well beyond that undertaken here. At least one of the characteristics of any policy which conceivably might emerge from the Congress is predictable, however, and it suggests that each of the difficulties discussed thus far is likely to be encountered. Political reality and common sense combine to support a prediction that a federal policy would necessarily be couched in flexible terms. Some aspects of such a policy might, of course, be susceptible to rigid statement but the political and other diversities in a country of more than continental size are simply too great for any requirement of absolute uniformity. Congressional recognition of the need for such flexibility is currently reflected in a number of the statutes which seek to influence state-local governmental structure. Section 204 of the Intergovernmental Cooperation Act of 1968, in a delayed congressional reaction to Oregon's difficulties with HEW, permits the head of any federal department or agency to waive the "single state agency" requirement imposed by any grant-in-aid statute upon finding that the "provision prevents the establishment of the most effective and efficient organizational arrangements within the State government" and that "the objectives of the Federal statute authorizing the grant-in-aid program will not be endangered by the use of such other State structure or arrangements." Similarly, the congressional effort in the same statute to strengthen general-purpose local governments at the expense of special districts is expressed only as a direction to prefer the former in the award of grants, not as an absolute restriction of eligibility. So, too, Section 204 of the Demonstration Cities and Metropolitan Development Act of 1966, in requiring comment upon applications for certain grants by an areawide agency designated to perform metropolitan or regional planning, specifies only that the agency shall *"to the greatest practicable extent,* [be] composed of or responsible to the elected officials of a unit of areawide government or of the units of general local government within whose jurisdiction such agency is authorized to engage in such planning" (emphasis added).

A good deal of flexibility is to be anticipated also with respect to the time within which state and local governments would be required to come into compliance with federal policy. The very large differences which almost certainly would exist among the states concerning the changes necessary to conform to a federal policy, the varying procedures for effecting change, and the political differences among the states would have an obvious impact upon the pace at which they might reasonably be expected to progress toward compliance. It would be naive to suppose that Congress would or sensibly could establish a certain date by which all state and local governments would either come into compliance with federal requirements or lose the financial assistance currently provided under the grant system. There is, indeed, considerable experience which teaches that even when Congress does lay down seemingly mandatory requirements for grant-in-aid eligibility, varying amounts of time are allowed to different recipients to meet those requirements because of the awareness of federal administrators that not all congressional objectives can be achieved simultaneously. State and local governments have often continued to receive grants, even though not in compliance with a congressional requirement, as long as they are making progress toward eventual compliance.

The partially open-ended quality of any policy likely to be adopted by the Congress, taken together with the inevitable hostility of some, and more likely many, state and local officials to at least aspects of such a policy, means that the administrative burdens of implementation and enforcement are certain to be highly demanding. It is because of these burdens that the obstacles to implementation and enforcement through the existing system of functional grants takes on such significance. Emphasis upon these obstacles ought not, however, to be taken as denying the possibility that a federal policy favoring reorganization of state and local government and other reforms designed to strengthen their capacity for responding to urban problems could in some measure be promoted through the leverage provided by the system of functional grants. Federal grants-in-aid have had in the past, and there is no reason to doubt that in the future they will have, a significant impact upon the structure of state and local government and the ways in which they exercise their powers. The conditions established for eligibility under the Social Security Act have undoubtedly contributed to centralizing and professionalizing the administration of welfare programs. The growth of metropolitan planning agencies and of metropolitan councils of government has been stimulated by federal incentives. The pace of southern school desegregation, however disappointing in absolute terms, has increased dramatically since the enactment of Title VI. In each of these areas, as well as numerous others which might be cited, there is considerable difficulty in assessing the actual impact of the federal programs. Inevitably, the "strings" attached to federal grants reflect trends which exist independently of the federal programs and which find other expressions. The extent to which these trends would have developed without federal support normally involves guesswork at best. A further difficulty in assessing the impact of federal grant programs arises from the need to distinguish between form and substance. Hundreds of plans have been written and codes adopted since the federal government began to promote such activities in the 1950s, but whether the plans have affected development or whether the codes have contributed to improving the quality of the housing supply—or, indeed, whether the codes represent more than a compilation of previously enacted requirements—are aubjects upon which we have precious little information. Even with all these doubts, it is difficult to escape the conviction that federal grant programs have had some impact upon state and local governments. The impact may not be as great as popularly believed but it seems likely that they have at least accelerated trends which might otherwise have developed more slowly and, on occasion, perhaps provided the support without which some changes would never have been made. It is such an impact which might reasonably be anticipated were the federal government, through the existing grant system, to attempt to implement a policy of reform in state and local government, though how substantial the impact would be is likely to depend in large measure upon the content of that policy.

THE REUSS PROPOSAL

An alternative approach to federal stimulation of the reform of state and local governments has been suggested by Congressman Reuss of Wisconsin. Instead of using the leverage of the existing system of functional grants, the Reuss proposal would condition a state's participation in a federal revenue-sharing plan upon the adoption of a "Modern Governments Program" to be developed in each state with financial support from the federal government. The program would set forth proposals and a timetable for a variety of reforms in state and local governments with respect to matters such as governmental organization, tax structure, debt restrictions, the special problems of interstate metropolitan areas, and personnel. Once adopted, the program would be submitted to one of four regional coordinating committees composed of representatives designated by the governor of each of the states in the region. The coordinating committees would review the programs, make suggestions for their improvement, and ultimately certify those which reflect "sufficient creative State initiative" to qualify the state for participation in the revenue-sharing plan. A similar review and certification would be provided by the Advisory Commission on Intergovernmental Relations. States whose programs had been appropriately certified by a regional coordinating committee and the ACIR would qualify for participation in the revenue-sharing plan established under the proposal. The failure of a state to carry out any of the reforms called for by its program would have no effect on its right to receive revenues under the plan, but the regional coordinating committees and the ACIR would be required to report annually to Congress and the President on the progress made by each participating state in carrying out its program, presumably to provide the basis for a judgment whether the scheme had induced reform adequate to justify retention of the revenue-sharing arrangement.

There are a number of attractive aspects of the Reuss proposal. The most propitious time for inducing change, as Joseph Pechman observed in commenting upon the Reuss proposal before a subcommittee of the Joint Economic Committee, is when new money is introduced into the system. Initially, governments are most likely to be engaged in or at least receptive to planning at such times. Whether or not they would do so of their volition, however, it seems particularly appropriate that they do that which is necessary to the accomplishment of the federal objective in adopting the revenue-sharing arrangement. Second, tying reform to revenue-sharing rather than to the existing grants may provide a more appropriate sanction for recalcitrant states. The sanction of withdrawing funds available under existing grants-in-aid is an imposing one—perhaps too imposing to be credible. Functional grants now account for nearly twenty percent of state-local revenues and for some functions, such as welfare, a far larger percentage of state-local expenditures comes from the federal government. Withholding federal funds would thus have so great an impact upon state and local budgets as to seriously deter use of the sanction by federal officials, a phenomenon which is currently observable in a number of grant programs. And, even if the executive branch were inclined in a particular instance to impose the sanction, the political pressures from Congress and elsewhere not to do so would be enormous. Of course, any attempt to deprive a state of its share under a newly adopted revenue-sharing plan would also involve political and other difficulties, but since the impact upon the state would be significantly less, the sanction would be somewhat more realistic. Even so, there may well be a problem of "overkill" even under the Reuss proposal.

A further advantage of the Reuss proposal is that it moves responsibility for implementation out of grant administering agencies, where that would frequently be a matter of peripheral interest. Some would add, as a related advantage, that the delegation of responsibility to regional coordinating committees and the ACIR for determining the adequacy of state programs avoids the risk of federal domination. My own view is that this is the major weakness of the Reuss proposal. Although the manner of selection and role of the coordinating committees is likely to increase the salability of the proposal, that is accomplished at a very considerable price. To begin with, it seems likely that a regional coordinating committee, though operated on the basis of majority rule, would in practice simply support the position of the governor of each of the constituent states with respect to his state's program. Each member of the coordinating committee, politically responsible in his own state and lacking either political or program responsibilities in any of the other states in the region, would be likely to conclude that his primary interest lies in deflecting adverse action with respect to his own state's program. In consequence, a process of log rolling, in which each state's representative implicitly trades favorable action by him on the programs of other states for a similar vote on his state's program, is likely to emerge. Exceptions might arise occasionally if governors were to attempt to use the veto power of the coordinating committees to push through reform proposals which they lack political strength to impose by more conventional political means. If one indulges the heroic assumption that governors' positions are likely to be desirable ones, the coordinating committee's review might at times serve a useful purpose. In general, however, it seems safe to predict that the coordinating committee would not play a significant role.

The concurrent review and veto power of the ACIR would not be subject to the same infirmity, though questions might also be raised as to whether it is well constituted to serve a policing function. The central difficulty with this aspect of the Reuss proposal, however, is not the failure to provide for an adequate reviewing agency, a function which the ACIR might satisfactorily perform or for the performance of which another agency might be designated, but rather the failure to develop or to provide an appropriate process for developing a federal policy with respect to the structuring and reform of state and local government. Although the Reuss proposal suggests some elements of such a policy, partly by identification of the areas in which the states would be expected to develop reforms and partly by suggesting the direction which reform should take, the main thrust of the proposal is toward state initiative and state determination of a reform policy. Such a conception, of course, accords with our traditional notions concerning the federal role with respect to state and local government, but it is a fair question whether it is adequate to deal with the dilemma with which this paper began.

CONCLUSIONS

The question how state and local governments shall be strengthened is closely linked with the question "for what purposes ought they be strengthened." Determination of "desirable" governmental organization or the content of other reform proposals requires attention to the substantive policies which one wishes to promote. Policies with respect to metropolitan governmental organization, for example, depend heavily upon whether we seek only economic efficiency or whether increased equity among the inhabitants of the region is also a goal. And if the latter, we need some definition of what is meant by "equity." If redistribution of income is to be achieved through national or state budgets, whether by transfer payments or support of selected services, issues concerning metropolitan governmental organization and taxing authority are likely to look very different than they would if redistribution is to occur to some extent at the metropolitan level.

The question how state and local governments are to be reformed and reorganized is not, in short, one with respect to which Congress can properly remain indifferent. A judgment that strengthened state and local governments are necessary to enable the nation to cope adequately with its urban problems must rest ultimately upon some conception of the roles which state and local governments are expected to perform in the years ahead. Responsibility for defining these roles appropriately belongs with Congress and the President as the most broadly representative of our political institutions.

REFERENCES

George Washington Law Review (1968) "Comment: Title VI of the Civil Rights Act of 1964–implementation and impact." 36 (May): 824-1022.

HAMILTON, A. (1885) "Report on manufactures to the House of Representatives, December 5, 1791." In The Works of Alexander Hamilton, Volume 3, New York: G. P. Putnam.

HELLER, W. (1966) New Dimensions of Political Economy. Cambridge: Harvard University Press.

University of Pennsylvania Law Review (1968) "Tondro, urban renewal relocation: problems in enforcement of conditions in federal grants to local agencies." 117 (December): 183-227.

BIBLIOGRAPHY

CAHN, E. S. and J. C. CAHN (1968) "The new sovereign immunity." Harvard Law Review, 81: 929.

MICHELMAN, F. I. and T. SANDALOW (1970) Government in Urban Areas. St. Paul, Minn.: West Publishing.

REUSS, H. (1967) "Revenue sharing as a means of encouraging state and local government reform." In Revenue Sharing and its Alternatives: What Future for Fiscal Federalism? Subcommittee on Fiscal Policy of the Joint Economic Commission. Washington, D.C.: Government Printing Office.

Royal Commission on Local Government in England (1969) Report of the Royal Commission. London: Her Majesty's Stationery Office.

TONDRO, T. J. (1968) "Urban renewal relocation: problems in the enforcement of conditions on federal grants to local agencies." University of Pennsylvania Law Review, 117: 183.

U.S. Advisory Commission on Intergovernmental Relations (1964a) Impact of Federal Urban Development Programs on Local Government Organization and Planning. Washington, D.C.: Government Printing Office.

——— (1964b) Statutory and Administrative Controls Associated with Federal Grants for Public Assistance. Washington, D.C.: Government Printing Office.

——— (1961) Governmental Structure, Organization, and Planning in Metropolitan Areas. Washington, D.C.: Government Printing Office.

U.S. Commission on Civil Rights (1967) Southern School Desegregation, 1966-67. Washington, D.C.: Government Printing Office.

WEIDNER, E. W. (1955) "Decision-making in a federal system." In J. McMahon (ed.) Federalism: Mature and Emergent. New York: Doubleday.

PUBLIC EXPENDITURES

How the Pie Gets Divided

8

Determinants of Local Government Expenditures

GAIL WILENSKY

☐ INTEREST IN the relationship between social and economic characteristics and state or local government expenditures is a relatively recent phenomenon. Before 1950, the subject had been generally ignored. Since that time, the number of these "determinant studies" has reached almost epidemic proportions. Roy Bahl (1968a) has compiled a list of sixty-six studies on state or local government expenditures available as of the spring of 1967 and does not claim to have completely exhausted the supply. Since 1967, expenditure studies have continued to flourish (Kee, 1968; Osman, 1968; Pidot, 1969; Sharkansky, 1967; and so forth).

No attempt will be made to review all or even most of the studies. The interested reader should refer to the excellent and recent review of this literature by Bahl (1968a). The purpose of this paper is to consider the usefulness of these investigations for the policy maker. To do this, it will be necessary to review at least a few of the studies. The intent of the review is only to provide

AUTHOR'S NOTE: *I am very grateful to Alan Ginsburg and Gunter Schramm for their helpful comments and suggestions.*

some indication of the types of variables which have been examined, the way in which they have been examined, and the general nature of some of the findings. Brief consideration is then given to some of the limitations of determinant studies. The remainder of the chapter is concerned with three issues which have been raised in determinant studies, the likelihood that the present set of studies can provide or have provided answers to these issues, and where relevant some other directions which might be taken.

To anticipate the conclusion, it is my opinion, although not uniquely mine (Morss, 1966; Bahl, 1968a; Ginsburg et al., 1969), that the present set of studies do not, and in their present form cannot, provide relevant answers to many of the crucial issues which most concern public finance economists and policy makers. This is not to say that the determinant studies have been useless. At the time they were initiated, little or nothing was known about the association between economic or demographic characteristics and governmental expenditures. While this information has been of limited use, primarily because it frequently tells us little about the underlying structural relations, the studies have provided us with a first approximation of the important variables. Furthermore, for some purposes the information gained from these studies may be sufficient. In particular, as long as there are no or few changes in the structural relations, predictions based on the results of determinant studies should be reasonably accurate.

STATISTICAL TECHNIQUE

The statistical technique most commonly used in the determinant studies is a single equation, cross-section, multiple-regression analysis in which variations in expenditures are explained in terms of variations in social and economic characteristics. The fact that they are cross-section studies means that they are concerned with differences in expenditures relative to differences in socioeconomic characteristics or determinants at a given time.

In a regression analysis, the regression coefficients reflect the effect on the dependent variable—per capita expenditures—which results from a small change in a particular independent variable—density or some other social or economic characteristic—assuming all of the other independent variables are held constant. Diagrammatically, we can depict this relationship in the following way:

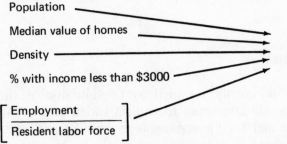

DETERMINANTS

Population

Median value of homes

Density

% with income less than $3000

$\begin{bmatrix} \text{Employment} \\ \hline \text{Resident labor force} \end{bmatrix}$

EXPENDITURES

Per capita expenditure
on some service

These forces can be summarized in a regression equation of the following form:

$$\text{EXPEND}_K = a_0 + \sum_{i=1}^{n} a_i x_i$$

where EXPEND_K = per capita expenditures on the Kth functional category, e.g., police

a_0 = constant term

x_i = value of the ith determinant of expenditures, e.g., density

a_i = regression coefficient or weight for the ith variable

n = number of expenditure determinants

A few studies have departed from this generalization. Bahl and Saunders (1965) and Kee (1965) explained changes in expenditures over a specified period in terms of changes in socioeconomic characteristics during that same period. While this approach allows for some description of a pattern of change over time, it is basically still a cross-section analysis. Fredland et al. (1967) used both cross-section and time-series data to explain variations in state expenditures. Wood (1961) used a factor analysis rather than a regression analysis and Pidot (1969) used a principle components analysis to reclassify socioeconomic characteristics into a small group of independent "underlying" variables.

DETERMINANT STUDIES

Determinant studies[1] have been done on state, state and local, and local government expenditures. Some have been concerned with total expenditures, others with expenditures on specific functions and still others with both. A few of them are highlighted below.

STATE AND LOCAL STUDIES

Fabricant (1952) is usually regarded as the initiator of the determinant approach. He attempted to explain variations in 1941 expenditures by state and local governments in terms of *per capita income, population density,* and *urbanization.* He found that all three were important for both total operating expenditures and expenditures on specific functions, and that income was the most important of the three.

Fisher (1961, 1964) repeated the analysis with 1957 data and then extended it by including more socioeconomic variables. He found that the "Fabricant three" explained less of the variation and that the prevalence of low per capita income was important. As Fisher recognizes, however, these results may not be reliable because of the high correlation between the level and the distribution of income.

Sacks and Harris (1964) added federal and state aid as independent variables. They found that both were important and that the Fabricant three were less important. They concluded that the declining importance of the three was due to the increasing importance of intergovernmental aid, a conclusion which is contrary to the results of a more recent study by Bahl and Saunders (1965).

As a result of these studies, it appears that per capita personal income and intergovernmental aid are important in explaining variations in state and local government spending. Density and urbanization are also important but less so than in early studies. Unfortunately, this information is not very useful for policy purposes. To know that federal or state aid is a significant expenditure determinant would be valuable. However, as will be argued in a later section, the results established so far are unreliable. More importantly, analyses of state and local governments combined represent decisions made by very heterogeneous groups. If we are to understand the forces which influence expenditures, we must disaggregate further and analyze the decision-making units themselves.

LOCAL GOVERNMENT EXPENDITURE STUDIES

Most of the attention has been focused on local government expenditures. These include studies done at the county level, the metropolitan level, the city level and in cases where school districts are independent units, the school district level. Studies done at the county level (Adams, 1965; Vieg, 1960) and those which focused exclusively on education (Renshaw, 1960; Shapiro, 1962; Miner, 1963; Bishop, 1964) are ignored in the section below.

In an early study, Hawley (1951) used a large number of socioeconomic variables to explain variations in per capita 1940 city expenditures for seventy-six United States central cities. Among the variables he found important were population and housing density in the central city and population size, percentage of population incorporated, and housing density in the rest of the metropolitan area. All of them were positively related to expenditures. Among the variables he found unimportant were population size and growth, number in the labor force and area in the central city, population density and growth number in the labor force, and area in the rest of the metropolitan area. Clearly the most interesting finding is that the size and the percentage of the population in the metropolitan area outside of the central city is important in explaining the central city's per capita expenditures while the central city's own population is not.

Brazer's (1959) analysis of 1951 municipal and metropolitan expenditures is still the most comprehensive and authoritative study to date. Because of this, his findings are presented in more detail and then some of the results of later studies are related to them. His work utilized five data samples: a large sample of 462 cities; three smaller samples of cities in California, Massachusetts, and Ohio; and a small sample which included the forty largest cities. The forty largest cities' sample included expenditures by overlapping governmental units in addition to expenditures by the cities themselves. Regressions were run for specific expenditure functions such as police, fire, sanitation, and so forth and for total general operating expenditures.

In the 462-city analysis, income was found to be a significant factor for all specific expenditure functions. Intergovernmental revenue was significant and positively related to expenditures for all functions; population density was significant for all functions except recreation, and positively related to expenditures for all functions other than highways. Except for police expenditures, population was unimportant. Population growth and employment in manufacturing trade and services were important for one or two selected functions but in general were not significant.

Bahl (1968b) did a study similar to Brazer's using 1950 and 1960 data for 198 central cities and in general found the same variables to be important.

There were several interesting findings in Brazer's analysis of cities in the three separate states relative to his findings for the larger sample of cities. The first is that, in each case, more of the variation is explained when the analysis is confined to cities within a single state. This, in itself, is neither surprising nor disturbing; it merely suggests that the structure of financial relationships between a state and its local governments is important. What is disturbing, although not surprising, is that the variables which were important in one state were not always important in another state nor were they always the same variables that were important in the interstate sample. In Massachusetts, population was important for all expenditures except highways; density and per capita income were important for only three or four selected functions and intergovernmental revenue was unimportant. In Ohio income and population generally were important, while population density and intergovernmental revenue generally were not. As Brazer notes, however, the importance of population may be the result of the smallness of the sample and the nature of the population size distribution and therefore may be spurious. In California, population *growth* was consistently important; income, intergovernmental revenue, and employment in three major sectors were important for selected functions. Scott and Feder's (1957) study of total 1950 city expenditures for 196 California cities was generally consistent with Brazer's findings, although they found property value and retail sales to be more important factors than income.

The primary advantage of limiting the analysis to cities within a state is that it eliminates much of the statistical "noise" due to differences in functional responsibilities, and to historical or geographical peculiarities. It also apparently limits the generality of the findings. This is unfortunate, but if the determinants of local expenditure are, in fact, specific to a particular state or region, this information is useful.

Several people have gone one step further and confined their analyses to local government expenditures within a single metropolitan area. Sacks and Hellmuth (1961) looked at 1956 data for villages and cities in the Cleveland metropolitan area. The traditional variables did not account for much of the variation in expenditures, probably because of the large differences in program responsibilities between villages and cities. When the analysis was confined to villages, assessed property valuation was important, while measures of population were not, although there was some problem of reliability because of intercorrelations among the variables. Bollen (1961) analyzed 1955 expenditures for local governments in the St. Louis metropolitan area. He found that assessed property valuation and an index of service quality were important explanatory variables while population and population density generally were unimportant. He did find, however, that dwelling unit density was negatively related to refuse collection. Sunley (1968) analyzed 1960 city expenditures in four separate metropolitan areas. He found that per capita income, the prevalence of low income, employment relative to resident labor force, and distance from the central city were the most frequently important variables. He also found that the regression equation for suburban expenditures in one metropolitan area did not provide good estimates for suburban expenditures in other metropolitan areas.

A final set of studies involves the analysis of expenditures by overlying units of government. In his study of the forty largest cities and their overlying units, Brazer found that population density, the ratio of city population to metropolitan population, and intergovernmental revenue were important for most functions. By way of comparison with his larger sample, it is interesting to note that per capita personal income was generally unimportant. Kee (1965) analyzed 1957 expenditures for thirty-six central cities and their overlying governmental units. Although the specific expenditure functions analyzed were not exactly the same as Brazer's, the results were quite similar. One exception was that income was found to be an important variable. Pidot (1969) analyzed expenditures for the eighty-one largest metropolitan areas. His study is unique in that he used a principal components analysis to create six independent measures which were then assumed to describe basic characteristics of a metropolitan area. He found that the "degree of metropolitan development," the "level of general wealth," and an "index of size" were important and in general were positively related to expenditures. He also found that state aid was significant for most expenditure functions and that while federal aid was significant for some functions, it was less important and less consistent in its effect.

On the basis of the studies outlined above and several others not specifically mentioned, some generalizations are possible. Per capita personal income or a variable which is related to income, such as property valuation is generally important in explaining expenditure variations. Intergovernmental revenue is frequently important. So is population density which in almost all cases is positively related to expenditures. In most studies and for most expenditure functions, population is unimportant in explaining variations in per capita governmental expenditures. Some measure of the nonresident population likely to make use of the city's facilities is frequently important and generally positively related to the city's expenditures. The most frequently used socioeconomic characteristics can explain more of the variation in some functions, such as police and fire than they can for other functions, such as sewerage and recreation. Finally, limiting the analyses to intrastate or intrametropolitan local government expenditures increases the amount of explained variation and illustrates the importance of regional, historical, political and administrative influences peculiar to a particular locality.

CONCEPTUAL AND METHODOLOGICAL LIMITATIONS

While determinants studies have clearly added to our knowledge about local government expenditure patterns, conceptual and methodological limitations have made it difficult to interpret some of the results and have restricted the usefulness of the findings. Not all of the limitations are major ones, nor are they confined to determinant studies.

INTERCORRELATIONS

One of the assumptions of a regression analysis is that the independent variables, in this case the social and economic characteristics, are truly independent, i.e., are uncorrelated with each other. In practice, this is rarely true. If there is a substantial amount of intercorrelation, which has been the case in some of the determinant studies, the results may be highly questionable or even meaningless altogether. Examples of variables which are correlated with each other are income and education; state aid to education and assessed property value; and population, density, and area.

Two ways of circumventing this problem have been used. One is to use a single variable as a proxy for both itself and for one or two other closely related variables. This is generally acceptable as long as the results are interpreted accordingly. A second way to eliminate intercorrelation is to use a principle components analysis. This technique creates a small number of new variables which are linear combinations of the old variables. While the new components have convenient statistical properties, it is not always obvious what the components represent (see Pidot, 1969). It is also not very helpful if one is interested in the effects of a specific variable such as, for example, state aid.

A good way to reduce at least some of the intercorrelation is to choose variables on the basis of a logical theoretical framework. Regrettably, the need for such a framework has not always been recognized. As several observers have noted, the search for variables in determinant studies often appears to have been almost casual and guided by convenient data availability rather than logic.

CROSS-SECTION VERSUS TIME-SERIES ANALYSES

As already mentioned, the existing studies have been based on cross-section data. An alternative would be the use of time-series data. In theory, the choice between cross-section and time-series analyses depends on whether one is interested in the aggregate pattern of, say, public spending at a given point in time or in the trend in public spending for a particular community over time. In practice, the choice frequently depends on the kind of data that are available. Since income and other socioeconomic variables are usually not available for local governments on an annual or a semiannual basis, determinant studies of necessity have been restricted to cross-section analyses. Policy makers, however, are likely to be at least as interested in why expenditures change as they are in why expenditures differ. And although it is done frequently, drawing dynamic inferences from static differences can be misleading. Essentially we must assume that what is true among different communities at a given point in time will be true for a particular community over time. Obviously little else can be done about this problem until adequate time-series data are made available.

OPERATING VERSUS CAPITAL EXPENDITURES

The use of cross-section data raises the question of whether to use total expenditures or operating expenditures. Whenever possible the usual procedure has either been to eliminate capital expenditures and explain only operating expenditures or to run separate regressions for operating and capital expenditures. This procedure is necessary because of the lumpiness of most capital expenditures —once built they may remain in service for years without requiring replacements or additions. The smaller the political unit, the more nonrecurring various types of capital expenditures are likely to be.

The exclusion of capital expenditures raises serious problems. At least two assumptions must be made in order to justify the exclusion. First, the ratio of capital to operating expenditures must be the same regardless of the scale of output of municipal services; New Haven has the same ratio of capital to operating as New York City. This is a questionable assumption. Substantial economies of scale may be available for at least some municipal services. And while the economies of scale may affect both operating and capital expenditures, the effect will not necessarily be proportional. In a study of sewerage treatment plans, for example, Robert Smith (1968) presented data which show that ratios of capital to operating and maintenance costs were 1.5, 2.2, and 8.5 for plant capacities of one, ten, and one hundred million gallons per day. Although both fixed costs and operating costs per unit of output decline as the size of the plant increases, they decline at different rates. The second assumption is that for a given amount of output, capital and operating expenditures are nonsubstitutable. For many services, this also is an unreasonable assumption. General city accounting tasks, for example, can be handled by a small army of clerks and accountants or by a few operators and a computer installation. If substitution is possible, the inclusion of only operating expenditures gives a distorted relationship between the socioeconomic determinants and actual output. The practical effect is that socioeconomic characteristics which are in fact important determinants may not appear to be so, whereas others that appear to be important may represent no more than suurious relationships.

The use of time-series data is not likely to solve the problem because of the longevity of some capital structures. It would also raise so many other problems, such as variations in the population growth rate, inflationary price changes and relative price changes, that the results would be difficult or impossible to interpret.

The problem is not readily solvable. One could correlate measures of capital expenditure with the regression residuals but the existence of correlation would only confirm the fact that capital expenditures are important. We already have indirect evidence of the problem in that typically we are able to explain more of the variation in expenditure functions where capital is less important (e.g., police) than we can for other functions (e.g., sewerage).

ESTIMATION OF A SINGLE EXPENDITURE EQUATION

A major limitation of the existing set of determinant studies is that they only estimate a single relationship between per capita expenditures and various social or economic characteristics rather than estimating the underlying structural relationships which, in fact, determine expenditures. In other words, we can view expenditures as being determined by a set of behavioral relations; some defining supply relations, some defining demand relations, some defining political relations. While we can reduce the set of equations to a single relationship between expenditures and socioeconomic characteristics by mathematically manipulating the equations, once this is done, there is no way of calculating the separate effects of the variables on each of the equations.[2] What this means is that the regression coefficients only reveal the *net* effect of variables which may influence *both* the cost of and the demand for particular expenditure functions. Furthermore, confining the analyses to per capita expenditures increases the ambiguity of the results since expenditures tell us little or nothing about quality or quantity of public services provided. Elsewhere, it is shown that even with the use of a highly simplified model, the coefficients of the standard social and economic variables are strkingly complex (Ginsburg et al., 1969). For present purposes, a simple example will suffice. Suppose, as has been found, density is positively related to police expenditures. We might expect that an increase in density increases the demand for police services since crime rates and traffic problems are likely to be influenced by the amount of contact and interaction between people. We might also expect that this effect might be partially offset by a reduction in per unit costs. At least within a given range, an increase in density may mean that an increased number of people could be served equally well by a given policeman. All we can say is that, apparently, the increase in demand is greater than the reduction in costs, if there is any reduction in costs. It should also be clear that the determinant studies can never say anything about a change in costs since the dependent variables are measuring expenditures rather than costs or outputs.

Formulating and estimating the structural relationships are not easy. But they clearly will be necessary if we are to gain anything more than an impressionistic understanding of the forces which influence public expenditures.

In order to assess the value of these studies it is instructive to consider three issues which have been raised in the determinants literature. The purpose is to see how much information they have provided or can provide and what else needs to be done.

GRANTS

Intergovernmental revenue has frequently been found to be an important determinant of local government expenditures. While this finding has important policy implications, the results are unreliable because of a misspecification of the variable and errors in the estimation procedure.

What has usually been done is to include federal aid, state aid, federal and state aid, or federal or state aid for a few specific functions as an independent variable, total state and local or local expenditures either by function or in total as the dependent variable. If the estimated coefficient is between zero and one, a dollar of aid is associated with less than a dollar of expenditures and the aid is said to be substituting for state or local expenditures. Similarly, if the coefficient is greater than one, a dollar of aid is associated with more than a dollar of state or local expenditures and the aid is said to be "stimulating" state or local expenditures. Some of the studies have found federal and/or state aid to be highly stimulative, some have found them to be slightly stimulative, some have found them to be substitutes for local expenditures. Part of this variation is expected. The effect of federal or state aid on one expenditure function may not be the same as its effect on another. But part of it is not. As Pogue and Sgontz (1968) demonstrate, the estimates are biased and causation cannot be inferred if the aid is determined by expenditures or if the aid and expenditures are jointly determined. This is clearly the case when the aid is in the form of a matching grant. In addition, the estimate is unreliable if the aid is determined by or related to a variable which itself is an important determinant of expenditures. This is likely to occur whenever aid is a direct or an inverse function of income or assessed property valuation. In this case, the problem is one of intercorrelation among the independent variables (determinants). Third, inclusion of aid as an independent variable and as part of the dependent variable may result in a significant correlation where no causal relationship exists simply because aid may be a large part of total expenditures. One or more of these problems have occurred in most determinant studies.

Although it is implied in the above paragraph, the major problem can be stated more explicitly. There are two basic types of grants: matching grants and nonmatching grants. Matching grants are analogous to price reductions. For one dollar of its own expenditures, the locality can purchase more than one dollar of a service. How much more depends on the matching ratio. Nonmatching (or block) grants can be either categorical or noncategorical: the grant must be spent on a particular function or it can be spent on any function.[3] Nonmatching noncategorical grants (sometimes referred to as unconditional grants) are conceptually the same as an equivalent increase in the community's disposable income. While the community may have to spend the grant dollars on a municipal function, it is also able to reduce its own taxes by the full amount of the grant. It is, therefore, effectively able to spend as much or as little of the grant as it wishes, just as it would be with an increase in income. Categorical nonmatching grants are also the same as an increase in income. Again, requiring the community to spend the grant dollars on a particular function does not insure that it will continue to spend what it would have spent. Since a matching grant is analogous to a change in price and a nonmatching grant is analogous to a change in income, it is clear that their effects should be estimated separately. This has rarely been done.

The effects of a matching grant depend on the matching ratio and on the community's responsiveness to a change in the price of the service being matched. Both of these are likely to vary. The matching rate for highways, for example, is different from the matching rate for welfare and the community's response to a change in the price of highways is likely to be different from its response to a change in the price of welfare. This means that the effects of a matching grant should be estimated program by program. To estimate the effect of a matching grant, we need to estimate the community's response to a price change for municipal services. If we can make this estimate, knowing the matching rate, we can calculate the effect of the grant. For nonmatching grants, we could assume that the effect is the same as an increase in income and calculate the effect of the grant on the basis of what we know about the relationship between income and the particular expenditure function. Alternatively, it might be assumed that even though the nonmatching grant is conceptually the same as an increase in income, the community may tend to spend more of the grant than they would from an equal increase in income. The effect of the grant on expenditures can be estimated directly as long as the grant is not directly or closely related to another variable which determines expenditures. Unfortunately, this occurs frequently. If there is a substantial amount of interrelationship or intercorrelation, there is little that can be done to get a reliable estimate of the separate effects of federal or state aid.

A side advantage to viewing grants in terms of being matching or nonmatching grants is that it places an outside boundary on results which can be regarded as being reasonable. Specifically, if the grant is nonmatching, then, the most that can happen is that the full amount of the grant is spent. In other words, "stimulative" nonmatching grants are unreasonable.

Some attempts to move in the direction suggested above have been made. Gramlich (1968) has estimated the effect of matching grants and unconditional grants, but not the effect for specific programs. He found that the effects of an unconditional grant were somewhat greater than the effects of equal increases in income. He also found matching grants to be moderately "stimulating" although this finding was determined by one of his assumptions (see Barlow, 1969; Gramlich, 1969).

A thorough understanding of the effects of grants has important implications for policy makers. If we can estimate how various types of communities are likely to respond to particular types of grants, a well designed system of grants would provide one way of altering expenditures from what they would be to what we think they should be.

ECONOMIES OF SCALE

The question as to whether or not there are economies of scale[4] in the provision of public services has long been of interest to policy makers. Although it is not the only consideration, it clearly is important in determining whether consolidation or metropolitan growth in general, is desirable.

Several of the determinant studies have reported finding no observable economies of scale. The basis of this conclusion was that population was either unrelated to expenditures or positively related to expenditures. By implication, observation of a negative relationship would have suggested economics of scale. Economies of scale exist when an increase in output is associated with a decline in the average cost per unit of output. This occurs whenever an increase in output allows for a more efficient combination of inputs than was previously possible. Some capital equipment, for example, may be extremely efficient but impractical unless output is very large. The apparent logic of associating a negative relationship between per capita expenditures and population with scale economies is the following: as population increases, output must be increasing and if this increase is associated with a reduction in "cost" (i.e., per capita expenditures), there is evidence of economies of scale (see Bahl, 1968a). However, this reasoning is fallacious. It seems clear that population is not a very good proxy for output and that per capita expenditures is not a good proxy for per unit costs. Furthermore, if there were scale economies present, we would not necessarily expect expenditures to decline.

On the contrary, they may well increase instead if it happens that the demand for the particular government service is elastic, i.e., decreased costs per unit of output could lead to increased per capita expenditures for that particular service.

To learn anything about the existence or absence of scale economies, it is necessary to estimate the long-run average cost functions for specific government services. The difficulties encountered in making such estimates are considerable. In addition to needing the costs of both inputs and outputs, it is necessary to solve the more basic problem of defining output. Defining output for sewerage or refuse collection is not too difficult. Defining output for services such as police, fire, education, and health is immensely difficult. Furthermore, it is necessary to take account of quality differentials.

A few attempts to estimate cost functions have been made. Hirsch (1965) estimated a cost function for residential refuse collection for St. Louis city-county cities and municipalities. Using the number of pick-up units as a proxy for refuse collected, he found no evidence of scale economies. Nerlove (1961) estimated a cost function for electricity and found evidence of significant scale economies. As mentioned earlier, Smith (1968) found evidence of scale economies in sewerage treatment plants. The studies mentioned above have made use of ex-post-statistical data. Will (1965) has used a different approach. His study is based on engineering specifications which are related to service level and service requirements. With the help of professional expertise, he estimated per capita standard service requirements for fire protection for thirty-eight cities having various population sizes. He then regressed the dollar value of the service requirements against population and found evidence of economies of scale. While most of the scale economies were realized by populations of 300,000, some existed for populations up to a million, the largest city in his sample.

Most estimates of cost functions thus far have been for services having more readily definable outputs. While the conceptual and empirical problems encountered in estimating cost functions for the remaining services are formidable, some attempts will have to be made if we are ever to establish the existence or absence of economies of scale. This is clearly one instance when we have to refer to the underlying structural relations if the results are to be useful for policy purposes.

EXPLOITATION

Several of the determinant studies have found a positive relationship between expenditures and measures of the nonresident
population likely to make use of city services. This has led to
speculation that central cities are being exploited by the suburbs or
that some suburbs are being exploited by other suburbs. The term
exploitation is used to mean that suburbs impose additional costs
on the central city or on other suburbs without providing full
payment for these costs. Since determinant studies have been
concerned with explaining expenditure variations, they obviously
cannot resolve this issue. We have some estimates of the effect of
nonresidents on the expenditure side although focusing on the
expenditures would exclude what is likely to be a major cost
imposed by nonresidents—the private cost of congestion. We do
not have estimates of the direct and indirect tax contribution of
nonresidents to the city.

There are two obvious types of exploitation which can occur.
Suburbanites may impose costs on the city in their role as worker-
commuters. Suburbanites may also impose costs by their use of the
city's facilities for shopping, entertainment, and so forth. In both
cases, the increased costs are likely to be in terms of roads, traffic
control, police protection, use of subsidized transportation systems,
etc. While there is no simple solution for estimating the direct and
indirect tax contributions of outside users, a brief discussion of
what we would like to estimate may be instructive.

Commuters are likely to contribute to the city's revenue in one
or more of the following ways: nonresident income taxes; user
fees; sales taxes, either directly or by the tax on output produced
by commuters in resident industry and commercial establishments;
increases in property taxes from resident industries and commerce
which employ commuters; and the positive differential (if only)
between per capita suburban versus central-city contributions to
and receipts from state taxes that are turned over to local governments on the basis of some per capita sharing formula.[5] One way
to begin approaching the problem would be to make use of
information provided by nonresident income tax returns. While this
information is not readily available in published form, information
regarding the number of nonresident filers, their income and their
tax contribution is sometimes available on a city-by-city basis. This
would provide the data for nonresident income tax and might also

provide for a crude way of estimating the portion of the sales tax attributable to commuters.

Other nonresident users may indirectly contribute to the city's taxes by increasing the value of city property and, therefore, increase property tax collections. They may directly contribute to taxes via the sales tax or via user charges. One way to estimate nonresident contributions to specific city facilities would be to estimate the ratio of nonresident to resident users of city provided or supported facilities (e.g., museums and zoos). While some of the facilities have user charges, most of them are, at least partially, tax supported. The allocation of the costs of the tax subsidies could then be allocated on the basis of the percentage shares of the nonresident, resident user groups (see Neenan, 1969).

A third and somewhat more subtle type of exploitation is also possible. Central cities have tended to have increasingly greater concentrations of low income, poorly-educated residents, both of which increase demands for welfare, public health, public housing, police protection, and so forth. To the extent that suburbs force or reinforce this concentration by zoning regulation or discriminatory practices in real estate transactions, suburbs can be regarded as exploiting the central city (see Brazer, 1961; articles by Noll and Riew, in this volume).

Increased use of user charges wherever possible would reduce some of whatever exploitation exists, but is unlikely to serve as a way of reducing all of it. Moreover, even if no monetary exploitation exists, the fact that nonresidents make use of a city's facilities is sufficient to cause nonneutrality in the allocation of resources to the public sector. We can expect that in deciding how much to spend on a particular municipal function, residents will only consider their own demands and not the demands of nonresident users. Furthermore, except when user charges are in effect, nonresident users have no way of making their preferences for services in another city effective, irrespective of whether or not they "exploit" the other city. The result will be an undersupply of services relative to the wishes of all who use a city's facilities. It is this fact, rather than the actual existence of exploitation which together with economies of scale argues for consolidation. The argument against consolidation is that it forces what may be very diverse groups to accept a common set of municipal services (see Brazer, 1962). While decisions regarding consolidation will not be made on economic grounds alone, we do not as yet know enough about the economic consequences of consolidation or the failure to consolidate to be able to offer useful guidelines to the policy maker.

CONCLUSION

The subject matter of the determinant studies clearly is relevant to the policy maker. Unfortunately, their findings in general, have not been. The results of determinant studies are suggestive. They indicate the general pattern of effects of various social or economic characteristics on expenditures. For some purposes, this may be sufficient. The approaches used so far provide at least some results which are helpful for forecasting although the use of time-series data would be better. Furthermore, the effects, at least of non-matching grants, could have been estimated although, so far, the distinction between matching and nonmatching grants generally has not been made and the results have, therefore, been unreliable. For most purposes, however, the present set of studies merely provides the groundwork for future investigations.

If we are to gain a thorough understanding of the forces that influence expenditures and the way in which they influence them, we will have to begin trying to estimate the structural relationships. This means we will have to define and estimate cost and supply functions, attempt to relate the supply price to the price which residents pay, and this price plus such factors as income and nonmatching grants to the residents' demand for services. Furthermore, we will have to account for the forces which influence nonresident tax contributions and the way in which they affect both the price which residents pay and the residents' demand for their own services. As is frequently the case, this advice is easier to give than to follow. The problems involved in formulating the structural relationships and in estimating them empirically are formidable. It implicitly requires the development of a "positive" theory of public expenditures. In other words, a theory which explains why expenditures are as they are rather than what they should be. This, in itself, would be a major advantage. At least some of the conceptual and methodological errors which have occurred could have been avoided by the use of a logical theoretical structure. In many cases, estimation of the structural relationships will require more intensive study of fewer areas than has occurred in the past since the detailed information which will be needed is unlikely to be readily available.

Understanding why expenditures are as they are, clearly is vital to changing expenditures to be what we think they ought to be. If we know the forces that tend to increase or decrease expenditures and think they are inappropriate determinants, we can attempt to change them by changing the characteristics of the municipality or by changing the expenditures directly.

One final point needs to be mentioned. Determinant studies implicitly assume that individuals' preferences determine the allocation of resources to and within the public sector. Some theories of political behavior are at least compatible with this view (see Downs, 1957). But, even if we do assume that individual preferences determine this allocation, we must recognize that the relationship is an indirect one. This is particularly true with regard to the allocation of resources *within* the public sector. While voters frequently have direct control over the overall level of local government expenditures, they rarely have direct control over the allocation of resources among expenditure functions. At best, individuals' preferences are likely to influence the within-sector allocation only after a considerable lapse in time. In the short run, the allocation is probably largely influenced by political or institutional decision-making processes.

More explicit recognition of the influences of political institutions is likely to be necessary if we are, in fact, to understand the determinants of public expenditures (see Meltsner and Wildavsky, in this volume).

NOTES

1. This section relies heavily on the summaries of the literature by Bahl (1968a), Hirsch (1968), and on a tabular summary, "Multivariate Studies of the Determinants of State and Local Government Expenditures in the United States," prepared by Robin Barlow for the Ford Foundation Workshop on State and Local Government Finance, Ann Arbor, June, 1966.

2. In technical terms, the problem is that regression equations provide estimates only of the reduced form relation. Under certain conditions, the parameters of the structural equations can be derived from the reduced form relation. Unfortunately, these conditions are rarely fulfilled.

3. Matching grants can be categorical or noncategorical, but, in fact, are always categorical. In other words, the matching is in reference to expenditures on a particular function, e.g., highways. Grants are sometimes referred to as being conditional or unconditional. In this case, attaching any stipulation to the grant makes it a conditional grant. Thus, a conditional grant would include matching grants, nonmatching categorical grants, nonmatching categorical grants requiring a minimum expenditure, and so forth.

4. For a more detailed discussion of economies of scale and other supply relationships, see Hirsch (1968).

5. Such a partial contribution by at least some suburbs to central-city revenues were recently found for the Detroit SMSA by William S. Neenan (1969).

REFERENCES

ADAMS, R. F. (1965) "Determinants of local government expenditures." Review of Economics and Statistics, 47 (November): 308-313.

BAHL, R. W. (1968a) "Studies on determinants of expenditures: a review." In S. Mushkin and J. Cotten (eds.) Functional Federalism: Grants in Aid and PPB Systems. Washington, D.C.: George Washington University Press.

――― (1968b) Metropolitan City Expenditures: A Comparative Analysis. Louisville: University of Kentucky Press.

――― and R. J. SAUNDERS (1965) "Determinants of changes in state and local government expenditures." National Tax Journal, 18 (March): 50-57.

BARLOW, R. (1969) "Comment on alternative federal policies for stimulating state and local expenditures." National Tax Journal, 22 (June): 282-285.

BRAZER, H. E. (1962) "Some fiscal implications of metropolitanism." In G. S. Burkhead (ed.) Metropolitan Issues: Social, Governmental, Fiscal. Syracuse: Maxwell Graduate School of Citizenship and Public Affairs.

――― (1959) City Expenditures in the United States. New York: National Bureau of Economic Research.

BISHOP, G. A. (1964) "Stimulative versus substitutive effects of school aid in New England." National Tax Journal, 17 (June): 133-143.

BOLLEN, J. [ed.] (1961) Exploring the Metropolitan Community. Berkeley: University of California Press.

DOWNS, A. (1957) An Economic Theory of Democracy. New York: Harper & Row.

FABRICANT, S. (1952) The Trend in Government Activity Since 1900. New York: National Bureau of Economic Research.

FISHER, G. W. (1964) "Interstate variation in state and local government expenditures." National Tax Journal, 17 (March): 55-74.

――― (1961) "Determinants of state and local government expenditures: a preliminary analysis." National Tax Journal, 14 (December): 349-355.

FREDLAND, J. E., S. HYMANS and E. L. MORSS (1967) "Fluctuations in state expenditures: an econometric analysis." Southern Economic Journal, 33 (April): 496-517.

GINSBURG, A. L., G. SCHRAMM and G. R. WILENSKY (1969) "Determinant studies: the need for new directions." Unpublished paper.

GRAMLICH, E. M. (1969) "A clarification and a correction." National Tax Journal, 22 (June): 286-290.

――― (1968) "Alternative policies for stimulating state and local expenditures: a comparison of their effects." National Tax Journal, 21 (June): 119-129.

HAWLEY, A. H. (1951) "Metropolitan population and municipal government expenditures in central cities." Journal of Social Issues, 7: 100-108.

HIRSCH, W. Z. (1968) "The supply of urban services." In H. S. Perloff and L. Wingo (eds.) Issues in Urban Economics. Baltimore: Johns Hopkins Press for Resources for the Future, Inc.

––– (1965) "Cost functions of an urban government service." Review of Economics and Statistics, 47 (February): 87-92.

KEE, W. S. (1968) "City-suburban differentials in local government fiscal effort." National Tax Journal, 21 (June): 183-189.

––– (1965) "Central city expenditures and metropolitan areas." National Tax Journal, 18 (December): 337-353.

MINER, J. (1963) Social and Economic Factors in Spending for Public Education. Syracuse: Syracuse University Press.

MORSS, E. R. (1966) "Some thoughts on the determinants of state and local expenditures." National Tax Journal, 19 (March): 95-104.

NEENAN, W. S. (1969) "Financing urban government." Unpublished paper.

NERLOVE, M. (1961) Returns to Scale in Electricity Supply. Palo Alto: Institute for Mathematical Studies in Social Sciences.

OSMAN, J. (1968) "On the use of intergovernmental aid as an expenditure determinant." National Tax Journal, 21 (December): 437-447.

PIDOT, G., Jr. (1969) "A principal components analysis of the determinants of local government fiscal patterns." Review of Economics and Statistics, 51 (May): 176-188.

POGUE, T. and L. G. SGONTZ (1968) "The effect of grant-in-aid on state-local spending." National Tax Journal, 21 (June): 190-199.

RENSHAW, E. F. (1960) "A note on the expenditure effect of state aid to education." Journal of Political Economy, 69 (April): 170-174.

SACKS, S. and R. HARRIS (1964) "The determinants of state and local government expenditures and intergovernmental flows of funds." National Tax Journal, 17 (March): 75-85.

SACKS, S. and W. F. HELLMUTH, Jr. (1961) Financing Government as a Metropolitan Area. Glencoe: Free Press.

SCOTT, S. and E. L. FEDER (1957) "Factors associated with variations." In Municipal Expenditure Levels. Berkeley: Bureau of Public Administration, University of California.

SHAPIRO, S. (1962) "Some socioeconomic determinants of expenditures for education: southern and other states compared." Comparative Education Review, 6 (October): 160-166.

SHARKANSKY, I. (1967) "Some more thoughts about the determinants of government expenditures." National Tax Journal, 20 (June): 171-179.

SMITH, R. (1968) "Cost of conventional and advanced treatment of waste water." Journal Water Pollution Control Federation, 40 (September): 1546-1574.

SUNLEY, E. M., Jr. (1968) The Determinants of Government Expenditures Within Metropolitan Regions. Ph.D. dissertation. University of Michigan.

VIEG, J. A. [ed.] (1960) California Local Finance. Palo Alto: Stanford University Press.

WILL, R. E. (1965) "Scalar economies and urban service requirements." Yale Economic Essays, 5 (Spring).

WOOD, R. C. (1961) 1400 Governments. Cambridge: Harvard University Press.

The City, City Hall, and the Municipal Budget

W. ED WHITELAW

☐ DURING THE LAST twenty years, there has been a large—even an enormous—number of studies attempting to find, identify, and explain the factors and forces shaping the municipal budgetary process. Although it is yet another comment on the expenditure and revenue activities of city governments, the present chapter is, hopefully, neither redundant nor unnecessary (for detailed analysis, see Whitelaw, 1968; forthcoming).

In the next section is a review of the major analytical descriptions of the municipal budgetary process that have appeared in the literature over the past twenty years. The section on the urban and federal scene contains a general description of urban development

AUTHOR'S NOTE: *The author is Assistant Professor of Economics and Research Associate with the Bureau of Government Research at the University of Oregon. This chapter is drawn largely from the author's doctoral dissertation (see References), the research for which was supported with funds from the Office of Economic Research of the Economic Development Administration in the Department of Commerce, Project OER 015-G-66-1. The author is indebted to the following individuals from whom he received numerous helpful suggestions on the dissertation: Professors Jerome Rothenberg, Edwin Kuh, Richard Eckaus, and E. Cary Brown at Massachusetts Institute of Technology, and Professor John F. Kain at Harvard University.*

and a few of the resultant municipal problems, a discussion of the constraints imposed by the federal government on municipal decision-makers, and an outline of some of the various intragovernmental constraints at the local level. Finally, a number of descriptive and prescriptive notions of the municipal budgetary process are present that constitute a step toward a theory of municipal finance.

ANALYTICAL DESCRIPTIONS OF THE BUDGETARY PROCESS

In this section is a summary of the various theories or conceptual views of the budgetary process with emphasis on those theories that are particularly relevant to municipalities. While the summary is designed to place this chapter in its proper perspective, it is not designed to be, nor is it, exhaustive.

Admittedly oversimplifying, we will divide the theoretical literature into two fundamental schools (Crecine, 1966). The naive position for the first school sees the sole source of influence internal to the municipal administrative structure, i.e., the primary determinants of government expenditures and revenues would be intragovernmental stimuli.

The naive position for the second school sees the sole source of influence external to the municipal government, i.e., the primary determinants would be economic, demographic, political, and/or social stimuli from the environment in which the municipal government is located.[1]

If one were to search for straw men defending each of these positions, he would do well to search for an "internalist" among organizational theorists and an "externalist" among economists, political scientists, or sociologists. In this subjective search, however, one would more likely run into a defender of a more sophisticated position who would concede the influence of other factors but who would consign them to positions of secondary importance.

Within the conceptual framework of organizational theory, the municipal budgetary process is viewed in terms of an internally negotiated, problem-solving sequence in an organizational environment (Cyert and March, 1963). The theory emphasizes a set of problems significantly different from those of economic theory;

specifically, virtually nothing is said about quantities, i.e., how output levels are set. An example of this qualitative emphasis is found in Argyris' (1959) view of the budgetary process wherein expenditure and revenue decisions are viewed almost exclusively in the context of interpersonal relations and their relevance to the enactment of change in the organization. The budget is viewed primarily as a means of applying never ending pressure to members of, or participants in, the organization.

Cyert and March (1963: 18) claim that a subdivision of organizational theory has derived its framework primarily from economics and originated historically in Ridley and Simon's work in 1938. Their work was directed almost exclusively toward constructing means of promoting and implementing efficiency in municipal government, and toward the organization theory subdivision that allegedly found its seeds in their study attempts to explain "how decisions are made in organization with special emphasis upon the processes of executive influence and the impact of organizational position on individual goals and perceptions."

To summarize, then, the emphasis of organizational theory, in the context of municipal government, is on the phenomena of bureaucracy, conflict among organizational subunits, coordination of departmental activities, and problems of the administrator dealing with an organization (March and Simon, 1958: 120-121, 123-125).

Abstracting, for the moment, from the numerous areas of debate, economists, political scientists, and sociologists tend to view the municipal budgetary process as one determined by environmental, economic, demographic, political, and social forces. In general, the municipal decision-maker is assumed to be, implicitly or explicitly, at least moderately rational as well as motivated to satisfy the needs and/or desires of the citizens of the municipality. The source of this motivation, i.e., an objective function, is discussed infrequently and is specified even more infrequently. To the extent that there is any discussion of the political procedure by which the needs of the voter taxpayers are transmitted to the municipal officers, a model is suggested which presumes that the flow of persuasion travels from the individual voters to the public officials (see, for example, Davis, 1964).

In those studies that might be described as economic by reason of the explanatory factors emphasized, the municipal officer is assumed to perceive the existence of needs through the economic and demographic environments within which his responsibilities lie, and then to attempt to satisfy those needs by providing municipal output of goods and services (Ando et al., 1965). An explicit attempt to apply the welfare maximization approach to municipal decision-making is embodied in Gramlich's macro-model of state and local finance.

The political scientists view the pattern of expenditures and revenues as evolving from the execution of the primary responsibility of municipal government which, in this admittedly abstracted view of the discipline, is the resolution of conflict, broadly defined, within the community. Dahl (1961; 1956) and the pluralists have emphasized the rationalist and individualist views of political structure and behavior. The pluralists see American society as fragmented into "hundreds of small special interest groups, with incompletely overlapping memberships ... and a multitude of techniques for exercising influence on decisions salient to them" (Polsby, 1963).

In order to explain why high-income groups in some municipalities appear to vote against their own self-interest, Banfield and Wilson (1964) have introduced a value premise into the individual voter's behavior that appears to bear some resemblance to an axiom of the organic theory of the state. They have postulated that voters of high socioeconomic status tend to vote on the basis of "public-regardingness" while those of lower socioeconomic status tend to vote in self-regarding ways.

A group of researchers (for example, Lynd, 1929), most of whom are sociologists and most of whose works preceded the recent spate of studies in this area, subscribes to what has been called the stratification theory of community power and, implicitly, of the municipal decision-making process. The basic proposition of the theory is that the pattern of the community power structure and of the municipal decision-making process is attributable to the community social structure and to the desires and interests of the "ruling elite," composed almost exclusively of members of the upper-income brackets of the community.

There are several important studies and topics that resist being slipped into the crude pigeon holes constructed for this taxonomic exercise. Rather than try to force them into categories, I will treat them separately.

Three individuals have developed a theory of political behavior by applying economic models to the structure and dynamics of the polity (Buchanan and Tullock, 1962; Downs, 1957). At the micro-political level, they assume the individual voter to be rational and motivated by a desire to maximize his self-interest in society, and the politicians (or government) to be rational and motivated by a desire to maximize political support.

In connection with the New York Metropolitan Regional Study, Wood (1961), with the assistance of Almendinger, developed a model of the local governmental process in which the influences of the static and dynamic characteristics of various environmental forces are linked simultaneously with the influences of the structure of government organization.

Since McNamara directed the Defense Department to adopt a planning-programming-budgeting-system approach to its annual budgeting, there has been a major controversy surrounding incremental versus comprehensive approaches to budgeting. In addition to debates on the normative issue of whether comprehensive, zero-base budgeting should be implemented, there is also disagreement on whether it is possible to implement it (see, for example, Weidenbaum, 1966). Avoiding the normative issue, there is evidence that at the municipal level, for medium-sized cities, rough approximations of zero-base budgeting are being used (Whitelaw, 1968).

THE URBAN AND FEDERAL SCENE

In the early development of the American metropolitan area, the municipal political boundaries usually coincided, almost by definition, with their respective, economically self-contained regions. Until 1920, the many municipal corporations were able to restrain the metropolitan population inside their boundaries only by expanding through annexation of the surrounding rural areas. Eventually, conflict arose between proximate localities and the process was halted.

Today there are few cities for which the local polity and economy coincide.[2] The typical pattern into which this divergence has evolved is a central city surrounded by suburban and ex-urban areas. As students of the urban scene are aware, the multiplicity and variety of governments serving metropolitan areas are enormous. Serving the Chicago metropolitan area in 1967, for example, were a total of 1113 governments, not counting the state and federal governments. These 1113 governments were composed of 6 county governments, 113 townships, 250 municipalities, 327 school districts, and 417 special districts.

Among the many factors influencing municipal revenue and expenditure decisions, none is so difficult to measure and yet so apparently significant as the set of constraints imposed by the federal system. Intergovernmental influences on any particular municipal government can come from all three levels of government—state and federal as well as other local governments. In general, there are three ways in which the state and federal governments can affect the budget policy of local governments: by delineation of the jurisdictional sphere within which the local units operate, by the provision of intergovernmental transfers, and by the impact of their respective fiscal decisions.

Constitutionally and financially, the city is inferior to the state. The municipal corporation is at once an artificial person created to serve as an agent of the state, and a governmental body serving the unique needs of the locality over which it has been granted jurisdiction (Banfield and Wilson, 1965). Even when it is serving the unique needs of the locality, however, it is still constitutionally subject to the dictates of the state.

There are various methods that states have adopted in order to grant powers to their cities. These methods range from special charters, which frequently approach and envelop the absurd in the detail of their requirements, to general charters, which in principle are designed to apply equally to all cities in the state (Banfield and Wilson, 1965: 65-66). The guidelines or, more accurately, the guideposts imposed by the state governments are prominent in the affairs of practically any municipal corporation in the United States.

The amount of direct state aid to city governments accounted for about 16.7% of total general revenues to the cities in 1955 and about 22% in 1968 (Department of Commerce, 1962). This amounted to more than $1.2 billions in 1955 and nearly $4.7 billions in 1968, and the latter was distributed over a large range of activities: school construction and operation, highways, welfare, health, airport construction, and parks and recreation.

In 1904, Professor Frank J. Goodnow observed, "As the city has no relations with the national government it is not necessary for our purpose that we make any study of the national administrative system." Although the city still does not exist in the eyes of the Constitution, the federal government plays a rather large role in the affairs of municipal governments—and the trend is in the direction of more federal influence.

The amount of direct federal aid to all city governments more than quadrupled between 1955 and 1968—from $202 millions to $941 millions—and this amount was distributed over the same activities listed above for state transfers plus low-rent public housing, urban renewal, and construction of sewage treatment facilities (Department of Commerce, 1962). Nonmatching federal grants undoubtedly affect the composition of the municipal budget since the relative costs or prices for various programs considered by municipal officers tend to shift in favor of those programs supported by grants. When federal aid is made available, subject to numerous conditions attached to acceptance, the nature of the federal influence is more direct and perhaps more distorting than when policy is effected through changes in relative prices.

In addition to its effect on local budgets through its direct participation, the federal government influences municipal financial decisions indirectly by its national activities in all three of Musgrave's (1959) conceptual branches—allocation, distribution, and stabilization. A tight money policy, for example, tends to have an initial, disproportional impact on residential and nonresidential construction which in turn constrains the tax base of the primary source of municipal revenues, the property tax.

If the complexities in the local budgetary process introduced by the federal system were limited to the three general areas of state and federal influence described above, the task of specifying a descriptive model would still be made rather difficult. There do exist, however, numerous other sources of influence attributable to the federal system that may effect variations in municipal budgetary policy. One of these is the interaction among local units, for rarely is there just one jurisdiction blanketing a particular area.

If the population distribution within a metropolitan area remained constant over time (and the overall level remained constant as well), and if the commercial and industrial patterns were unchanged, then perhaps there would exist a steady-state solution to the logistics problem of collecting and spending tax revenues in a metropolitan area that has been divided extensively into a seemingly random patchwork of governmental units. We doubt it. The hypothesis is hardly interesting, however, since in the process of posing it we have assumed to be constant the very factors whose changes give rise to the interaction—both conflict and cooperation —among overlapping jurisdictions.

> The economic units which comprise the city—the households and firms—are functionally and spatially differentiated. The organization of functionally differentiated economic units into an economy which satisfies the needs of the people has been the central topic of economic analysis. Spatial differentiation has been neglected, but herein lies the core of the metropolitan finance problems [Margolis, 1961].

The nature of the interaction among overlapping jurisdictions is subject to some debate. If the vertically and horizontally partitioned metropolitan governmental structure were able to promote cooperation among the various municipalities, counties, special and multiple function districts, then one might expect to see some tendency toward a market-like solution to the equalization of public benefits and costs.[3] Some observers suggest, however, that the prime movers in this tendency are private economic units. As they see it, intra-areal location decisions by households and firms are influenced by rational attempts to maximize public benefits while minimizing tax costs. There is no general agreement, however, whether the net effect of this competition is of negative or positive social benefit to the metropolitan area as a whole (Burkhead, 1963).

The concept of social benefit itself is subject to controversy. Since there is considerable suspicion, by this author at least, that one cannot conceive and deliver a social welfare function universally relevant to each actor on the urban stage, then to speak of a social benefit necessarily implies some presumption regarding the actor to whom the benefit accrues. The outcome of a rational choice among alternative municipal budgets will differ as the point of view differs.

From the citizen's point of view, matters are rather confusing. Not only does he act in a large number of locations, but even when he is in one location, he is faced by not one but many governmental institutions.

The extension of activities across jurisdictional lines makes it more and more difficult to relate benefits and taxes at the local government level. "In the modern metropolitan community, a family may reside in one jurisdiction, earn its living in one or more others, send its children to school in another, and shop and seek recreation in still others..." (Fitch, 1957).

From the local government's point of view, matters are equally confusing. "... to a considerable extent, the American local financial system still reflects the presumption that these various activities are concentrated in one governmental jurisdiction" (Fitch, 1957: 67).

Assuming the government is trying to satisfy some tangible needs, whatever its motivation, then for different functions it must satisfy the needs of different groups, subject to a variety of pressures. Not only are there myriad jurisdictions of a particular type within an area, but we also find a multiplicity of different types (Tiebout, 1956 and 1961). In those areas where the proliferation of units permits a closer approximation of individual tastes and government action, some observers (for example, Arrow, 1951) feel that precisely because of this correspondence, programs that are desirable in a broader social context are obstructed.

In the preceding paragraph, it was assumed that the municipal officers try to satisfy the tangible needs of the voter-taxpayers in the municipality in which their responsibilities lie. The actual means by which a municipal officer attempts to reach this goal is by the provision of goods and services through the expenditure of local revenues and intergovernmental transfers, subject to numerous technological and pecuniary constraints. A major source of complications in this process is the lack of an accurate and consistent method of measuring municipal output (Ridley and Simon, 1938). Although there is considerable difficulty measuring government output, municipal officers apparently find little difficulty changing its quality and thereby achieving "illusory economies in spending" (Pidot, 1965).

The municipal officers may face significant technological and/or pecuniary alterations in the factor inputs with which they attempt to produce municipal output. As a typical case in a central city, all other things being equal, the substitution of lower-class children for middle-class children in the public schools will result in a dramatic decrease in educational output if the unit of output is measured, for example, in terms of the number of pupils of a certain achievement (or performance) level per year. Similarly, for some constant level of health, air pollution appears to increase costs significantly. Furthermore, there is some suspicion that price increases are disproportionately concentrated in the local public sector (Eckstein, 1959).

TOWARD A THEORY OF MUNICIPAL FINANCE

In this section is described in a reasonably systematic fashion those factors that we feel may have an influence, over time, on the fiscal pattern of a municipality. Underlying the discussion are the behavioral assumptions that the municipal decision-maker is rational and attempts to satisfy the needs of the municipality. Furthermore, it is assumed that the needs of the community can be described by reference to specific economic and demographic variables in the community. Although our own notions of the community political process form a Madisonian view of municipal leadership, it is sufficient for the purposes of this chapter merely to imagine a political black box through which the needs of the community are perceived by, or transmitted to, the municipal decision-makers.

There is some doubt that a unique, general description of municipal budgetary processes exists, but even if it does, we would be neither willing nor able to perform such Sisyphean labor. In the discussion that follows, therefore, we will assume a municipality characterized by simple governmental structure, i.e., no special function districts and no independent school districts, and by little change in municipal boundaries, i.e., no major annexations.

It is important to emphasize the perspective from which this section is developed, namely that of the municipal officer or officers. Granted that there is no one individual or group who, in reality, makes all the decisions with respect to the municipal budget or even all the decisions regarding any one program in the budget, neither is there any single individual who could accurately represent an alternative perspective. This should not be interpreted to imply that this hypothetical officer is an autonomous decision-making unit for the municipal budget. Rather, it is a result of our goal in this section to describe the manner in which the municipal officers manipulate the fiscal instruments at their disposal in order to satisfy the needs of the voter-taxpayers in their community.

In order to suggest the importance of the budget constraint in municipal decisions and to describe, though simplistically, the role of welfare maximization at the municipal level, assume that community welfare, W, is a function of municipal output of goods and services, M, and of personal disposable income, Y, net of local taxes, T.[4]

$$W = w(M, Y - T)$$

where M and T are policy instruments available to the municipal officer who, it is assumed, tries to maximize this function subject to various taxing and borrowing constraints.

If the municipal officer tries to increase W (community welfare) by increasing M (municipal goods and services) and the increase in M involves increasing local taxes, then W is simultaneously decreased, a concept universally disliked during municipal elections and referendums. Financing any increased expenditures through borrowing results, of course, in the same trade-off. As Gramlich has argued, municipal governments "are generally required to balance their current operating budgets, and they must pay a default risk premium when they make use of credit markets." (Gramlich, unpublished mss. 1)

In order to understand the lags and timing in municipal expenditure and revenue decisions, it helps to know the number and length of the various stages in the municipal decision process. As an example, consider the municipal budget of Worcester, Massachusetts.[5]

On 1 December of calendar year T_0 (the fiscal year in Worcester corresponds to the calendar year), the department heads submit to the city manager, in addition to their budgetary requests for current or operating expenditures, a list of priorities for capital projects. This list orders projects according to their relative urgency and, outside of the two-period case depicted in the outline below, will contain projects to be initiated in T_1 or projects initiated in T_0 or before, and are to be continued in T_1 (Phelps, 1961). It is assumed that the city manager plans capital expenditures with respect to (a) priorities of proposed projects, (b) expected intergovernmental revenues, and (c) expected monetary conditions. In this framework, the priorities are determined by environmental economic and demographic forces in the community.

The outline below depicts the budgetary process for the current expenditures of the city government of Worcester.

FISCAL YEAR	STAGES OF THE BUDGETARY PROCESS FOR FISCAL YEAR T_1	
T_0	STAGE 1	(October, November, December of T_0) Departmental requests are developed within the individual departments and submitted to the city manager.
T_1	STAGE 2*	(1 December to 17 February of T_1). The city manager draws up the budget message and submits it to the city council.
	STAGE 3*	(17 February to 1 April of T_1). The city council reviews the budget and makes recommendations. The city council can neither increase nor reappropriate funds for a specific category. It can only decrease the funds. At the end of Stage 3, planned expenditures are authorized.
	STAGE 4	(1 April to 1 December of T_1). Planned and authorized expenditures for T_1 are expected.

* The actual expenditures occurring during Stages 2 and 3, i.e., from 1 December to 1 April, cannot exceed the expenditure requests from Stage 1.

Every expenditure is therefore considered on three occasions or, more accurately, during three different periods: once by the individual departments, once by the city manager and his staff, and once by the council.

In the past, empirical studies of municipal finance have dealt usually with current expenditures, rarely with revenues and never with capital expenditures. As suggested with the elementary welfare function above, the instruments used by municipal officers to effect community welfare include all municipal goods and services —capital *and* current—as well as revenues. If municipal officers plan, program, and budget to satisfy the needs of their community, then they should consider benefits *and* costs. To blunt criticism of this reform zeal, we have evidence that indicates the officers do respond to the needs of the community and they do consider, perhaps inadvertently, benefits and costs (Whitelaw, 1968: chapter IV).

In the remainder of this chapter, those factors most important in determining the level and composition of the municipal budget are described. Also discussed is the concept of ability and/or willingness to pay for municipal services, and, after describing the determinants of capital outlays, current outlays, and property tax revenues, two hypotheses concerned with general policy relation-ships in the budget as a whole are presented.[6]

ABILITY AND/OR WILLINGNESS TO PAY FOR MUNICIPAL SERVICES

The usual approach in models treating municipal expenditures is to view real, personal disposable income per capita as the appro-priate measure of the community's ability and/or willingness to pay (for example, see Brazer, 1959). Presumably, real disposable income per capita is included on the basis of the assumptions that the various municipal goods and services are superior goods,[7] and that per capita income is a long-run measure of the community's demand for, and willingness to support, the consumption of public goods and services. However, this measure is several steps removed from the more direct factors that influence the decisions of the municipal officers. As Renshaw (1960) suggests, short-run changes in per capita income should not be interpreted as permanent changes in the wealth of the community. In order for an increase in per capita income to indicate a lasting willingness to support an increase in school construction expenditures, it must persist.

In the short run, a lagged measure of the tax rate, which is clearly the more easily manipulated policy variable of the two factors that determine property tax revenues (the other is the rate of assessment), is a direct index of the community's willingness to spend more for municipal goods and services in general. A lagged value of the tax rate times assessed valuation of real and personal property is a measure of a constraining influence within which the officers must plan their budget. Real property tax revenues com-prise that portion of the budget constraint over which the muni-cipal officers maintain some direct control, and an appropriately lagged value of the property tax revenues is a relevant and accurate measure of the community's ability and willingness to pay.[8]

An additional factor in the specification of the municipal budget constraint is state and federal transfers. Other things being equal, an increase in intergovernmental revenues to a municipality will increase expenditures in all functional streams. Even though these revenues typically are accompanied by conditions to their acceptance, e.g., the money shall be used only for school construc-tion, they release local revenues for support of other functional streams or, alternatively, they are substituted for the local revenues

(Renshaw, 1960: 170-174). An increase in state aid may in fact reflect an earlier increase in personal income throughout the state (Miner, 1963: 52). This would support the view that changes in personal income may effect changes in municipal expenditures, although indirectly.

CAPITAL OUTLAYS

In the case of capital outlays, the priorities of proposed projects (established by the heads of departments in the example of Worcester, Massachusetts) are derived from estimates of the difference between the desired stock of capital and the actual stock. The larger the difference, the higher the priority. In education, for example, the gross investment in school plant and equipment for any year is assumed to be a function of the difference between the desired stock of school facilities and the actual stock, where the desired stock in turn depends on the current and expected level of the school-age population and the community's ability and/or willingness to pay for these facilities. If recent changes in the level of public school enrollment have been large, then the present students would be enduring crowded conditions and, without new construction, these conditions would be expected, at best, to remain the same. If our description of the decision process is accurate, then the municipal decision-makers would place a high priority on reducing the difference between actual and desired.

There are several reasons why the response of the municipal officers to changes in the index of need is spread over time. Figure 1 illustrates, roughly, a time pattern of the lag in the case of an increase in the index of need.

t_0	t_1	t_2	t_3	t_4
Latest Major Construction Outlay	Increase in Need Occurs	Recognition of Increase in Need	Initial Action in Satisfaction of Need	Outlays for New Facilities

FIGURE 1

First, there will be some lapse in time between an actual change in the need and recognition of that change by the municipal officers (Ando et al., 1963: 3-7). (This is a different question from the one concerning whether the tree that falls in the woods makes any noise. In this framework, the tree makes noise but some time elapses before it is heard.) Point t_1 represents the period in which the increase in need occurs. In the case of an increase in school-age children, this change would most likely occur over several years, e.g., over the period of t_0 to t_1. The time span between points t_1 and t_2 represents the period during which the needs are transmitted to the municipal officers.

Second, there will be a delay between recognition of the need and the taking of action (Ando et al., 1963: 3-7). The time span between points t_2 and t_3 represents the slippage that occurs between recognition of the need and the initial formulation of plans to satisfy the need.

Third, time will elapse between the initiation of action and the subsequent change in the magnitudes of revenues and expenditures (Ando et al., 1963: 3-7). The time span between t_3 and t_4 represents the various stages of the budgetary action process, i.e., the period during which legislative action is being taken to satisfy the increased need. These stages include, for example, plans, authorizations, and appropriations, and they terminate in t_4 with the outlays for the new facilities.

The last two reasons for slippage in the reaction to a change in need can be explained by institutional and technological factors. In a municipal organization, both custom and statute dictate that certain rules be followed in the budgetary process. An act approved by the state legislature of Massachusetts is illustrative of a statutory constraint:

> Any city, town or regional school district may apply to the commission for reimbursement . . . of any expenses incurred . . . for surveys made of school building needs and conditions, the contract for which has been approved by the commission. The said commission may, upon completion of the survey, certify to the comptroller for payment to the city . . . such amount . . . as it may deem proper, and the state treasurer shall forthwith make the payments so certified. . . .[9]

The primary technological factor contributing to a lag in capital outlays is the fact that such construction expenditures are lumpy. The durability and immobility of a school building, for example, will cause a particular district to have excess or insufficient capacity. Needs accumulate gradually, taking considerable time before they require, and are granted, new facilities.

In addition to the institutional and technological factors, there is undoubtedly some of the "psychological inertia" to which Koyck (1954: 8) referred in his analysis of distributed lags. The extent of the deviation in educational investment from a long-run average, for example, may be influenced by the municipal officers reacting to some form of budget constraint where the long-run proportion of the budget that any expenditure stream comprises can change only slowly over time.

In our own work in this area, the length of the lag which corresponded to the best performing models typically extended over seven or eight years, with those changes occurring within the past three or four years receiving the greatest weights.

In addition to the educational expenditure stream, the major components of total capital outlays for municipal governments are outlays for street, sewer, and water facilities. During the 1963-1968 period, for example, these four functional categories accounted for 55% of total gross investment by city governments in the United States. The remaining 45% was distributed in varying proportions among several functional streams: fire and police protection, hospitals, welfare, library, parks and recreation, cemetery, airports, housing and urban renewal, and general public buildings. Therefore, for the limited purposes of this chapter, I will discuss, in addition to school construction, only the construction of street, sewer, and water facilities.

The desired, or long-run equilibrium, stock of streets can be defined as that which would permit certain levels of present and expected vehicular activity to flow through the street system at some minimum vehicular and commutation cost (Friedlaender). The level of activity is a function of demand for vehicular services, the vehicular cost is a function of the frequency of street use and the quality of the streets, and the commutation cost is a function of the amount of congestion and an inverse function of the rate of flow. A major empirical problem arises in trying to find an accurate measure of demand in the form of intensity and frequency of utilization, e.g., ton-miles per unit time, which would provide an index of the desired stock of streets. Among the measures that have been used are vehicle registrations, vehicle-miles traveled, fuel consumption, and residential and nonresidential structures.

An additional problem in the estimation of the desired stock of streets is that posed by the spatial distribution of development in a municipality. Whether the new street construction is for conversion of unpaved roads and streets into paved ones, or whether it is for the widening and improvement of old arterials depends on, among other things, the pattern of dispersal of employment opportunities and the corresponding residential location choices. In any event, it seems reasonable that the nature of the spatial dispersion in the municipality would have an effect on the street construction outlays made by the municipal officers.

In general, then, the investment in street construction for any year is a function of the difference between the desired stock of streets and the actual stock, where the desired stock in turn depends on a spatial distribution variable, the present and expected stock of residential and nonresidential buildings in the community (as a measure of the level of demand for vehicular services), and on the community's ability and/or willingness to pay for these facilities.

Since the construction of both sewer and water facilities is characterized by trunk distribution systems that are linked at one end to a large treatment plant and at the other to myriad structures, it is not surprising that the analysis of the determinants of investment in these two functional categories is similar. The desired stock of public sewage and sanitation facilities can be defined as that which would permit adequate treatment and disposal of present and expected effluence and garbage from structures requiring public sanitation service. The desired stock of public water facilities can be defined as that which would permit adequate distribution of water suitable for residential and nonresidential use.

In the case of sewer and water service for residential structures, there are three major types of residential development, each of which imposes substantially different construction demands on the municipality (Kain, 1966). First, there is the residential development in which all sewage treatment and water supply are accomplished through the use of private facilities, private cesspools or septic tanks for sewage and private wells for water. Second, there is the type of development that occurs along existing trunk lines. For adequate sewage and water service to the residences in this type of development, the city need supply only house-to-trunk connections. Third, there is the type of residential development that requires public service in undeveloped areas. This final type of development demands trunk lines as well as house-to-trunk connections. From the standpoint of the municipal officers, these three are listed in order of increasing expenditures per unit of service (Kain, 1966: 43-44). If the above reasoning is sound, then a good measure of the residential needs for sewage and water facilities it total residential construction plus an index of the spatial distribution of that construction.

In addition to the needs of residential structures, municipal officers must also satisfy the needs of commerce and industry for sewage and water facilities. Although there is very little information on the specific water and sanitation infrastructure needs of industrial and commercial interests, what information does exist (for example, Stanford Research Institute) indicates that such requirements are uniform among those industries most likely to be found in the majority of municipalities in the United States.[10] A good measure of commercial and industrial needs for sewage and water facilities, therefore, is nonresidential construction.

In general, then, the investment in sewage and water facilities for any year is a function of the difference between the desired and the actual stocks of such facilities, where the desired stock in turn depends on residential, commercial, and industrial construction, a measure of spatial distribution, and the community's willingness to pay for the facilities.

CURRENT OUTLAYS

Current or operating expenditures by municipalities comprise the bulk of municipal expenditures and they have received practically all the attention in empirical studies of municipal finance. Given the limited space for this chapter, and assuming that the reader is at least moderately familiar with some of these other studies, we will describe only those models with which we have experienced the most success. It is important to note that the functional categories for which these models have been designed, i.e., education, welfare, and general health, constitute at least half of all municipal expenditures, including capital outlays.

For current expenditures on education, the major factors influencing the municipal officers are the school-age population, the community's willingness and ability to pay, the real salary scale the municipality faces in its annual contracting, and the stock of education facilities (for detailed studies, see Hirsch, 1960; Miner, 1963). The inclusion of the public school enrollment is based on the assumption that the fundamental variable from which the public activity of education derives is the number of individuals for whom public education is relevant.

For current expenditures on welfare, the major determinants are the number of persons in need of assistance. Examples of measures of the needy are the number of persons over sixty-five years of age and the distribution of per capita income in the community.

For current expenditures on general health (including public hospital) purposes, the major factors influencing the municipal officers are the number of individuals in need of hospital and general health care (e.g., persons over 65), the community's willingness and ability to pay for this care, the salary scale the municipality faces in its annual contracting, and the stock of facilities in which the general health services are offered. As the reader may have anticipated, two factors that provide large obstacles to the analysis are the degree to which hospital needs are satisfied by private facilities and services, and the relative responsibilities for the provision of public hospital services among municipal, county, and state governments.

PROPERTY TAX REVENUES

Compared with the number of studies on current municipal expenditures, there have been very few systematic examinations of municipal revenues.

From the brief discussion of community welfare at the beginning of this section, recall that the municipal officers may manipulate the policy instruments, M (municipal output of goods and services) and T (municipal revenues), so that the needs of the municipality are satisfied subject to the ability and willingness of the citizens to satisfy those needs through the municipal sector. As the needs of the municipality increase and are translated into expenditure requirements, then, unless they are offset by intergovernmental transfers or nonproperty taxes, municipal tax revenues will increase. Following this reasoning, the factors influencing the municipal officers in their property tax revenue decisions are those upon which the bulk of the municipal demands depend, namely, public school enrollment, real per capita income, and population.

In addition to the above factors based on need, there are two hypotheses associated with the relationships between the economic activity in a particular municipality and the supply of property tax revenues. The first hypothesis proposes that the greater the economic activity in the municipality, then the greater the tax base is likely to be, and, with a given tax rate, property tax revenues therefore will increase (Burkhead, 1963). This effect may also be transmitted through increases in per capita incomes, which should occur when economic activity increases.

The second hypothesis suggests that the greater the proportion of industrial and commercial land uses, of total land uses, in a municipality, the greater the incentive for the citizens of the municipality to impose higher tax rates (Margolis, 1957: 231-232, 235-236). This hypothesis is based on the view that as the industrial proportion of the tax base increases, the citizens find additional property tax revenues taking less from their incomes per dollar of revenue.[11]

GENERAL POLICY RELATIONSHIPS

In addition to the relationships for the specific functional categories, some of our work indicates that there may exist interrelations for the budget as a whole, such as the relationship between total current and total capital expenditures or the relationship among the various types of capital expenditures.

When a municipal budget is growing over time, it is reasonable to expect a long-run complementarity between capital programs and the operating programs that are subsequently developed to utilize the new capital goods. In the short run, however, because of the limited resources at the disposal of the municipal officers, a large capital program may cause a cut-back in the operating program, i.e., the two would be substitutes.

Another relationship that has been suggested by a few preliminary results is an interdependence among capital outlays in the major functional categories. First, it is suspected that there exists a complementary relationship among the capital outlays for street, sewer, and water facilities. Second, there appears to be substitutability between investment in the above set of functional categories and investment in public school facilities.

Regarding the alleged complementarity among capital outlays for street, sewer, and water facilities, large outlays for one of them will tend to occur with large outlays for the other two because of the similarity that exists among these three categories. This similarity is based on a priori notions of the determinants of outlays for these facilities, on the technological similarities of their construction, and on the institutional arrangements by which changes in the determinants are transformed into construction in the city.

As suggested above, the fundamental variable on which the needs for investment in street, sewer, and water facilities are based is total residential and nonresidential construction in the city. Furthermore, particularly in the case of investment in these facilities in response to residential development in new areas, it is typically quite rational to construct new streets, sewers, and water mains at the same time because of the technological complementarities that exist among them, e.g., streets, sewers, and water mains usually follow the same arterial patterns.

In addition to the above sources of complementarities, the departments in which the specific decisions for new facilities in these three categories are made, and the engineers who make these dicisions, are usually the same. Assuming that institutional and organizational arrangements have any effect in this area, then that effect would be toward more complementarity among the functional streams over which they have jurisdiction.

The feeling that substitutability exists between investment in the above three categories and investment in public schools is based on the assumption that municipal officers regard the capital outlays on sewers, water facilities, and streets as more postponable than the construction of new schools. Given the existence of an overall budget constraint, then an increase in the need for new school facilities, and the subsequent transformation of that need into new construction, will result in the postponement of outlays for new sewers, streets, and water facilities.

CONCLUSION

This chapter was begun with the hope that the effort would be neither redundant nor unnecessary, and there are several reasons why that hope is not vain. First, by characterizing the municipal budgetary process as a composite package of relatively simultaneous decisions, we have roughly approximated a theory of municipal finance. In so doing, we have suggested that descriptive models of the municipal budgetary process, if they are to be free of considerable error and bias, must include explicitly the fact that municipal decisions on revenues and expenditures are made simultaneously, i.e., there exists a budget constraint. Finally, we have summarized the specification of simple models of certain municipal functional streams in which the determinants are observable economic and demographic variables that are external to the municipal organization and budgetary process. It is important to note, however, that we have not urged exclusive reliance on such external factors. On the contrary, merging an internalist's or organizational theorist's arguments into this structure would add materially to both the empirical and conceptual values of the analysis.

NOTES

1. In terms of the amount of theoretical and empirical literature devoted to each of these viewpoints, this division distorts the relative weights one should assign. The "externalist" viewpoint has received much more of the attention than has the "internalist" view.

2. The exceptions are found almost entirely in Texas where relatively liberal annexation laws still exist.

3. For a rough approximation of this publicly initiated solution, see the Lakewood Plan for some municipalities in the Los Angeles area and the Metropolitan District Commission in the Boston Metropolitan Area.

4. In his macro-analysis of the budgetary processes of state and local governments, Gramlich has used a considerably more sophisticated welfare function. The version of his paper, "State and Local Governments and Their Budget Constraint," to which I had access is unpublished and is part of the Federal Reserve Board-Massachusetts Institute of Technology econometric project directed by Albert Ando, Frank de Leeuw, and Franco Modigliani.

5. Why Worcester, Massachusetts? The City of Worcester was the source of data for my dissertation; therefore, I am familiar with it. Furthermore, its budgetary stages are rather typical.

6. For the complete and rigorous specification of these models and for the results of the application of various econometric techniques to them, see Whitelaw (1968).

7. A superior good is defined as one for which the demand increases as income increases.

8. When Miner says, "A danger that must be avoided is to identify tax receipts as the measure of ability to pay" for public expenditures, and "Tax collections are not a measure of ability to pay," he has generalized, in error, from his experience with cross-section study where it is appropriate to regard the tax revenue as a residual item in the budget. In a time-series study, the same conclusion is appropriate only on nonlagged tax revenues. See Miner (1963: 75).

9. Section 6A, Chapter 645 of the Special Acts of 1948, General Laws of Massachusetts.

10. This classification would exclude the remote extractive industries and the large industrial complexes.

11. Elsewhere, I have used the ratio of nonresidential construction to residential construction as a measure of the industrial and commercial activity relative to residential activity. See Whitelaw (1968: Chapter IV).

REFERENCES

ANDO, A., E. C. BROWN and E. W. ADAMS, Jr. (1965) "Government revenues and expenditure." In J. S. Duesenberry et al. (eds.) Brookings Quarterly Econometric Model of the United States. Chicago: Rand-McNally.

ANDO, A., E. C. BROWN, R. M. SOLOW and J. KAREKEN (1963) "Lags in fiscal and monetary policy." Research Study Number 2. In Stabilization Policies. Commission on Money and Credit. Englewood Cliffs, N.J.: Prentice-Hall.

ARGYRIS, C. (1959) "Understanding human behavior in organizations." In M. Haire (ed.) Modern Organization Theory. New York: John Wiley.

ARROW, K. J. (1951) Social Choice and Individual Values. New York: Cowles Commission Monograph Number 12.

BANFIELD, E. C. and J. Q. WILSON (1965) City Politics. Cambridge: Harvard University Press.

——— (1964) "Public-regardingness as a value premise in voting behavior." American Political Science Review 58, 4 (December): 876-887.

BRAZER, H. (1959) City Expenditures in the United States. Occasional Paper Number 66. New York: National Bureau of Economic Research.

BUCHANAN, J. M. and G. TULLOCK (1962) The Calculus of Consent. Ann Arbor: University of Michigan Press.

BURKHEAD, J. (1963) State and Local Taxes for Public Education. Syracuse: Syracuse University Press.

——— (1956) Government Budgeting. New York: John Wiley.

CRECINE, J. P. (1969) "A computer simulation model of municipal resource allocation." Doctoral dissertation. Carnegie Institute of Technology. Now available as: Governmental Problem Solving: A Computer Simulation of Municipal Budgeting. Chicago: Rand McNally.

CYERT, R. M. and J. G. MARCH (1963) A Behavioral Theory of the Firm. Englewood Cliffs, N.J.: Prentice-Hall.

DAHL, R. A. (1961) Who Governs? Chapters 10 and 11. New Haven: Yale University Press.

——— (1956) A Preface to Democratic Theory. Chapter 1. Chicago: University of Chicago Press.

DAVIS, O., L. DEMSTER and A. WILDAVSKY (1966) "A theory of the budgetary process." American Political Science Review (September): 529-547.

DAVIS, O. (1964) "Empirical evidence of political influence on expenditure policies of public schools." In J. Margolis (ed.) The Public Economy of Urban Communities. Washington, D.C.: Resources for the Future, Inc.

Department of Commerce, Bureau of the Census (1962) Census of Governments: IV, 3. Also the annual report, City Government Finances in 1967-1968.

DOWNS, A. (1957) An Economic Theory of Democracy. New York: Harper & Row.

ECKSTEIN, O. (1959) Trends in Public Expenditures in the Next Decade. New York: Committee for Economic Development.

FABRICANT, S. (1952) Trend of Government Expenditures. New York: National Bureau of Economic Research.

FITCH, L. C. (1957) "Metropolitan financial problems." Annals of the American Academy of Political and Social Science (November).

FRIEDLAENDER, A. "The federal highway program as a public works tool." Unpublished draft.

GOODNOW, F. J. (1904) City Government in the United States. New York. (Quoted in Banfield and Wilson [1965]: 73.)

GRAMLICH, E. "State and local governments and their budget constraint." Unpublished paper sponsored by the Federal Reserve Board and directed by Albert Ando, Frank de Leeuw and Franco Modigliani.

HAWLEY, A. J. (1951) "Metropolitan population and municipal government expenditures in central cities." Journal of Social Issues 7: 100-108.

HIRSCH, W. Z. (1963) "Quality of government services." In H. G. Schaller (ed.) Public Expenditure Decisions in the Urban Community. Washington, D.C.: Resources for the Future, Inc.

––– (1960) "Determinants of public education expenditures." National Tax Journal (March).

HOLLINGSHEAD, A. B. (1949) Elmtown's Youth. New York: John Wiley.

HUNTER, F. (1953) Community Power Structure. Chapel Hill: University of North Carolina Press.

KAIN, J. F. (1966) "Urban form and the costs of urban services." Harvard Program on Regional and Urban Economics. Discussion Paper Number 6 (October).

KOYCK, L. M. (1954) Distributed Lags and Investment Analysis. Amsterdam: North-Holland Press.

LYND, R. S. and H. M. LYND (1929) Middletown. New York: Harcourt, Brace & World.

MARCH, J. G. and H. A. SIMON (1958) Organizations. New York: John Wiley.

MARGOLIS, J. (1961) "Metropolitan finance problems." In Public Finances: Needs, Sources, and Utilization. National Bureau of Economic Research. Princeton: Princeton University Press.

––– (1957) "Municipal fiscal structure in a metropolitan region." Journal of Political Economy (June).

MINER, J. (1963) Social and Economic Factors in Spending for Public Education. Syracuse: Syracuse University Press.

MUSGRAVE, R. A. (1959) The Theory of Public Finance. New York: McGraw-Hill.

NETZER, R. (1961) "Financial needs and resources over the next decade: state and local governments." In Public Finances: Needs, Sources, and Utilization. National Bureau of Economic Research. Princeton: Princeton University Press.

PHELPS, C. (1961) "The impact of tightening credit on municipal capital expenditures in the United States." Yale Economic Essays I, 2 (Fall).

PIDOT, G. B., Jr. (1965) "The public finances of local government in the metropolitan United States." Unpublished doctoral dissertation. Harvard University.

POLSBY, N. W. (1963) Community Power and Political Theory. New Haven: Yale University Press.

RENSHAW, E. F. (1960) "A note on the expenditure effect of aid to education." Journal of Political Economy 67 (April): 170-174.

RIDLEY, C. P. and H. SIMON (1938) Measuring Municipal Activities. Chicago: International City Managers Association.

SACKS, S. and T. HELLMUTH (1961) Financing Government in a Metropolitan Area. Glencoe, Ill.: Free Press.

SHAPIRO, H. (1961) "Measuring local government output." National Tax Journal (December).

SIMON, H. A. (1957a) Administrative Behavior, 2nd edn. Introduction, chapters 4-5. New York: Free Press.

––– (1957b) Models of Man. Chapter 14, "A behavioral model of rational choice," reprinted from February, 1955 Quarterly Journal of Economics.

Stanford Research Institute. Costs of Urban Infrastructure for Industrial Development as Related to Sizes of Urban Centers in Developing Countries. Preliminary draft, prepared for U.S. Agency for International Development (Contract AID/csd 802): Chapter III, 17-20.

TIEBOUT, C. M. (1961) "An economic theory of fiscal decentralization." In Public Finances: Needs, Sources and Utilization. National Bureau of Economic Research. Princeton: Princeton University Press.

––– (1956) "A pure theory of local expenditures." Journal of Political Economy LXIV (October): 416-424.

UYEKI, E. S. (1966) "Patterns of voting in a metropolitan area, 1938-1962." Urban Affairs Quarterly 1 (June): 65-77.

WARNER, W. L. et al. (1949) Democracy in Jonesville. New York: Harper & Row.

WEIDENBAUM, M. L. (1966) "Program budgeting–applying economic analysis to government expenditure decisions." Business and Government Review 7 (July/August): 22-31.

WHITELAW, W. E. (forthcoming) "The determinants of municipal finance: a time-series analysis." Annals of Regional Science.

––– (1968) An Econometric Analysis of a Municipal Budgetary Process Based on Time Series Data. Doctoral dissertation at Massachusetts Institute of Technology, discussion paper under the Harvard Program on Regional and Urban Economics (September).

WILDAVSKY, A. and A. HAMMANN (1965) "Comprehensive versus incremental budgeting in the Department of Agriculture." Administrative Science Quarterly 10 (December): 321-346.

WOOD, R. C. (1961) 1400 Governments. Cambridge: Harvard University Press.

BUDGETARY REFORM AND THE RESTRUCTURING OF PUBLIC EXPENDITURE DETERMINANTS

10

PPB for the Cities:
Problems and the
Next Steps

SELMA J. MUSHKIN

☐ CITIES HAVE BEEN increasingly in trouble since soon after the end of World War II. Among the responsible factors were the burgeoning of urban population; the consequent growth in demand for services; the ironic fact that as the general prosperity of the nation spiraled, the demarcation between the "haves" and the "have-nots" deepened, so that welfare needs grew much heavier; the complexity of modern living itself, which most cities had not been designed to accommodate; and finally, the long-due eruption of the neglected problem of racial discrimination. All of these problems were accruing at the same time that the tax base of cities was deteriorating. The stark fact was finally evident: too many people needed too many things—and most of the cities didn't have enough resources to satisfy the needs. In other words, cities had been caught up in a cycle of poverty not unlike the vicious circle in which some families are trapped as poverty is transmitted from one generation to another.

As the crisis has grown since the end of the 1940s, city officials have become very familiar with strained budgets in the face of the need to choose among pressing programs. But for many years, the processes of choice were weighted in favor of unquestioning continuation of program policies and practices of the past, with only very small incremental changes from year to year. In fact, high tribute is due to most local elected officials and to city managers that they were able to keep their cities functioning at all, given not only the lack of money, but also such assorted lacks as analytical staff work on behalf of the decision makers, documentation of program formulation, hard questioning of purpose and of methods for providing program objectives, and analytical devices whereby preferred ways to fulfill the objectives could be judged.

In the past several years, a new tool has been offered to city governments to deepen and broaden the processes of choice in program and budget decisions, which, with the crisis of the cities as a backdrop, have acquired a new urgency. The process of planning-programming-budgeting—known as PPB—is an integrated system intended to provide the information needed to clarify the range of choices as well as the meaning and present and future consequences of the choices in the resource allocations that the decision makers must consider. Thus, for local officials—mayors, city and county managers, city councils, and department heads—the process supplies the possibility of assembling a maximum of information in a manner and at a time when the hard choices must be made on program and the concomitant budget allocations.

PPB: WHAT AND WHY

We referred above to PPB as a "new" tool, and that is accurate as far as cities go. Otherwise, it and its precursors have been around for some time. Both the British and the United States Armed Forces used it in a certain form ("operations research") in World War II. In 1949, the Hoover Commission recommended a "performance budget" based on such principles. In 1954, David Novick of the RAND Corporation developed the system further in association with the United States Air Force. When Charles J. Hitch, RAND economist, became comptroller of the Defense Department in John F. Kennedy's administration, he brought the idea along, expecting to implement it over several years, but Secretary Robert S. McNamara decided to put the system in operation at once. In August, 1965, President Johnson announced that PPB, which had worked well in the Defense Department, would be applied throughout the major executive agencies of the federal establishment.

The most recent application of PPB has been in states, counties, and cities. Beginning in January, 1966, the State-Local Finances Project of the George Washington University held conversations with various states and localities about implementing a PPB experiment in their jurisdictions. In July, 1967, letters of agreement were signed between the George Washington University project and five states, five counties, and five cities to undertake a year's pilot demonstration of installing PPB systems; inevitably the operation was dubbed the "5-5-5 Project." The five states participating in the demonstration were California, Michigan, New York, Vermont, and Wisconsin; the counties, Dade, Florida; Nashville-Davidson, Tennessee; Los Angeles, California; Nassau, New York; and Wayne, Michigan; the five cities, Dayton, Ohio; Denver, Colorado; Detroit, Michigan; New Haven, Connecticut; and San Diego, California. At about the same time, application of such a system was being discussed independently by officials in New York City and Philadelphia. And, by early 1968, a survey indicated that some twenty-eight states had reported beginning steps toward an implementation of PPB systems.

The system is fundamentally designed to link program planning, via analysis of potential program effectiveness, to budget decisions and thus to underscore that, by the budget allocations made, important decisions are reached on programs. It has certain characteristics that distinguish it from other modes of assembling information on behalf of public officials, as the following excerpt from a 1967 report (Hatry and Cotton: 38-39) of the State-Local Finances Project indicates:

> The operation of a PPB system can be viewed as a continuing dialogue between the chief executive and the agencies which are the public producers of goods and services. . . . This dialogue is cyclic. The chief executive, reflecting the demands of the electorate, generates broad objectives and guidelines. The operating agencies, in conjunction with analytical staffs, translate these broad objectives into more specific operational objectives and structure information to display the options for pursuing the objectives. The chief executive, upon reviewing the array of options, makes broad policy decisions. In accordance with these policy decisions, specific operational decisions are made within the agencies. The formal "multiyear program and financial plan" is prepared on the basis of the decisions made. The document gives a multiyear, government-wide perspective of approved programs. The entire cycle is repeated at each planning interval.

Systems analysis is primarily focused on the first and second portions of the cycle—interpretation of broad objectives and guidelines, and the structuring of information in such a way that the decision maker can see and evaluate the options available to him. Government can be viewed as a consumer, and PPBS as a system aimed at helping government behave more rationally as a consumer. Loosely speaking, the rational consumer of economic theory knows the price and utility of every good that he might purchase. Ideally, the results of systems analysis would inform the government decision maker of the price and utility of all potential governmental programs.

However, that is an overstatement of the likely contribution of systems analysis to rational decision making. Systems analysis *will* help government officials to see more clearly the true price and value of programs; it *will not* be able to measure or estimate accurately the cost and value of all possible programs in the context of the total government operation. The realistic goal of systems analysis is the improvement of the decision-making process, not the creation of a mechanism that automatically produces ideal decisions.

The above statement is, of course, an idealized version of how the system could work. In the experience so far of various governments working with PPB, there have been many slips between the cup and the lip. Nevertheless, most people who have been actively involved in applying the system acknowledge the advantage of its two special distinguishing characteristics: (1) the assembling of alternative ways of meeting specific objectives and (2) the obligation of discovering what the specific objectives really are—or should be.

Both of these items have had ramifying results, described in various ways. For example, one federal executive involved in applying the system is quoted as saying that the PPB implementation forced some agencies, possibly for the first time, to examine what their objectives were, not in generalities, but as achievable specific results. And Alice M. Rivlin (1969), in describing her experience with the system in the Department of Health, Education, and Welfare, mentions two main factors that hampered the analytical effort, even in federal agencies—lack of staff and lack of information—and remarks that the lack of information was the more basic difficulty. But she adds:

It may be that the most important result of the PPB effort in HEW so far has been the discovery of how little is really known, either about the status of the Nation's health, education, and welfare, or about what to do to change it. . . .

I remember being astonished when we first started . . . [a child health study] that doctors could produce no evidence that children who saw doctors regularly were healthier than children that did not. They all believed it (and I did, too) but they did not have any statistics to prove it. I was equally astonished to discover that educators have little or no evidence that children who get more expensive education (newer buildings, higher paid teachers, more teachers, etc.) learn more than children who get less expensive education. They believe it (and so do I) but the available statistics do not prove it, nor does presently available information give any solid clues about what kinds of schools are best, or whether particular educational materials are more effective than others.

Charles Hitch (1961), in speaking at a Brookings Institution symposium a few months before he was appointed to the Defense Department staff and started the PPB ball rolling, defined the combination of analysis and objectives as follows:

The key to progress, I am convinced from my own experience, is modesty in our objectives. It is perfectly apparent that when we have incommensurable and ill-defined objectives . . . the *optimum optimorum* will be illusive. Fortunately, to be useful, economists do not need to find that Holy Grail. The practically important conclusion is that A is better than B, or even that A is better than B *if*. . . . If economists can learn to do that much well, we can begin to make some sense of expenditure decisions.

In the government expenditure area, economists can also suggest improvements in the institutional environment. Our traditional budgetary practices seem designed to defy economic rationality. . . . A large percentage of the energies of the best people in government are consumed in fighting for large budgets rather than in making the best use of those they have.

And making the best use of what funding is available is exactly what the PPB system can assist a government to do. It sets in motion a process through which, as a routine of government, the question is asked: How better—i.e., more effectively and efficiently—can available resources be allocated among competitive uses?

The newness of PPB arises not from the individual techniques that have been developed as parts of the routine, but from their integration into a system and their procedural application to government decision-making. These techniques have zeroed in on the preparation of a series of documents that are the major tools of the system. They may be enumerated as follows: (a) an output-oriented statement of objectives and program structure, (2) program analyses and memoranda, (3) the multiyear program and financial plan.

In the preparation of these documents, the PPB system requires that the ultimate goals or objectives of each activity for which a government budgets money will be clarified and specified; that contributing activities will be gathered into comprehensive categories or programs to achieve the specified objectives; that a continuous process of examination will be carried on as to the effectiveness of each program, as a first step toward improving or, if necessary, eliminating it; that any proposals for new programs or for old program improvements will be analyzed to see how effective they may be in achieving program goals; that the entire costs of each proposal will be projected not only for the first year but for several subsequent years; and that a plan, based in part on the analysis of program cost and effectiveness, will be formulated for budget implementation.

Statement of Objectives and Program Structure

The program structure reflects decisions on the goals and objectives that are being sought through the government. Essentially it represents a classification of activities and programs in accord with the goods and services produced to achieve the defined objectives. Formulation of basic objectives, goals, missions, or purposes becomes an initial phase of implementation of a PPB system. Such goals or objectives are not value free, since they reflect the value judgments of those who represent and are responsible to the public. Thus, there is no single perfect model of a program structure that encompasses all functions. Even for a single set of defined objectives, there are many ways of classifying programs.

Nevertheless, a program structure is intended to highlight basic objectives and to display the competing and complementary activities involved in carrying out such objectives. It provides the framework for analyzing programs and for the preparation of a multiyear program and financial plan. Essentially the definition of objectives of a governmental jurisdiction provides answers to such questions as the following:

What needs doing and for whom?

Why is each activity currently performed being performed?

How does the activity relate to what we want to achieve?

The program structure itself displays a hierarchy of activities. At the topmost levels are the broad categories that reflect the programs designed to achieve the fundamental objectives of the government. The second levels display narrower groupings serving more limited objectives. As the categories become progressively narrower in scope at the lower levels, the structure sets forth the complementary and/or substitute approaches to the fundamental objectives at the highest levels in the structure. The lowest levels of any structure would be composed of activities that are intended as specific means for moving toward the larger objectives.

An array of expenditures by purpose contrasts sharply with a traditional line-item budget that may show the following items but gives little indication of why the expenditures are made:

Salary and wages

Contractual service

Supplies and material

Equipment purchases by type

Employees' retirement

Utility purchase

Workmen's compensation

Repairs and maintenance

Rentals

Such a listing conveys little information about why employees are on the public payroll, why supplies are purchased, or what groups in the population benefit from the outlays. If, in addition to or in place of such line-item information, a display shows the amounts expended in relation to objectives, as is intended by a program structure, both the public and the officials concerned with program policies have before them information on expenditures according to purposes being sought. The relative amounts in the display illuminate the priorities that have in fact been assigned to each objective, whether or not such a ranking of priorities was intended.

Implementation of a PPB system does not require that for programming purposes expenditures be grouped in terms of output-oriented program objectives rather than in terms of the items bought. Some jurisdictions, however, have altered their budget formats in accord with their program structures. And such budget formats clearly give a meaning and use to the work done on structuring.

Whatever the historic or conceptual basis for government production or provision of services, in some cases, such services have come to be viewed in terms of relative levels of expenditure or processes of government, rather than as end products. As one observer has commented, an assessment of the workload performance measures customarily used by many local governments somewhat justifies the hackneyed caricature of the government employee who sees the public purpose as one of processing forms and shuffling paper.

The following listing (Hollinger, 1969: 82-84) of the categories and subcategories of the Los Angeles County program structure and budget illustrates the emphasis given to basic program purposes by an objective-oriented program structure budget.

LOS ANGELES COUNTY PROGRAM BUDGET
1968-1969 APPROPRIATIONS

[Includes general county, special county, and special district funds under control of Board; dollar amounts in millions]

Programs and program elements	Salaries and employee benefits	Services and supplies	Other charges	Fixed assets	Capital projects	Total	Percent
I. Personal safety:							
(A) Law enforcement	$81.1	$10.4	$1.9	$1.3	$2.6	$97.3	
(B) Judicial	57.8	13.6	(¹)	.4	2.0	73.8	
(C) Traffic safety	5.7	7.1	9.0	.2	.4	22.4	
(D) Fire prevention and control	29.6	8.5	.1	1.8	.8	40.8	
(E) Safety from animals	1.7	.1	(¹)	(¹)	.2	2.0	
(F) Protection of and control of the natural and manmade disasters	.1	23.0	35.3	31.8		90.2	
(G) Prevention of food and drug hazards, non-motor-vehicle accidents and occupational hazards	4.6	1.0		(¹)	.3	5.9	
(H) Unassignable research and planning							
(I) Unassignable support							
Total personal safety	180.6	63.7	46.3	35.5	6.3	332.4	(22.0)
II. Health:							
(A) Physical health	145.5	133.0	5.4	2.1	5.5	291.5	
(B) Mental health	9.0	14.8		.2	.1	24.1	
(C) Drug addiction prevention and control	9.3	2.0		.2	.4	11.9	
(D) Environmental health, included under IV, D–H.							
(E) Other							
(F) Unassignable research and planning							
(G) Unassignable support							
Total health	163.8	149.8	5.4	2.5	6.0	327.5	(21.8)

(continued)

Programs and program elements	Salaries and employee benefits	Services and supplies	Other charges	Fixed assets	Capital projects	Total	Percent
III. Intellectual development and personal enrichment:							
(A) Preschool education							
(B) Primary education	.4	.5		(1)	.1	1.0	
(C) Secondary education	.3	1.1		(1)		1.4	
(D) Higher education	.4	.2		(1)		.6	
(E) Adult education							
(F) Public libraries, included under VI, C.							
(G) Museums and historical sites, included under VI, C.							
(H) Vocational education other than III, E., included under V, B.							
(I) Other							
(J) Unassignable research and planning							
(K) Unassignable support							
Total intellectual development and personal enrichment	1.1	1.8		(1)	.1	3.0	(.2)
IV. Satisfactory home and community environment:							
(A) Comprehensive community planning	2.2	.3		(1)		2.5	
(B) Provision of satisfactory homes for dependent persons	4.8	5.4		(1)		10.2	
(C) Provision of satisfactory homes for others	8.5	.7		(1)	(1)	9.2	
(D) Maintenance of satisfactory water supply	4.2	2.8	.7	2.1	.8	10.6	
(E) Solid waste collection and disposal		1.3				1.3	
(F) Maintenance of satisfactory air environment	4.5	.8		.2	.1	5.6	
(G) Pest control							
(H) Noise abatement							
(I) Local beautification	1.1	1.6	(1)	(1)		2.7	
(J) Intracommunity relations	1.3	.3		(1)		1.6	
(K) Homemaking aid and information							
(L) Other							
(M) Unassignable research							
(N) Unassignable support							
Total satisfactory home and community environment	26.6	13.2	.7	2.3	.9	43.7	(2.9)
V. Economic satisfaction and satisfactory work opportunities for the individual:							
(A) Financial assistance to the needy	82.9	452.7	.1	.6	.4	536.7	
(B) Increased job opportunity		3.2				3.2	
(C) Protection of the individual as an employee							
(D) Aid to the individual as a businessman, including general economic development	.1	.8		(1)		.9	
(E) Protection of the individual as a consumer of goods and services	1.6	.1	(1)	(1)	.2	1.9	
(F) Judicial activities for protection of consumers and businessmen, alike	(1)	.1		(1)	(1)	.1	
(G) Other							
(H) Unassignable research and planning							
(I) Unassignable support							
Total economic satisfaction	84.6	456.9	.1	.6	.6	542.8	(35.9)

(continued on next page)

(continued from preceding page)

Programs and program elements	Salaries and employee benefits	Services and supplies	Other charges	Fixed assets	Capital projects	Total	Percent
VI. Leisure time opportunities:							
(A) Outdoor recreation	10.4	5.9	.2	.2	8.6	25.3	---
(B) Indoor recreation	2.6	.5	.1	.1	2.1	5.4	---
(C) Cultural activities	11.3	9.5	---	.2	3.1	24.1	---
(D) Leisure time activities for senior citizens	.2	(1)	---	(1)	---	.2	---
(E) Other	.3	(1)	---	(1)	---	.3	---
(F) Unassignable research and planning							
(G) Unassignable support							
Total leisure time opportunities	24.8	15.9	.3	.5	13.8	55.3	(3.7)
VII. Transportation, communication, location:							
(A) Motor vehicle transportation		17.5	65.3	1.9	---	84.6	---
(B) Urban transit system							
(C) Pedestrian							
(D) Water transport							
(E) Air transport	.5	.1	---	(1)	2.9	3.6	---
(F) Location programs							
(G) Communications substitutes for transportation							
(H) Unassignable research and planning							
(I) Unassignable support							
Total transportation, communication, location	.5	17.6	65.3	1.9	2.9	88.2	(5.9)
VIII. General administration and government:							
(A) General government management	2.3	.4	---	.1	---	2.8	---
(B) Financial	23.1	4.2	---	.1	(1)	27.4	---
(C) Unassignable purchasing and property management	11.6	6.9	.1	.2	.1	18.9	---
(D) Personnel services for the government	3.4	.4	---	(1)	.1	3.9	---
(E) Unassignable EDP	.1	(1)	---	(1)	.2	.3	---
(F) Legislature	2.2	.4	---	.1	---	2.7	---
(G) Legal	2.6	.2	---	(1)	(1)	2.8	---
(H) Elections	2.9	2.5	---	(1)	(1)	5.4	---
(I) Other	14.5	20.6	.9	.8	14.3	51.1	---
Total general administration and government	62.7	35.6	1.0	1.3	14.7	115.3	(7.6)
Subtotal all programs	544.7	754.5	119.1	44.6	45.3	1,508.2	(100.0)
Less costs applied						33.3	
Total all programs						1,474.9	
Appropriation for deficiencies						1.7	
Reserves						21.8	
Estimated delinquencies						14.0	
Grand total, requirements						1,512.4	

¹ Denotes figure less than $50,000.

SOURCE: L. S. Hollinger (1969) "Changing Rules of the Budget Game: The Development of a Planning Programming Budgetary System for Los Angeles County." In *Innovations in Planning, Programming and Budgeting in State and Local Governments*. Washington, D.C.: Government Printing Office.

Program Analysis: Costs and Benefits for Alternatives

Basically, a PPB system is a unifying and comparing process. Optional programs are compared in terms of: (1) their consequences and effectiveness and (2) their costs—both those that are immediate and those that are implicit for subsequent periods as a result of the immediate action.

Analysis involves a reduction of complex problems into their component segments so that each segment may be studied. Questions of fact can be subjected through this process to the test of observed experience. Those aspects of the problem that involve value judgment can be separately identified and the basis of judgment made explicit.

In local governments, program analysis has often been divided into two components: (1) the important defining of the problem, (2) the comparing of cost-effectiveness of options that have been identified as feasible. The problem definition statement, termed "issue paper," is intended to represent the first stage of any program analysis and to stop short of assessing the costs and effectiveness of alternative ways of meeting the identified governmental objectives (State-Local Finances Project, 1968). For local governments, the problem definition statement has proved to be an especially important process. In-house staff groups with some training, but without extensive analytical experience, have related what they knew and understood to the process of working out an issue paper. Experience has indicated that some issue papers can be completed within a relatively short time, enabling interagency conversation to be started in problem-definition terms. When the issue paper is used as a combined training-doing exercise, it has given added meaning and relevance to learning. The carrying out of a "good" cost-effectiveness analysis often involves more than a short-term study. Initial questions tend to be broadened and to illuminate the unknowns more than the knowns. The requirements of analysis are quite severe: wisdom to avoid pitfalls of superficiality or of sophisticated manipulation of inadequate data; imagination to go outside the bounds set by the usual approaches to a program; ability to combine the range of knowledge from the various academic disciplines and administrations and from relevant basic research findings; and, possibly most important, judgment about the political feasibility. Such analytical efforts demand an expenditure of talent that is at present likely to be beyond the capacity of most city governments.

But cost-effectiveness analysis of a more rudimentary type can be pursued through improvements in staff documentation to aid policy officials. Components that represent a simplified procedure of a cost-effectiveness second stage of analysis can be blocked out where more complex analytical talents are not available. Metropolitan Dade County furnished a set of guidelines for preparation and display of program analysis based at least in part on the work of Martin Jones (1968). Such guidelines can assist city analysis staffs in their cost-effectiveness work; they call for: (1) the gathering-in of relevant facts that are available; (2) new collection of data when feasible; and (3) the clear identification of qualitative considerations that are important, together with judgments about those factors and the reasoning that underlies the judgments set forth. Constraints by way of special interest group concerns, legal barriers, and administrative feasibility are all identified, and the guidelines call for descriptions of those concerns in capsule format.

Such directions are helpful if they are not converted to rigid rules. It is *not possible* to pursue analysis by rigid routine; the very essence of option generation and assessment is the unroutine.

The Multiyear Program and Financial Plan

A PPB system makes long-range fiscal planning a routine for government. Analysis of programs that provides the information for decision includes an examination of effectiveness and of cost implications for a future period. Programming is thus given a longer time perspective. The multiyear program and financial plan is of particular importance at two major points in the planning cycle: (1) the initial phase of program planning, and (2) the time of adoption.

A tentative summary of initial individual program decisions provides the basis for an examination of the fiscal feasibility of overall decisions. In many instances, these tentative sets, when summarized, may indicate a requirement for funding that is out of reach in the current budget or even in the projected budgets of subsequent years. However, such a tentative summary does not necessarily imply that programs should be planned only to the extent that their revenue sources are currently authorized. A government jurisdiction may choose to plan for a level of expenditures exceeding that which can be raised under current revenue provisions. The question of revision of revenues is thus clearly posed.

At the time of adoption, a second role is played by the multiyear program and financial plan. The revised plan actually adopted becomes the base or baseline display of the decisions that have been made. It provides planners at all levels with a perspective of what the government is doing and where it is heading.

A BEGINNING BY BEGINNING

"PPB," in the words of the chief administrative officer of the County of Los Angeles, "is a welcome addition to the budgeting scene for local governments" (Hollinger, 1969: 75). As the fifteen jurisdictions in the 5-5-5 Project neared the end of the pilot study year, they prepared final reports that covered the period from July, 1967 to approximately July 31, 1968 (State-Local Finances Project, 1969). (Some of the reports included activities overlapping the latter date.) The assessments of the initial work done coincided, in general, with the Los Angeles County statement that the experiment had shown that "the concepts and techniques of a PPB system have a very real potential for providing a more systematic and explicit flow of information needed in the decision-making process of county government. . . . Our future efforts are directed toward bringing such explicit analyses into the decision-making processes of local government" (Hollinger, 1969: 78-79).

Thus, in an essentially very brief period of years, PPB techniques have developed and spread from RAND, to the Department of Defense, to federal civilian agencies, and to states, cities, and counties. In this present discussion, we are primarily concerned with the five cities that participated (Dayton, Denver, Detroit, New Haven, and San Diego) in the 5-5-5 demonstration, and also with the five counties (Dade, Nashville-Davidson, Los Angeles, Nassau, and Wayne), since each of these is heavily urban, and two of them—Nashville-Davidson and Miami-Dade—were, at that time, the only existing metropolitanwide governments in the nation.

The ten local jurisdictions, together with New York City and Philadelphia, came to be called the "adventuresome dozen" (Mushkin, 1969: 167). Each faced the same questions: what is to be done, how, and with what effect? Each drew comfort and support from the activities of the others. For each, there was necessity to adapt the PPB idea to the specific political climate, personalities, and governmental structure of the locality. And, although all of them were aware that this short-run exercise on which they were launching actually concerned a complex, long-run effort in many ways, the excitement of adventure made these beginning steps seem plausible. The press, sensing the importance of this beginning, gave encouragement through editorial coverage. In some quarters, however, there was, not surprisingly, caution about the impact of the system.

The progress that was made during the demonstration year encouraged many other local governments to begin PPB system installation. In 1968, 1,134 city and county governments were surveyed; 356 responded to the survey questionnaire. Of these, between fifty and sixty reported that they were taking steps to implement PPB systems (Mushkin et al., 1969).

The largest concentration of local governments reporting active steps to begin the implementation was in California. This is possibly due to the discussions on PPB in 1966 between the State Municipal League and California's own programming and budgeting system, as well as the interest of the aerospace industry in applying aerospace technology to civilian public services. Interactions among governments that appear to encourage PPB are also indicated by responses from other geographic areas. For example, in Michigan, where the participants in the 5-5-5 demonstration included an across-the-board group of jurisdictions—the state, a city (Detroit), and a county (Wayne). Reports of beginning efforts toward implementation came from three cities: Ann Arbor, Lansing, and Wyoming. More recently, active support has been given by California to local PPB efforts by the provision of training programs for local employees. And in New Jersey, early in the implementation phase, state technical assistance to localities' programs was initiated. (See Appendix for a summary of beginning steps as reported by local governments, exclusive of the original initiation of PPB in the local governments in the 5-5-5 Project.)

APPROACHES TO PPB

A review of the steps toward initiating PPB implementation indicates two general approaches. The first approach was taken by most of the local governments. The second is possibly best represented by New York City's activity. For purposes of contrast, the approaches may be labeled (1) incremental and (2) systematic.

However, the difference between the two approaches is more a matter of gradual shading than of sharp contrast. The systematic approach has, for example, been applied at times flexibly and opportunistically, and the incremental approach has been planned as "installation by installments."

In most of the local governments, the initial incremental effort has had the following characteristics:

(1) The chief executive of the jurisdiction gave his formal endorsement, but not a strong supporting hand.

(2) The PPB budget and the analytical staffs for implementation were small.

(3) Experienced analytical staffs were not recruited.

(4) Implementation was heavily dependent on training of staff already working for the government.

(5) The effort was supported and nurtured by the career professional with due caution, and steps taken were faltering.

(6) Line-agency cooperation became essential, but was difficult to attain on what was almost a voluntary basis.

(7) The timetable for implementing the procedures was a long one.

In the systematic approach, typified in New York City, the mayor determined the installation of PPB. Recruitment for implementation and placement of central staff was also essentially determined in terms of the program the mayor decided upon. The city's commitment to PPB was relatively large. Resources allocated probably approximate the total of those of all other local governments combined. The approach has had the following characteristics:

(1) A strong commitment was made by the chief executive.

(2) Sizable staff resources were provided for analytical efforts, and considerable emphasis was placed on recruitment of new talent.

(3) Some experienced staff was recruited, and new staff was trained on the job.

(4) A link was made with RAND of New York to carry out analytical studies in at least four areas: housing, fire prevention, police, and health and hospitals.

(5) The staff was largely new to the city government and enjoyed the excitement of a new adventure.

(6) The timetable for the showing of a payoff was brief.

Clearly, the commitment of the chief executive is an important determinant in any approach. On this commitment depends, at least in part, the size and scale of the effort and the relationship with departments of the city government. Small-scale incremental efforts become heavily dependent on career professionals and their relations to each other. Philadelphia, for example, started down an incremental path toward PPB implementation that contrasted sharply with that of New York City. Whereas, New York's PPB was initiated by the mayor and moved forward with his active involvement and support, Philadelphia's was originated by the professionals in government and was furthered largely through their cooperation at central and agency levels. Gradual development of the system was sought step by step with the build-up of staff capability through periodic training and orientation. The first products were structural—the development of an output-oriented classification of programs, subprograms, elements, and subelements. To assure that the structures developed by the agencies would have an overall rational relationship to each other, general agreement on a broad basic program framework for the entire city government was reached initially between the Department of Finance, which housed the central staff for the PPB effort, and operating agency officials. From its beginning, Philadelphia's PPB system was grounded in agency cooperation.

Analysis in Philadelphia was, however, undertaken on a hesitant, trial basis. First directed to the venereal disease program in the Health Department, it was subsequently broadened selectively and for a long-run developmental program. In contrast, New York City started down the path of program analysis with a large effort to demonstrate short-run payoff from analysis. Accordingly, at least at the outset, much emphasis was put on maintenance of effectiveness of programs at reduced costs.

A city government's adaptation of PPB is likely to be extensive because a city is a whole system, not an agency. Since it is a taxing unit as well as a spending unit, decisions on tax rates and tax levies have to be made along with decisions on program levels. The multiyear program and financial plan, formulated as a summary document in the routine of a PPB system, takes on new attributes in the setting of city financial policies. It necessarily becomes a document from which to judge taxing and borrowing levels and, in turn, constraints on spending. To a much greater degree than is required of federal agencies, cities must translate program plans directly into an analysis of the consistency of that plan with the revenue options available, and take account of state restrictions on the city's deficit financing, debt limits, and authority to tax.

The cities provide services directly to people. There is no intervening buffer layer between the decisions on program and the recipients of the services. In the cities, specific form is given to hard problems of coordination of activities and programs that represent integral parts of a package of public services to serve the same or interrelated purpose, frequently for the same families or target groups. Interagency consideration of citywide program objectives within a PPB process provides a base for communication on activities of common concern among the departments and bureaus. Moreover, analysis of policy issues serves further to identify options that cross departmental lines. In contrast, federal agencies with public service responsibilities in such areas as housing, public health, education, and welfare are primarily grantor entities channeling funds to cities and states, rather than providers of services.

The "close-to-home" government is open to all the pluralistic pressures that are generated in a political system. The government is "exposed" and in most cases, highly accessible. Differences in interest between those who pay the taxes and those who receive the services are fairly clear. These pressures make the information that the analytical techniques of PPB can generate ahead of time—ahead, that is, of program decision—even more important than it is in a less exposed type of jurisdiction. PPB does not lighten the burden of program decisions. In a way, it makes the burden heavier because the range of options must be displayed along with total costs and possible effects. At the same time, by opening the possibility of assessment ahead of the time of decision, it permits the after-decision consequences to be eased by information about anticipated response.

Cities, furthermore, are at both the receiving and doing ends of intergovernmental relations. To plan their programs, they depend to varying extents on federal and state fund offerings and meet concomitant restraints, both on types and levels of services and on administrative arrangements, imposed by federal law or state statute. Within this complex pattern, a city, as an immediate agent for joint intergovernmental action, brings together the national, state, and local efforts and faces the problem of estimating composite federal-state-city program effects, not in aggregate terms, but in the specifics of impact on the people who live and work in the city. Federal direct grants to a city are almost exclusively concerned with physical plant (through urban renewal, mass transportation, public housing, waste-treatment works) and thus reinforce the practice in many cities of separate capital budgeting. Analysis of total cost for purposes of evaluating program options helps to assure that the combined current and capital outlays are considered. Such combined costing does not necessarily mean a change in budget format, but it may alter the focal point of decision on whether a project should go forward or not.

Thus, program analysis for a city is a harder task than it is for other levels of government. This is so because of: (1) the particularity of the city; (2) the directness of the relationship between the city's residents and those charged with policy responsibility; and (3) the combined task of spending and taxing when revenue resources are severely constrained by authorized taxing and borrowing capacity, and by intergovernment tax competition.

ISSUES AND PROBLEMS

A number of main problems face cities when they undertake to begin the process of implementing PPB systems. Some of these are discussed briefly in the sections that follow.

Structure or Analysis? The Relative Emphasis

As indicated earlier, many of the local governments that have worked with PPB in the past two years have concentrated on its structural aspects. Others, in common with New York City, preferred to begin with analysis of programs. The choice is largely one of initial emphasis, for both structural process and analysis have their place in the system.

The structural aspects of PPB serve to set a framework for subsequent analysis of programs through identification of objectives and the hierarchy of programs and activities undertaken to move toward their achievement. The program structure serves also as a starting point for the development of multiyear program and financial plans that display the changes in program outputs along with projected changes in expenditures and that pose the revenue issue ahead of time.

While from the outset it was clear to the "adventuresome dozen" that the initial work on program structures would have to be considered as experimental, with changes made as analytical efforts were subsequently undertaken, the task of defining objectives, of structuring programs in a hierarchy, and of identifying (even in a routine way) outputs of public services was formidable. The structuring exercise set in motion the process of inquiry about the purposes of the city: i.e., what is a city government trying to do, for whom, and why? This is an inquiry that kindles an awareness of government as a producer or provider of products for the people. It is no small wrench for government officials to begin asking questions about the products for people—for example, opportunities provided for learning, changes made in health status, the housing conditions that are fostered, protection provided for persons and property, the extent to which an advance from poverty can be achieved, the recreation and cultural opportunities furnished. Officials who have long been accustomed to the routine of comparing expenditure levels and manpower inputs into public services, when suddenly required to examine the quality and quantity of public products in terms of end products rather than processes or inputs, discover many lacunae in knowledge about such products, their nature, and their magnitude. The yardsticks for that type of measurement are often not in the battery of statistics which city statistical offices or agencies have formerly produced.

Most of the local governments engaged in PPB structuring were faced with a common problem. Defining broad objectives and developing a program structure for the largest categories of their activities were relatively easy, but the governments found it far more difficult to group elements and subelements of their programs in a meaningful way. Especially at those levels of government where the public products are provided directly to the citizenry, the PPB aspirants run head-on into the problem of identifying in a fundamental way the results that lead to employment or use, for example, of such services as those of social workers or of the police.

Whatever the focus in structuring, it is clear that the central core of a PPB system is analysis of cost and effectiveness of alternative ways of satisfying specified city objectives. With analytical talent in short supply, the energies devoted to refinement of structure can detract from analytical efforts. While New York City concentrated its PPB work on program analysis, with a number of analytical studies made by the staff of its Bureau of the Budget buttressed by a joint effort with the RAND Corporation, the city undertook subsequently to institutionalize PPB work through preparation of a program structure and setting up program definition statements.

In discussing this phasing of the work, the problem posed by a city official was "the real danger that we will lose the substance in the process of installing the form.... We have from the beginning operated ... on the assumption that it was analysis which was of primary importance and that we could establish the formal trappings of PPB only after we had established a capability for, and a credibility of, an analytic approach" (Mushkin et al., 1969: 191).

Without substantial investment of staff resources, analytical efforts are necessarily very limited and the payoff in new program approaches and in basic questioning of current processes of resource allocation is small. Moreover, analytical efforts that are not to be confined to "another study" status have to be used—used, that is, in a process of program and budget decision. From Dade County comes the following (Grizzle, 1969: 68):

> It would appear that analysis can play a useful role ... by developing information to be used in evaluating whether or not a new program ought to be funded. Analysis should be a useful method of conducting studies when "yes" can be answered to the following questions:
>
> (1) Does the problem require a decision?
>
> (2) Can the analysis be finished before the decision must be made?
>
> (3) Would more information make a difference to whoever will make the decision?

Bottom Up—Top Down

The relative emphasis on structure and on analytical efforts is closely related to a second question faced by most of the local governments: should a governmentwide approach be adopted in program structuring and analytical studies, or should the main responsibility be assigned to the departments and perhaps in turn to the agencies within departments? The close interdependence of a city's activities suggests that a large measure of responsibility should be given to a central staff reporting to the mayor or city manager in initial work on a governmentwide structure and for the analytical studies carried out. However, this was not always the process followed; some governments experimented with implementation or analytical studies in a few departments.

A balance is needed. Central staff work isolated from departments jeopardizes the potential of joint consideration of the goals of a city. Such consideration can become an occasion for interdepartmental communication. Only through team approaches to analytic studies can the range of program options considered be made to cover the entire responsibility of the government and not of the departments—yet be responsive to the policies of the departments. Reports from several of the governments participating in the 5-5-5 Project underscored the usefulness of a governmentwide effort in sharpening department personnel's awareness of the interaction between programs of different departments (State-Local Finances Project, 1969).

Dependence upon line agencies for PPB implementation has both its pluses and minuses. For one thing, a substantial educational effort is required to gain agency acceptance of goal formulation that can lend itself to measurement and cost-effectiveness analysis. But this kind of educational effort also serves to institutionalize analysis as the approach to program formulation. The question-raising proclivities of the more analytical approach become, in the course of agency implementation, a routine of the agency response to budget preparation and reviews of proposed legislation. Moreover, a department's consideration of its purposes and examination of its programs and activities in relation to those purposes set the stage for specific policy and program design. Departments that find program reassessment useful have a reason for follow-up in analytical studies and program evaluation. In any case, since the departments have the specific program knowledge, they are the agencies that must be held accountable for program assessment and program execution. Accountability can only be fully realized through active agency participation at the stage of policy review and formulation.

However, assigning major or exclusive responsibility to departments invites a continuation and even a strengthening of the separateness of functional agency performance. It also invites a narrow view of agency missions and encourages the tendency to derive objectives to fit the activities carried out rather than to determine those objectives and identify, if indicated, gaps in program. Complementarities among activities and program interdependence, moreover, are lost from view. And program options come to be restricted to those already within the purview of the agency.

Effective use of PPB as a coordinating and informing mechanism depends upon active participation of departments in task forces or groups that represent the various functional departments as well as the staff agencies. Clearly, a system for program planning and resource allocation is needed that can take account of interprogram effects. Action by functional agencies to improve health care or educational achievement levels, or to control alcoholism and drug addiction, or to lower juvenile delinquency rates yields fragmented attacks on closely interwoven multiple disabilities of families. And on the environmental level, the interactions are even more marked. An educational agency, for example, will look to educational programs as possible methods of approach and ignore preventive steps such as maternity care and feeding, or infant health services. An agency concerned with solid waste disposal may ignore the impact of such waste disposal on water supplies, on air quality, or on recreational services.

The problem of achieving interagency cooperation is a familiar one. But the beginning of efforts to implement a PPB system offers an occasion for interagency dialogue. While analytical efforts require assessment of all programs that contribute significantly to the same purpose, regardless of organizational structure, the optional methods for interagency cooperation—such as task force arrangements (with or without outside experts), "lead agency" assignments, review of individual agency submissions by central staff—all have been used. Each has its virtues, each its drawbacks. The PPB process alters the questioning and the questions; it does not alter the problem of the need for separate organizational units in a government in which everything is related to everything else.

Analysis of What?

Cities are complex integrated systems, but they are not closed systems. In an open system, steps taken to remedy a problem—for example, to reduce unemployment and underemployment or to eliminate slums—may in fact enlarge rather than reduce the problem. New entrants into the city attracted by recently created jobs or housing may make the deterioration worse than it was before. Our basic knowledge about behavioral response is rudimentary. Moreover, while cost-effectiveness literature is now voluminous, when one turns to the specific issues in a specific city, the needed data are often lacking, the analytical concepts have not been sufficiently developed, outputs may have been inadequately researched, and information about production functions may be deficient. Often, even when conceptual work has been done, it has not been "translated" into a form that lends itself to practical application.

What, then, can a city undertake by way of analysis of relative costs and effectiveness of program options to satisfy defined objectives? A number of local governments have begun on an analytical process that may be far from sophisticated, but is, nevertheless, a process being carried out in a determined spirit of inquiry that does make a difference in governmental services.

The following listing of some of the issues on which at least a beginning toward analysis has been made may give some perspective to the work presently underway as the result of the early PPB implementation.

Nursing home care

Supplemental food programs

Fire protection

Swimming opportunities for children in model city areas

Leisure time activities for residents of deprived areas

Emergency ambulance service

Library services

Police services

Cultural needs in a suburban area

Criminal rehabilitation

Mental health

Juvenile delinquency

Drug abuse and addiction

Employment opportunities in a target area

Bus transportation routes to job opportunities

Waste collection

Waste disposal

Infant and maternal health care

Even rudimentary exercises of problem definition, when systematically carried out, can provide more information than policy officials had prior to the undertaking of beginning steps to PPB implementation. The director of the Administrative Analysis Division of Nashville-Davidson (Horton, 1969: 103) put it this way: "All jurisdictions should undertake the exploration of the PPBS concept because few urban areas cannot afford not to benefit from the short-haul end products of this type of exploration."

The starts that have been made by the various city governments suggest great variation in the quality of the analysis undertaken. The scope of the problems selected was frequently unduly restricted. Yet expansion of the scope may enlarge the areas of study beyond either the capacity of the persons undertaking the inquiry or the basic research knowledge that can make for a workable investigation of the problems.

Selection of issues for analysis is clearly a major phase of the process of PPB implementation. There are probably thousands of issues within a city government that would lend themselves to analysis, but up to now the number of such potential analytical studies has not been determined. Even federal agencies, with vastly greater resources at their command than cities have, have had to restrict their analytical studies to a few selected issues. At the city level, where the persons engaged in analytical undertakings are usually few in number and, for the most part, inexperienced, selectivity becomes even more important. Issue selection requires a clear view of: (1) the major policy problems on which decisions are urgent, and (2) the possibility of completion of cost-effectiveness studies in time to contribute to the decisions.

Even a small start toward the question-raising—that is, the analytical process—may move in the right direction. For example, an identification of the number of families actually receiving food stamps out of a total number of possible eligible families can clearly pose a policy question in a local government. This question may have been blurred earlier by previous practices of recording expenditures or even by changes in numbers of those receiving stamps, without considering the relation of such numbers to the number of families in comparable economic circumstances that did not have access.

Requirements for Intergovernmental Comparative Information

Another of the problems concerned program structures. Should they be the same or comparable from government to government?

The importance of local government performance in producing public services and delivering them is increasingly being recognized as the essential component in the achievement of federal program objectives. The pressures for information about results of programs and their combined levels of financing—federal, state, and local—and especially for data that are comparable from one place to the next have grown. The response of the federal government and also of the states is in a few instances that of encouraging uniformity among local governments in program categories and output indicators. The temptation to encourage replication at the state or local level of a federally-designed program structure is strong. A uniform national and state structure would tend to produce, as a by-product, operational statistics consistent over the nation, and would open up the potential of program analysis within common definitions of public purposes and measures of outputs.

There is a repetition of this striving for uniformity from the state to the city, with the county raising similar, but not exactly identical, problems. The federal government cannot legislate program standards for the state. County governments, however, derive their program responsibilities from the states, and their program objectives are closely related to the requirements of state law that the counties are delegated to perform.

The temptation to issue directives and impose requirements that can yield uniformity and comparability endangers the gains that are possible to local government when it undertakes its own exercise of setting goals and of comparing programs relative to effectiveness and costs, both as to expenditures and opportunities foregone. Especially important for local governments and the states in a federal system is an independent search for alternatives within the processes of program analysis. In fact, the routine of analysis requires such a search, and opens the way for the inventive public employee to inject new approaches into the program assessment. An inventive program proposal, as a minimum, would be assured a more prompt consideration as an inventive alternative in the course of analytical study of an issue.

Data needs of a higher level of government are large and comparability is required, but the steps toward such comparability in data collection will have to be complementary to, and not a part of, PPB systems. Some steps are being taken by the U.S. Bureau of the Census toward comparability of data. Such intergovernmental cooperative efforts in gathering statistics, buttressed by work on defining in a uniform way small program elements and collecting information on source of funds for each of the elements would help to fill the data gaps. Similar cooperative efforts between state and federal agencies have been carried out for mental hospital data and data on educational finances.

Staffing and Staff Development

PPB, viewed as a policy-informing instrument, is primarily a staff product and process. Questions then arise: What size staff? What staff skills—at what price? What provision for staff training and when, using what resources, and at what cost? In another context (Mushkin et al., 1969), we have reviewed the experience of the local governments that participated in the 5-5-5 Project. This experience suggested no pat answer to these questions. The size of the government, the range of its program responsibilities, the time schedule set for implementation are all factors that will affect the decision on staff size. Among the observations in this regard that emerged from the 5-5-5 demonstration are the following:

(1) Rather surprisingly, a real advance in informing policy officials during the process of PPB installation can be made by a small, perhaps three-man, staff that can devote full time (or almost full time) to PPB.

(2) Among the personnel employed by most local governments, there are usually some people who have general analytical ability. This capacity can be further developed through training programs to yield a rudimentary analytical competence. Most governments also have certain quantitative specialists in such departments as engineering, public works, and health who can be drawn into the PPB effort.

(3) One of the surprising findings of the PPB survey made in 1968 was the extent of exposure of local government officials to orientation sessions in PPB. Well over a third of the 356 cities and counties responding to the survey questionnaire reported that staff personnel had attended such sessions. And we can reasonably surmise that the numbers exposed to PPB concepts have grown considerably since that survey date.

(4) Approaches to developing staff capacity for PPB implementation varied: consultant firms were employed for training, unversity services were used, and in-house training programs were

developed. The best training appears to have been derived through the process of doing, i.e., the combining of learning with doing tended to yield results superior to other methods. Because the timetable for PPB installation in most local governments is likely to be long, the build-up of staff capacity in phase with implementation calls for sequential training programs specially designed to advance doing-on-the-job. In fact, few substitutes exist for learning by doing-on-the-job—a learning exercise that requires leadership and the type of analytical capacity in a leadership position that could yield the learning capacity for additional staff.

(5) Few local governments have salary scales that permit recruitment of highly-experienced analysts. Thus, substitute methods must be used in developing staff capacity. Consultant, university, and industry support is needed, along with assistance in implementation that comes with on-the-job training. How best to get the required support from consultants, university, or local industry remains largely an unresolved problem of implementation. Some state governments are engaged in a process of making use of consultants or universities that may eventually offer models for the cities. The International City Managers Association's program of training in systems analysis and the National League of Cities Urban Observatory program for university research on city problems—both financed by grants from the U.S. Department of Housing and Urban Development—are also importantly showing the way.

(6) The beginning steps taken by the earliest participating local governments mean that a body of knowledge has been accumulating and will be available to other city governments, especially regarding staff development. Materials that can be used for training are increasingly becoming available. Personnel in the cities that first initiated PPB systems have been willing to offer their services for training their counterparts in other cities.

Next Steps

A stocktaking at this time suggests at least four major further efforts that are needed to foster PPB systems in the cities and to lower barriers to implementation: (1) new (or enlarged) programs of education to increase the number of analysts available for work in the cities; (2) precise definition of requirements for evaluation of intergovernmental programs; (3) instituting a programmed yet flexible mode of communication among governments, and among governments and universities; and (4) enlisting direct federal government support of PPB implementation in local governments as well as in the states.

Expanded Education Programs

A number of education programs are presently providing the curriculum choices and instruction that produce the type of quantitative knowledge to be applied in analysis of public service problems. For the most part, such education has been afforded by graduate schools of business and by graduate economic faculties in various universities. More recently, a few schools of administration have been created to merge study of business administration with education for specific administrative fields—for example, the administration of hospitals, schools, and government. New programs can be found at the University of Rochester, University of Michigan, and University of California at Los Angeles. A number of other universities, through their faculties of engineering, economics, political science, and business administration have been cooperating with the federal government in training federal employees in analytical methods.

The focus thus far has been on students enrolled for graduate degrees. But the pay scales of local governments for analytical staffs cannot attract, by and large, those with graduate degrees in a market where demand far exceeds the supply. Moreover, customary hiring practices in the few local governments that recruit from graduate schools appear to suggest a heavy dependence on schools of public administration for filling the type of central staff position that would be involved in program planning and budgeting in state and city.

The possible options for recruiting are numerous—if governments can be induced to use them. For example, a government's customary recruitment sources could be altered, or it might be induced, to train persons for tasks of analysis. Pay scales could be amended to make graduate degree students available to local government at levels competitive with industry and the national government. And, actually, there is no reason why analytical competence cannot be developed as part of undergraduate education, at least in certain key schools, if governments would demand it forcibly enough. Economic education ought to give greater prominence and importance to applied analysis, and more schools need to build into their regular curricula at least part of the methodology of the economics discipline required for governmental analysis and evaluation of programs.

In general, it appears desirable to urge that undergraduate courses in cost-effectiveness analytical techniques become part of public administration education, and that graduate schools of business and of public administration develop, either in a joint educational program or separately, a curriculum that can develop cost-effectiveness analytical competence with an emphasis on public service careers. This may entail new faculty to match the new curriculum or at least faculty retraining.

Additional Requirements for Intergovernmental Program Evaluation

Several of the newer federal grant programs call for program evaluation as part of the grant administration process. Of these, the most important for the cities are the Model City programs originated in the Demonstration Cities and Metropolitan Development Act of 1966. The law calls for evaluation of the effects of the federally aided activities and provides technical assistance directly or by contract to city demonstration agencies. The 1970 United States government budget calls for more than a tenfold increase in the range of specialized technical assistance for these evaluation and related research efforts.

Other federal grant programs that have importantly stimulated local interest in application of analytical techniques and encouraged work on analysis include Title I of the Elementary and Secondary Education Act of 1965 and the Omnimbus Crime Control and Safe Streets Act of 1968. An even greater incentive for analysis was embodied in the Juvenile Delinquency Prevention and Control Act of 1968, which sets forth an assessment of relative costs and effectiveness of the programs proposed for achieving rehabilitation of juvenile delinquents as a criterion of project selection.

The current financial bind of the cities creates an immediate response to the means of qualifying for federal assistance—and if evaluation is a precondition, then steps to evaluation are taken by the cities. In fact, there is some reason to speculate that the anticipation of federal requirements for application of cost-effectiveness analysis has encouraged some local governments to start down the path of PPB implementation.

The growing awareness of federal officials that management of social programs needs to be decentralized gives impetus to encouraging the analysis and evaluation of programs on the local level. For example, the U.S. Bureau of the Budget's assistant director for program evaluation recently wrote (Carlson, 1969: 16):

Thus, to be responsible, decentralization must assure that the means for wise use of funds exists and that elected officials at all levels are given the tools they need for both effective and responsible action.

We may expect in the months ahead that an increased use will be made of evaluation requirements in federal grants-in-aid—or requirements that demand of communities an assessment and quantification of the effectiveness of the federally-aided programs in achieving the objectives set. Revision of federal grants-in-aid to require program evaluation could perhaps be achieved faster by administrative regulation than by statutory change. If such administrative regulation is not in accord with congressional authority, appropriate amendment of the Intergovernmental Cooperation Act of 1968 could be designed that would apply to all grant programs. Grants might be enlarged to finance the evaluation studies, or, within the allocations made from sums appropriated, evaluation costs could be counted as a necessary expense of administration. For the seven federal programs that carry specific authorization for the use of between one-half of one percent and one percent of the program funds for evaluation, the question has already been posed of use of a part of the funds for state and local evaluation (Carlson, 1969).

Along with requirements for evaluation, provision could be made by federal officials to facilitate and encourage exchange among communities on completed evaluation studies. And to help build local competence for evaluation studies, technical assistance could be provided by federal agencies, with federal personnel undertaking to train local staffs and develop analytical competence.

Evaluation studies by cities and states that assess the results of programs have become vital to the national government in its undertaking to achieve its purposes through financial assistance to states and cities. The problem of managing federal responsibilities in areas such as urban renewal, community development, crime control, juvenile delinquency protection, and education through decisions taken by states and communities clearly requires that the localities institute procedures for evaluation to provide the data needed by the national government. But such evaluations, if they are to have an impact on the decisions of the local governments, must be built into a machinery for the decision-making that is specific to each government's own process. Essentially, this means the development of processes of program planning and budgeting in state, city, and county.

A small start has been made through "the Federal Technical Assistance Team" to undertake visits by federal agency staffs to states and communities under the leadership of the U.S. Bureau of the Budget and the Department of Housing and Urban Development. The visits, even if only a brief week in duration, can be used as a communication chain on analytical studies that have been carried on throughout the nation, and to encourage, by examples of rudimentary studies, the application of analytical techniques. Such studies undertaken jointly by local personnel and federal analysts become a local staff training-and-doing exercise.

Federal Technical Assistance Teams have gone to several states and to at least one city—Denver. The immediate purpose of the visits has been an improved use of federal planning grant assistance. Thus, the teams have concentrated on review of a state or city government PPB performance, to the end of recommending improvements in organization and process. Such performance team visits need to become available on request to a far larger number of governments. The procedure requires a firmer institutional arrangement with clearer mandate and separate funding. Expert analysts from the functional agencies can provide the specialized skills that are of special interest to the requesting governments and might well become a part of the Federal Technical Assistance Teams.

The team visit is one method of local staff training and transmission of analytical study findings. Other methods that can be applied as a substitute for or in conjunction with such team visits are indicated below.

Communication Channels Between Governments and
Between Universities and Governments

Start-up problems on PPB implementation in cities have by now been documented and the comparative experience has been reported. But, as local governments advance in their efforts, there is more need for organized communication channels on program structures, on program analyses, and on evaluations completed. Open channels are indicated between universities and governments, between levels of government (federal-state-local), and between cities.

Universities are engaged in research that has an application to governmental problems. They have the expertise for bringing together specialists who can contribute to the understanding of city problems and ways of meeting them, but the results need to be translated by the universities so that the research can be understood and judged by the layman. Publication vehicles usual for university scholars are not, in the main, customary reading materials for city officials, and the language of scientific journals is not always clear to laymen.

In addition, a reverse flow of communication from city to university could serve an important research purpose. The problems as perceived by the city are many and offer a field of research study that can provide the hypotheses for investigative study. Many researchers are seeking problems to investigate; cities with the problems seek information and research findings.

Several recent institutional arrangements have been designed to develop or foster new communication channels. The Urban Observatory project under the direction of the National League of Cities specifically aims at generating partnerships between cities and universities. The concept calls for creating coordinating organizations that: (1) can focus university resources on the problems of metropolitan governments, and (2) undertake research on matters of particular concern to local communities. The Urban Institute, established in 1968 as an independent nonprofit research organization, has the basic mission of exploring the whole complex of urban problems as an aid to federal policy formulation. It constitutes a de facto channel for communication between national and local governments and between governments and the university research community by: (1) building on university research already completed, (2) undertaking major studies indicated by existing deficiencies in knowledge, and (3) working with cities to improve their management.

Communication among governments and universities can be furthered by other similar or additional institutional arrangements. Clearinghouse activities on research are, in fact, being fostered by several federal agencies, each of which is concerned with special areas. Such organizations as the Council of State Governments, the International City Managers' Association, the National Association of Counties, and the National League of Cities are also involved in the transmission of research findings and demonstration experience.

The vehicle that will serve, specifically, as an effective communication channel may vary according to the types of activities considered desirable as effective aid to local governments. Among the activities that would presumably be useful are the following:

(1) Compilation of materials on program structures for citywide programs and for specific broad program areas—for example, education, health, transportation.

(2) Comparisons of program structure formats through brief reviews of the structures and their use for (a) analytical purposes, (b) public program displays, and (c) budget formats.

(3) Compilation of analytical studies concerned with the same or similar public program issue.

(4) Development of case materials on the basis of analytical studies completed for staff training purposes.

(5) Comparative reviews of analytical studies.

(6) Design of experiments that can help answer questions illuminated by the analyses carried out.

(7) Assessment of the special characteristics of the locality in which the analysis was undertaken and the limitations on general use that result.

(8) Compilation of evaluation studies.

(9) Development of case materials based on those studies.

(10) Comparisons and assessments of evaluations carried out.

(11) Assembly of materials on procedures and methodology used in multiyear programs and financial plans.

(12) Evaluation of revenue and expenditure projections.

(13) Evaluation of problems in application of multiyear program and financial plans.

(14) Improvement in information on advance estimates of intergovernmental aid, both federal and state.

Not all of these activities need to be carried out through the same organizational arrangement. But each would facilitate the implementation of PPB systems in cities. What we are suggesting here is a system of exchange of information on: (1) PPB documents produced by states and local governments and (2) university research that bears on the analyses of public issues.

FEDERAL GRANTS FOR PPB

The steps that are now being taken by the federal government to encourage states and localities to improve their program planning and evaluation and to assure more effective use of federal program grants are inadequate. They fall short of the type of incentives required to achieve the improved program planning in the city that will assure that national program objectives are met. Various proposals are being advanced to free up the federal regulations and conditions on state and local use of federal grant funds. The thrust of the current administration's fiscal federalism goes further to provide federal aid to states and localities without federal strings of any sort. A necessary counterpart of such free funds is assurance of program management methods in states and localities that can derive the largest gain in program results for the dollars spent. States and localities require better program planning and management tools if they are to function effectively as partners in a federal system. Such program planning and management would be furthered by development of PPB systems in state, city, and county. Each government as part of its "management by objectives" processes would formulate objectives and plan in accord with its priorities responsively to federal grant incentives.

New approaches to encourage implementation of PPB systems in states and localities are needed. They should provide incentives for application of analytical techniques that would furnish continuing federal support for central staff work on program analysis or review of such analysis. A specific categorical grant that would allocate funds by formula and, accordingly, provide continuing and assured support could achieve the improvement in program planning that is sought. With a specific set-aside of a part of the new categorical grant for training of personnel in program analysis, the required staff build-up will be encouraged as well.

APPENDIX

For those localities that indicated a beginning on PPB implementation beyond a planning phase, summaries are presented below on status of work, based on the responses to both questionnaires and such additional remarks as they included. The summaries were reviewed by the governments, and revisions were made as they requested. The summary includes information from Orange County (California), which had not responded to the questionnaire but at our request provided materials to be included here. It does not include material from the cities and counties in the 5-5-5 project. Summaries for 20 cities are listed first.

Anaheim (Calif.)—Operating agency personnel oriented on PPB; staff assigned to perform cost-effectiveness studies on new programs; governmental objectives not fully established, but criteria of effectiveness selected for all major program categories; the question of whether to implement one agency at a time or by successive steps for the entire government under consideration. A systems analyst has been hired from the aerospace industry to "bring a 'systems approach' to municipal operations."

Fremont (Calif.)—Governmental objectives established; some program analysis underway; emphasis on improving budget request back-up through use of projections and analytical material; limited staff orientations carried out, but no training program planned. "Time and money" are the major problems.

Fresno (Calif.)—Some work done on defining governmental objectives; comprehensive cost data for program alternatives being included in program studies; staff training program planned but not yet underway.

Pomona (Calif.)—Central PPB unit established in City Administrator's Office working through Systems Task Force, which coordinates all management information systems; staff personnel attended training sessions; orientation sessions held for departmental personnel and City Council; budget format changed to conform to major goals and objectives; program structure implemented for 3 departments, full program structure for 1969-70 budget; cost-effectiveness studies started in a few departments; improved budget request back-up in all departments; specific objectives and criteria of effectiveness for each department being developed; have full City Council direction to implement as rapidly as possible.

Hartford (Conn.)—Executive budget submitted to Council expressing human and material resources within a program format and including a narrative statement of program objectives; comprehensive cost information being developed for use in program studies; multiyear fiscal projections prepared; no personnel training planned. The remaining effort is devoted to better budget request back-up.

Tampa (Fla.)—Program structure planned; work largely in the area of performance analysis within some departments and emphasis on better budget request back-up; no training program planned. Obtaining adequate cost and work program data presents a major problem.

Savannah (Ga.)—Central PPB unit established; departmental goals set; work underway to refine program structure; output measures selected for all programs and subprograms; program analysis being carried out in all departments to some extent, with some interdepartmental analysis; all agencies have received both initial orientation and detailed instruction on PPB; budget request back-up for new or changed programs includes discussion of program alternatives. A major problem is the lack of time and money to institute a training program. Multiyear projections are hoped for, beginning 1970.

Peoria (Ill.)—Governmental objectives established, and a program structure constructed including measures of output for programs and subprograms; some departments have received both orientation and detailed instruction on PPB; budget request back-up is more analytical, includes future-year costs and effects, and discussion of program alternatives for some department submissions. The major problem is succinctly stated: "time."

New Orleans (La.)—Program structure being planned; output measures selected for programs and subprograms; some program analysis being done, both within departments and interdepartmentally; some departments and agencies have received both orientation and detailed instruction on PPB.

Portland (Maine)—Some work done on a program structure; within budget preparation process, development of program goals emphasized as first step; departmental and divisional objectives being developed and refined each year; improved program statistics on past and present workloads being required of each department; program budgets for a limited number of division accounts included within regular city budget but do not replace regular budget format. Bringing capital and operating budgets together is identified as the major problem.

Ann Arbor (Mich.)—In 4 major operating departments—Public Works, Parks and Recreation, Traffic, and Water—output measures developed for programs and subprograms; program structures prepared for each department; major efforts directed toward building up and developing accurate and usable information through new reporting systems (information currently available inadequate as basis for PPB); some departmental personnel have received orientation and detailed instruction in PPB; budget and information collection forms being prepared and written as program learning texts to be used as training tools, as well as reporting systems.

Lansing (Mich.)—Program structure being planned; most work being done in program analysis and in more analytical budget request back-up; interdepartmental program analyses being planned; and budget requests of some departments include cost-effectiveness information and future-year cost data; no personnel training program planned.

Wyoming (Mich.)—An objective-oriented program structure developed, with output measures for programs and subprograms; department and agency staffs have received both initial orientation and more intensive instruction on PPB; certain budget submissions include analytical back-up material, including future-year costs, cost-effectiveness information, and program alternative discussion. The major problem is seen as training personnel and establishing routines.

New York City—The main thrust has been on program analysis, carried out across department lines where necessary; an analytical approach to programs in the departments is emphasized; institutionalization of the PPB system by development of a program structure with criteria of effectiveness underway, and information gaps being defined; some programs being documented with multiyear projections that include comprehensive cost data. A training program for departmental staff is underway.

Lancaster (Pa.)—Program structure developed, with output measures for programs and subprograms; program analyses being prepared both within departments and interdepartmentally; departments have received orientation on PPB but no detailed instruction as yet; budget request back-up of an analytical nature required for all department submissions, including cost-effectiveness information, future-year costs, and discussion of program alternatives. The biggest problem is constitutional restrictions on the city's budget procedures and techniques.

Philadelphia (Pa.)—Program structure established and incorporated into a budget format; a training program for central and agency personnel conducted, using an outside consultant firm; orientation for key agency heads carried out; work initiated on program analyses that cross departmental lines; effort being made to incorporate analytical work into the budget process. Restraints on funds are identified as barriers to development of program options.

Wichita Falls (Texas)—No training program on PPB for departmental personnel; program structure covering selected departments being developed; cost-effectiveness study is within departmental organization; budget request back-up is more analytical, including future-year cost information and discussion of program alternatives for some departmental submissions. Independent effort on long-term projections (20-25 years) being made; availability of funds for research and study of better methods is identified as a major problem.

Danville (Va.)—Training program on PPB planned; program structure work planned covering all departments to be used in budget format; some program analysis being carried on within departments. Training staff and department heads is identified as a major problem.

Tacoma (Wash.)—Steps being taken to initiate a limited training program for personnel; more analytical budget back-up being sought; major problems are inadequate financing of the PPB system and state limitations that constrain development of program options.

Beloit (Wis.)—Program structure developed, involving a change in the budget format; output measures developed, by program and subprogram, in some departments; some studies of cost effectiveness being made; orientation for department personnel has been conducted; training program planned. The small size of the city impedes necessary data gathering and makes for direct citizen confrontation in program formulation.

SOURCE: State-Local Finances Project of the George Washington University (1969) *Implementing PPB in State, City and County.* Washington, D.C.: George Washington University (June).

REFERENCES

CARLSON, J. W. (1969) "Federal support for state and local government planning, programming, and budgeting." In U.S. Congress, Joint Economic Committee, Innovations in Planning, Programming, and Budgeting in State and Local Governments. Ninety-first Congress, First Session. Washington, D.C.: Government Printing Office.

GRIZZLE, G. (1969) "PPBS in Dade County: status of development and implementation." In U.S. Congress, Joint Economic Committee, Innovations in Planning, Programming, and Budgeting in State and Local Governments. Ninety-first Congress, First Session. Washington, D.C.: Government Printing Office.

HATRY, H. P. and J. F. COTTON (1967) Program Planning for State, City, County Objectives. Washington, D.C.: State-Local Finances Project, George Washington University Bookstore.

HITCH, C. (1961) "The uses of economics." P. 104 in the Brookings Institution Research for Public Policy. Washington, D.C.

HOLLINGER, L. S. (1969) "Changing rules of the budget game: the development of a planning-programming-budgeting system for Los Angeles County." In U.S. Congress, Joint Economic Committee, Innovations in Planning, Programming, and Budgeting in State and Local Governments. Ninety-first Congress, First Session. Washington, D.C.: Government Printing Office.

HORTON, R. A. (1969) "Planning, programming, and budgeting in metropolitan Nashville-Davidson County, Tennessee." In U.S. Congress, Joint Economic Committee, Innovations in Planning, Programming, and Budgeting in State and Local Governments. Ninety-first Congress, First Session. Washington, D.C.: Government Printing Office.

JONES, M. V. (1968) A Regional Project Evaluation Model. Bedford, Mass.: Mitre Corporation (August).

MUSHKIN, S. J. (1969) "PPB in cities." Public Administration Review, 29 (March/April): 167.

———, H. P. HATRY, J. F. COTTON et al. (1969) Implementing PPB in State, City, and County. Washington, D.C.: George Washington University (June).

RIVLIN, A. M. (1969) "The planning, programming and budgeting system in the Department of Health, Education, and Welfare: some lessons from experience." P. 917 in U.S. Congress, Joint Economic Committee, The Analysis and Evaluation of Public Expenditures: The PPB System (volume 3). Ninety-first Congress, First Session. Washington, D.C.: Government Printing Office.

State-Local Finances Project (1969) PPB Pilot Projects Reports from the participating 5 states, 5 counties, and 5 cities. Washington, D.C.: George Washington University.

——— (1968) PPB Note 11. Washington, D.C.: George Washington University.

11

Implementing PPBS:
A Practitioner's Viewpoint

DONALD J. BORUT

☐ THE INTRODUCTION into an organization of a new system which changes existing patterns of behavior and existing relations may well produce anxiety, fear, hostility, and resistance. A prime consideration in the implementation of such a system must be the structuring and modification of behavior required to make the new system functional to both the participants themselves and the overall objectives of the institution.

A planning-programming-budgeting system, fully implemented, should have a radical impact on the procedures, relationships and activities of an institution. It is, therefore, essential that serious consideration be directed toward accommodating the personal concerns of employees. More specifically, it is in the processes by which the behavioral consideration are met that PPBS emerges as an operable and exciting technique for improving the performance of an organization.

Unless the several technical elements comprising a PPB system are modified, tempered and carefully structured to accommodate the particular behavioral characteristics of a given organization, its probability of success is radically reduced, no matter what the performance criteria. A prime element, then, of this systems approach must be the development of mechanisms which inter-relate participant needs with technical and administrative activities.

As an administrator responsible for the introduction of a planning-programming-budgeting system in the government of a municipality of 110,000 population, I have become increasingly aware of the need to establish participant interest and enthusiasm in the benefits that the system will produce, not just for the city as an institution, but for employees personally. This consideration has become central to all decisions made on the project and provides the focus for this paper.

Ann Arbor, Michigan, like many municipalities in the 1950s and 1960s, has grown at a rapid rate with significantly increased capital and operating demands, extensive growth, and new program requirements. The existing budgetary and administrative process within the city, while creative and responsive to these changing demands, nevertheless required a reorientation in order to better utilize municipal resources and to better comprehend the long-range objectives toward which the several programs were aimed.

Ann Arbor is in the process—the second year of a five-year effort—of introducing a PPB system. The evolution of the system and the modification of its several elements to respond more appropriately to the particular needs of Ann Arbor may be instructive, not because of their success, for certainly that assessment is premature, but, rather, because it reflects what practitioners are forced to keep foremost—that is, keeping some system going while change is being made.

SYSTEM MODEL

A normative concept or model of the planning-programming-budgeting system proposes a rational procedure for developing a program structure, for modifying organizational information systems, and for applying several analytical processes to increase the rationality of operations of the organization. This model begins with planning considerations and identifies administrative and management procedures which assist in both the operations and control functions. Unfortunately, the implementation of a PPB system following this model could result in serious practical problems, thereby reducing success in meeting the system objectives.

While it is not my role to develop and describe a model for PPBS (see Mushkin, this volume), it may be useful to consider the several elements of the system from the perspective of a practitioner, and to consider how these elements may be modified to facilitate implementation.

PPBS has been defined as the composite of three elements— program structure, analytical techniques, and information systems —which together provide a means of rationalizing, defining, and achieving program goals. A fourth element, behavioral modification, structurally less defineable, is, I will contend, an equally important element.

PROGRAM STRUCTURE

A program structure provides a means of hierarchically describing the activities of an institution beginning with the output objectives or goals and working back to activity components of the organization: these may be defined as programs, program elements, and activities. Viewing an institution in this manner not only forces a recognition of structural relationships, but also permits the actors to see how their participation and activity meaningfully relates to the overall objectives of the organization. In layman's language, the program structure is the skeleton of the system. It also permits the individual and subunit to identify with the overall goals and objectives of the organization.

Municipalities are proud of their ability to model their formal organizational structure—a process model that is intuitively accepted—but the organizational structure only relates organizational inputs to other inputs and to nondefined outputs. A program structure is designed to link defined outputs to inputs. The particular nomenclature used to produce a program structure need not follow a particular model. In fact, the nomenclature should be based on the terms generally used and understood by the organizational participants themselves. Introducing a new approach involves enough new considerations without introducing new terminology when existing language may be perfectly adequate to describe the hierarchical program structure of the organization. If people speak of objectives and programs and generally agree as a result of past perceptions about what is meant, there is good reason to avoid the inevitable confusion which would result from changing the descriptor guideposts.

A program structure, besides requiring actors in the institution to relate their actions to organizational goals, should reinforce and clarify perceptions of the interrelationship of activities, not confuse them. Tampering with nonessentials in developing this structure merely provides a focus for participants to criticize and find fault with the overall process. Reassurance is as important a consideration in the introduction of PPBS as any of the most sophisticated analytical techniques.

INFORMATION SYSTEMS

Managing the activities of any institution depends primarily on information and data which can be used to understand the system, evaluate performance, make required modifications or adjustments, and continue the action processes of the organization. Informational feedback acts as a thermostat and without it, managers can only operate on hunches or nonverified philosophies. Thus, the ability to modify performance requires accurate descriptors of performance as a principle factor. Municipalities, like most institutions, are not lacking in the generation of information in the form of daily, monthly, quarterly, and annual reports. Periodic reports have a way of being introduced, and whether relevant or not, they are generally perpetuated, institutionalized, and ultimately formalized to a point where the generator of the report has a stake in its continuation. Few, if any, reporting systems are introduced that eliminate other information systems; rather, they supplement, complement, and duplicate previous efforts—often losing sight of the prime purpose for the information: to assist some person in knowing what is being done or to what degree something is being effected.

Information systems are fundamental parts of a meaningful description of the activities of an institution, just as is the program structure. Information is the necessary substance for analytical tools designed to improve and modify the process and practices of the institution.

The notion that information should be generated only if it is necessary for a defined purpose is not novel—in fact, it is obvious.

Achieving the obvious is more difficult. As an element in the overall PPB system, information and reporting techniques, particularly the method by which they are introduced, becomes extremely significant. How information is produced and how people use it provides one of the most effective focal points in the introduction of the total concept of PPB simply because it is a point of relevance to large numbers of participants and provides rapid feedback to administrators.

ANALYTICAL TOOLS

The ways in which organizations arrange inputs and combine them to achieve defined outputs or goals are as numerous as the organizations that exist. Certain "standard" procedures and methods have been developed and applied to many organizations, yet even these must be modified and tempered by the particular organization. Demands, personnel configurations, resources, and political considerations all vary, necessitating the modification of model systems to meet particular needs. Traditions, personal idiosyncracies, and so on, usually play an even greater role in determining the way an organization carries out its activities than do objective, external reasoning.

Procedures can be changed and modified through the application of a number of formal approaches which in the nomenclature of PPBS are described as analytical techniques or tools. They include methods by which problems can be defined and recognized, and means by which practices and procedures can be understood and modified to better achieve desired effects. Systems analysis, the catchall for a multitude of activities, is perhaps best thought of as an *approach* to a situation or discipline. The approach can involve varying degrees of sophistication but is presumed to be objective. It is the application of ordered, rational thinking to the methods by which complex organizations systematically relate resources to objectives.

Today, the systems approach, systems analysis, and the various techniques harbored under these titles take on a greater degree of importance in public institutions than previously, partly because of recent alleged success in the public sector. The proliferation of consultants in this field is a sign of the growing importance of such techniques along with the dearth of practitioners in public organizations.

BEHAVIORAL MODIFICATION

The fourth rubric or element of the overall system encompasses behavioral considerations; that is, the means by which people are considered in implementing the technical elements of PPBS and the process by which these elements are utilized to effect organizational performance. It is the method by which this system is introduced that makes it relevant and workable. How often have we heard managers lauding the virtues of some technique they read about or learned about at a conference, but failed to introduce in any practical way! To be relevant, the PPB system must not only improve, in some general way, the effectiveness of the institution, but also, in a felt way, it must assist individuals both in changing given practices and, more importantly, in helping them to appreciate and understand that they are personally assisted by the approach or system.

This element, in fact, will be the central issue of this paper. Only through the ability of managers to effect real changes in behavior can the system be made relevant.

COMMUNITY PROFILE:
ANN ARBOR, MICHIGAN

Ann Arbor has an estimated 1969 population of 110,000 and is working to meet the problems which face all rapidly growing independent cities—housing, racial tension, responsive law enforcement, adequate planning, changing community needs and demands, and adequate resources and responsibility. A unique characteristic of Ann Arbor, one which both generates problems and provides unusual resources to deal with them, is the presence of the University of Michigan. The university community requires unusually sensitive and responsive municipal services in such fields as police protection, housing inspection, human relations, and planning. As a complex and nonmonolithic organization, the university requires and demands special municipal capabilities.

In 1960, the city had a population of 67,340 of which five percent was nonwhite. Between 1960 and 1969, it grew in land area by over thirty-three percent to 21.78 square miles; increased in assessed value from $216 million to $414 million; and moved from a 1960-1961 general fund budget of $4,043,531 to a 1969-1970 general fund budget of $10,707,162. In 1960, Ann Arbor had 548 full-time city employees—a number which increased to 849 by 1969. Ann Arbor has an average per household income of $13,350 (Sales Management Magazine, June 10, 1969)—a figure which reflects both relative affluence and a population which demands a highly responsive municipal government. Growth and change are the basic characteristics of Ann Arbor. The desire to anticipate and meet new demands and the conflicts inherent in an older community developing at a rapid rate provide a strong impetus for an innovative municipal organization.

Several additional considerations may be useful in putting Ann Arbor's PPB effort in perspective. Physical growth, for example, is seriously inhibited by antiquated state annexation requirements. This generates a high degree of uncertainty in long-range planning, particularly in utility development planning. A high level of interest in human relations and quality human and physical services made known by an articulate population requires a more highly visible governmental operation than is generally found. This is further accentuated by the political characteristics of the community-partisan ward elections with strong and active political organizations and a great deal of citizen participation.

The crucial question, and the one that motivates the city's PPBS effort, concerns how well the city resources are being utilized and how well the organization is providing the services for which monies have been allocated. That dollars are appropriated for new efforts is one means by which the city council and the public can be satisfied that the government cares. Over time, however, the measure must be how effective that funded service is.

In 1956, the city adopted a compromise council-manager form of government, with a large span of control for the manager, seventeen departments, partisan ward elections for ten councilmen and a mayor elected at large with the power to appoint two department heads. By 1964, the actual operation of the government had been tempered and modified by charter amendment to more closely follow a traditional council-manager form of government.

Several efforts were undertaken at this stage to generate mechanisms for closer control over the operating activities of the city. Capital planning and programming, utilizing a capital improvement plan and budget, provided controls and constraints over the more capital intensive efforts of the city. The operating activities, however, continued to reflect the independent and often uneven planning and programming of the different department operations.

In 1965, a reporting or crude management information system was developed to provide regular departmental and top administrative review of ongoing programs to help in determining program status and, hopefully, to provide for a means of reallocating resources within a budget year. This system was introduced with the best of intentions (by the author), but the fact that it was planted on top of an existing reporting system and was not perceived by the departments as relevant to their actual operations proved to be a significant drawback. Because data produced was of only minimal relevance to the departments, over time, its accuracy and usefulness became marginal. Equally important, the amount of data generated was so vast and the techniques for assimilating it so limited that it was useless.

After reviewing these efforts, it became abundantly clear that major steps would have to be taken not only to improve reporting procedures but also to gain greater control over departmental programs to insure that they were in fact achieving the general, stated objectives set forth in the line-item budget. While the political process offered a relatively effective check on the quality of the most visible services provided, the city administrator recognized that as the city and its services grew, internal procedures had to be modified to achieve increased internal coordination and control over operations.

In 1966, a concept known as **PPBS** was being discussed in professional circles and was "in" like miniskirts and MacNamara. Just what **PPBS** entailed or how it might assist the administration was not altogether clear at the time. To many, it meant cold efficiency, the application of business techniques to municipal government, systems analysis, the use of high-speed computers —administrative nirvana. Whether all these notions were true or even relevant was less significant than the myths we (and I include myself) believed. To be uncomfortably honest, we backed into PPB and only after the initial floundering began, were we able to gain some perspective and articulate a consistent plan for the implementation of the system and a feel for expected benefits. It is difficult to justify bumbling into something as significant as the introduction of a modification of a management system; nevertheless, I would argue that immersion under fire identified, more rapidly than any other approach could, the significance and importance of the personnel problems and the apprehension generated by a potential systems change.

The prime impetus for the introduction of a PPB system in the city of Ann Arbor was the desire on the part of the city administrator to develop a management reporting system to provide him with realistic yardsticks for measuring the activities of the several city departments—control on outputs as well as inputs. The growth of the city, the proliferation of programs, the city council, and the general public demand for "better" services all gave credence to the need for more responsive descriptors of what the city was in fact doing, not just how much it was spending. In addition, the city administrator, having lived with the organizational goliath built into the 1956 charter in which seventeen department heads reported directly to him, was prepared to carry out some rational modification of the structure. There was also an interest, though not totally articulated, in attempting to relate long-range operating programs and problems to the ongoing efforts of the municipality, in much the same way that the capital budget and program had been operating for capital expenditures.

Finally, the administrator was anxious to make the budget a more useful tool, useful in the sense that it was a clear *line-item* budget and that it contained program statements from the several departments which could be described as contractual commitments by the department heads to provide stated services.

With these stated and implied objectives set out, PPBS appeared to offer a potential approach for implementing the several operating changes and relating them to the long-range planning consideration, while simultaneously providing greater control over the city's operations. Clearly, we did not have a considered perspective of the total implications of PPBS, nor had we considered the impact which these "simple" changes would have on the overall organization.

It must be reiterated that, at this early stage, few in the city recognized the potential or the broad implications of the steps to be taken. It was only after the implementation process began that broader implications of the system became apparent. It could be argued that before a system of this kind is introduced, the total strategy for implementation should be spelled out and articulated. Even if this were possible, it is not necessarily desirable, since the discovery process and resultant participation in change may prove to be a highly functional element of the system. This is not to say that those introducing a system of this kind should disguise the intent of their efforts by being totally nondirective, but, rather, that the means of implementation should grow to meet the demands and requirements of the particular institution. Rather than trying to anticipate *all* problems, a better strategy is to respond to them as they arise.

IMPLEMENTATION IN ANN ARBOR

In the spring of 1967, the city administrator made the decision to begin the implementation of a PPB system. The generally accepted objectives of the effort were to: (1) improve the budgeting process so that it would better describe the activities of the city, (2) introduce a more administratively relevant daily and monthly reporting system for the city, (3) provide a better understanding of the municipal activities for the purposes of reorganization, and (4) relate annual programs and program efforts to long-range objectives of the city.

Concern for the budget format and the dearth of adequate reporting information in that document, were the catalysts that permitted PPB system approval. It was decided, however, that responsibility for this task should not be delegated to the controller, chief financial officer of the city. Even at this early stage, PPBS was seen as far more than a method of better refining the existing budget document, although this was an important motivation for introducing the system. The city controller and his staff have an orientation, as they should and must, toward dollar *control* (i.e., to insure that expenditures are made on those items for which they are allocated and to see that there are adequate funds to cover expenditures). The quality or quantity of services—outputs—is not of primary concern to the controller. Instead, prime responsibility for the introduction of the system was placed in the office of the city administrator and delegated to his assistant. City personnel were not sufficient to introduce PPB. As a result, a staff configuration was developed which brought together a variety of skills including departmental supervisors; graduate interns from the University of Michigan's Institute of Public

Administration (now the Institute of Public Policy Studies); and systems personnel from the Community Systems Foundation, a nonprofit consulting firm with previous experience in the city, with the assistant to the city administrator as the program director.

This staff mix was significant. It insured that any program changes would be carried out by those who would ultimately be responsible for its ongoing implementation, namely, department personnel. Second, it brought onto the team, technically-trained individuals with a systems orientation and background, skills lacking in the city's own operation. Finally, the use of young interns provided both manpower and aggressive participation not wedded to some preconceived notion of municipal operations. In reality, we created an independent group working out of the city administrator's office with close ties to department heads and supervisory personnel within the several participating departments.

Once the commitment had been made to proceed, a more realistic rationale and method of operation began to be articulated and the full implications of the system began to be identified. If all four elements of the system were brought to bear on the city government simultaneously, it was apparent that the total organizational structure would have to be modified and the orientation of planning as then seen would need to be elevated to a position of primary importance. Change and control would have been obvious by-products, both of which were sure to generate understandable resistance from departmental personnel.

In developing an implementation procedure for the PPB system in Ann Arbor, the administrative staff reviewed what appeared to be the normative and logical model, namely, to begin by developing a program structure. Such a structure would start with a definition of citywide program objectives or outputs, and through the development of a hierarchical order of elements, would work back to a description of all activities comprising the inputs of the city. From a normative perspective, this approach appears both logical and necessary before proceeding further. Two problems became obvious as the program-structure-first approach was pursued: (1) city personnel had never been forced to articulate "program objectives" and define, in a close-ended manner, precisely what they did. To even a casual observer, it was clear that many activities overlapped departmental lines, and these lines were fixed only in a loose sense; (2) an attempt to create or describe citywide "program objectives" would require large amounts of time and commitment from all department heads, most of whom were extremely busy just in dealing with normal day-to-day crises faced by their staffs.

The development of an operating plan or a descriptor of operating activities—program structure—did not appear to be conducive to generating support from the operating manager; rather, he was more likely to see it as a nuisance.

While the definition of program objectives may appear to the outside observer as a basic and vital factor, necessary before any program can be undertaken, to the manager immersed in the day-to-day activities of his department, the exertion of this kind of effort is more likely to be seen as an academic exercise, something nice to do but not at all relevant to the "crises performance" against which he has traditionally been measured. Given this, to begin the implementation of a PPB system with an extensive commitment of time to the definition of objectives appeared to be highly disfunctional and not designed to generate enthusiasm.

An alternative, one which seemed to counter the program-structure-first approach, was to begin at the opposite end of the system, concentrating on inputs (personnel, expenses, material, etc.). These were concepts with which the operating people had traditionally worked and were most comfortable. In light of the interest in generating and insuring support for the PPBS effort, a decision was made to begin implementation by describing and developing measures of what was currently being done (current departmental outputs) and systematically relating these to current input measures. By beginning with the definition of current activities, personnel would be required to speak to the activities with which they were most familiar, activities which were of prime concern to them and reflected their relevant world.

It is worth noting that no matter how much effort goes into the development of a new plan or into changes in procedures, a fundamental concern of operating managers will remain the status of existing activities. Streets need to be paved and park lawns must be mowed. It is not possible to one day stop concentrating on daily programs and the next day begin a new approach; rather the new approach must be phased in and must reinforce ongoing activities in meaningful and relevant ways.

A rudimentary or even crude data development operation which builds upon daily operational considerations appeared to have a greater possibility of generating support early in the process than a similar expenditure of effort in defining and creating a program structure. Furthermore, because the program objectives and outputs of the city government had never been rigorously defined, the probability of generating interest in this latter exercise seemed remote; there was no obvious way such an effort could produce immediate rewards, that is, reinforcement.

Another constraint mitigating against the utilization of a program structure as a first step in implementing PPBS was the fact that from the date of its inception, the regular budgeting process was within three months of beginning. Since the introduction of changes in the budgeting procedure was a prime motivation for introducing the system, some type of successful changes had to be reflected in the upcoming budget.

Even the most dramatic efforts on the part of city personnel and consultants could not produce a total program budget within the three months required. It did seem possible, however, that by selecting four major operating departments, certain changes could be effected during the budgeting process, specifically, better descriptions of departmental output. Furthermore, by selecting two departments that appeared to understand the implications of the system, and two that were having difficulty with programs, a functional competition could be generated through which the latter two departments would improve their management procedures.

The decision was therefore made to select four major operating departments—Public Works, Parks and Recreation, Parking and Traffic Engineering, and Water—and to utilize the PPB group and department staff to systematically describe their current activities in output terms and to use these descriptors in the budget for proposed departmental programs in the following fiscal year.

The net result of this initial effort was the development of some crude measures of program output, measures which structured, for the first time, *all* of the department activities. In brief, the PPB staff and departmental personnel systematically examined everything the departments did, describing every unit of output as an element of a program. The output measures were clearly unpolished. Many of the supervisors had never thought through their own responsibilities and such efforts were unpleasant and disruptive for many in spite of early orientation meetings. Much of the defining of outputs ultimately was completed by the PPB group and not by the department personnel, although the latter had to approve it in the final analysis.

The initial tendency of the staff personnel was to describe programs in terms of inputs. For example, descriptors of street maintenance output was the amount of material used to patch holes or the manhours required to keep the holes filled, rather than measures which reflected service received by the public. These efforts had to be reviewed, reworked, and new measures created. The process was barely completed at the end of the three months.

BUDGET PROCESS

The PPB effort received a reality test during the budget hearings for the fiscal 1968-1969 budget. While it is not the purpose of this paper to review the budget process in Ann Arbor, a brief discussion may be useful in understanding its relationship to PPBS.

Like virtually all municipal budgets, Ann Arbor *in practice* operates on the system known as incremental budgeting; that is, the preceding year's allocations become the status quo or platform from which the current budget begins.

The city controller in Ann Arbor is the budget officer and is responsible for the mechanics of developing a document in which projected revenue estimates are balanced with department expenditure requests. At the start of the budget process, he calculates expected revenues and provides blank forms to all departments on which they must prepare their requests. The general guidance offered is that requests should be held down. These requests are reviewed by the controller and his staff and reduced, based on some system known only to controller types, but assumed by department heads to reflect the general magnitude of the increase over the preceding year. The budget product developed at this point is a document which more or less balances revenues against expenditures.

The reduced department budgets are then reviewed by the controller with each department head and then again with the city administrator. It is at this stage that the advocate process (budget game) is presumed to operate. The department heads plead their case and the city administrator argues for reduction on behalf of the balanced budget. If a department head is persuasive, that is, has program data which are difficult to refute, cuts may be restored. If, on the contrary, the back-up data and department presentations are weak, one of two things may occur: either the reduced figures are retained, or the city administrator switches positions and becomes an advocate for the requesting department, that is, the advocate system breaks down. It is this last alternative that weakens the advocate model. It is necessary, however, because several department heads are not able to defend the interests of their department in an effective manner. In fact, without this transfer of roles, that is, the disruption of the advocate procedure, programs for which the city administrator will ultimately be responsible may not be adequately defended and funded.

Once the final administrative hearings are held, the controller and administrator make final adjustments; that is, the controller, who generally estimates revenues low, "finds" additional dollars and the budget is balanced for presentation to the city council.

Ann Arbor's municipal elections are in April, which means the budget document for the fiscal year beginning July 1 is the very first responsibility of each new city council. It is on the budget that new councilmen cut their teeth. Traditionally, the councils have tended to do one of two things with the budget: (1) make minimal changes at the margins, that is, cut a few new positions, add some police, or (2) propose some totally new program which requires additional modification of other expenditure requests to provide the necessary funds. In few cases, if any, however, has the final budget deviated significantly from the budget proposed by the city administrator. (The only real changes are those generated as a result of negotiations with unions necessitating additional wage and salary increases after the budget has been printed for council review.)

This budget process may best be characterized as a period of pressure, uncertainty, and last-minute preparation.

Up to the time of the program budget effort, all budget materials were presented in line-item form with a narrative discussion describing, in very general terms, what each department intended to produce in terms of service. Some output measures were provided, but in no instance were they tied to inputs; that is, the specific costs of producing the service. (For the novice in municipal budgeting practices, let me point out that this procedure is by no means unique; in fact, the budget process in Ann Arbor is perhaps in the class of the more sophisticated.)

A prime consideration in implementing the PPB system, as previously stated, was to make improvements in the budgetary process, particularly the addition of output—service and program data—to the budget document. In this first effort, the PPB staff worked with the four operating departments to develop descriptors for stating what service levels would be met during the year and relating estimated unit costs to projected output or service levels.

An early and hard lesson learned through this effort and one, I might add, that required a second attempt before it was driven home, was that the budget process "ain't no time" to introduce a radically new system.

The budget process is a time of crisis and pressure and, during such times, people, be it department heads, controllers, or city administrators, all revert to previous and internalized behavior as the preparation deadline approaches. In other words, people under pressure are not apt to be amenable to changes in procedure; rather, they prefer to continue with procedures which are most comfortable *even* if they know it may not be in their best interest.

For example, we found that during the advocacy process of the administrative budget hearings, the department personnel rarely argued their cases from the program data they and the PPB staff so laboriously prepared. In more than one instance, however, this data provided the justification needed to restore cuts and avoid further cuts. It was only when the city administrator, switching to a departmental-advocate role, asked the PPB staff for their comments that these program statistics had the desired effect of reinforcing line-item requests.

Some additional observations may be useful at this point, even though they are based on the more refined efforts carried out during the budgeting process the following year. Better program data is a two-edged sword. It can be used to justify the necessity of additional funding *and* it can be used by a sophisticated advocate to verify the validity of budget reductions. In a word, greater information means greater control for the stronger of two adversaries. Control, therefore, may very well be seen as an insidious threat and any effort to refine program data may, from the perspective of a weak advocate, be a dangerous consideration. Department heads may prefer not to have program data, and at the critical moment, when a reduction is to be made, stand on their "reputation as professionals" or possibly even play on the sympathies of the administrators or council. In the budget game, few rules are sacrosanct.

Those persons willing and committed to expend their own and their staff's time in the laborious task of generating accurate output data for the budget presentation are necessarily the department heads who recognize how such data can be used to reinforce their budget requests. Furthermore, the more tightly service levels and cost data are tied together, the greater the probability that, should budget reductions be imperative, the department head will be given the discretion to make the cuts in his own budget.

In tying input projections to outputs, a far greater degree of accuracy and detail is required than is the case with a simple line item budget presentation. Again, for the sophisticated staff member, this merely reinforces and strengthens his position. It also means, however, that the time expended in preparing the budget increase significantly from the preparation of a line-item budget.

A final point driven home to the PPBS group during the initial budget effort was the tendency to produce so much data and in such detail that it could not be easily assimilated by decision makers. Thus it was difficult to use during the budget process. It is, in fact, this difficulty in simplifying and reducing PPB information that may insure the primary status of a line-item budget in the municipal government process for many years to come. People, materials, and equipment are concepts easier to grasp than foreign, complicated output measures.

BUDGET PREPARATION

It would indeed be an overstatement to claim that the initial implementation phase of the PPB system was a success, especially when measured by the usefulness of the data generated and the perspective gained by the departmental participants. For the PPB staff, however, the experience was extremely enlightening because it put the relatively less important requirement of technical expertise into perspective with the much more important ability on the part of the staff to communicate and build basic understanding and support from the department personnel. In other words, the technical process of carrying out change appeared to be less important than the means of generating genuine understanding of, belief in, and excitement for the new approach.

Following the introduction of the 1968-1969 budget, the PPB staff returned to the same operating departments and again, working with the department personnel, discussed, reviewed, and analyzed their programs. The focus was on the direct benefits more refined descriptors of program output could provide to individuals operating in the system and for the department as a whole. This process actually varied from department to department, depending on the particular needs, understanding, and skills of the personnel involved. In one department, it became obvious that the supervisors, resistant to the intrusion of young "experts" with little technical knowledge of their department, needed to be convinced at the outset that the time expended on the project could ultimately assist them in scheduling and controlling their work. In another instance, the supervisors were highly supportive and the PPB staff directed its efforts to generating understanding and interest in the modification of data collection and reporting by the first level supervisors and their staffs.

These varying activities were ultimately focused on the preparation of the 1969-1970 budget. In considering the 1968-1969 procedure, the absence of uniformity in the output descriptors, their relation to line-item costs, and the apparent lack of understanding of the process, the PPB staff developed a unique set of data collection forms which proved to be a kind of programmed learning text for budget preparation.

To begin with, the department supervisors, working closely with the department heads, were asked to set forth the services and the levels of service they believed *could* and *should* be achieved during the next fiscal year. These estimates were not based on cost; in fact, they were specifically told not to consider dollars but rather, to identify what outputs they as responsible officials believed were necessary. These proposals and projections varied in their degree of completeness and relevance, and on their basis in reality, as later tested. Perhaps most important, this approach *forced* the departments to consider seriously in program terms, the services being provided the public. It was a new perspective, a different way of looking at job responsibilities.

The second phase of this new procedure involved the assigning of costs to the individual programs and program elements first described. The budget manual—program text—actually laid out a step-by-step methodology for determining these costs beginning with the calculation of manhours at different skill or job levels, including fringe benefit costs and the projection of material and equipment requirements for each and every element. The process provided checks which insured the calculation of all inputs.

Although the procedure was broken down into elemental steps, it was both a time consuming and an intellectually intense experience for those involved. Presumably, department personnel in the past have intuitively made similar if much less precise calculations in preparing line item budgets. However, when all inputs must be identified, particularly the assigning of all personnel time to program elements, the procedure proves to be extremely laborious and time consuming. With little previous experience or data to apply as a basis of comparison, many of the service level projections costed out at astronomically high figures, a fact which proved to be most frustrating for the individual who needed to recalculate each of these programs. This was not a totally unrewarding experience, though it was often perturbing. Personnel were literally forced to think through the actual relationships between services measured in quantifiable units and inputs reduced to dollars.

Paralleling the budget process, the PPB staff began to pull together the several program elements and develop them into a hierarchical set of departmental programs which, when completed, graphically described the interrelationship of each department's activities, that is, a program structure built upon the data and programs generated by the department staffs themselves, rather than vice versa.

The results of this second effort were more positive than the first attempt, although they also suffered from several of the problems generated during the previous year.

The line-item budget preparation closely paralleled the program process. Since the line-item document continued to be the control document, all program figures had to be translated into line-item terms not only at the outset but each time a change was made in the line-item document. This conversion from program to line-item figures proved to be horrendous and generated a great deal of hostility and confusion.

The behavioral or commitment issue was more effectively considered during the second round, particularly within the four departments. It is worth noting, in fact, that several of the department heads actually honed the new information and used it with an incredible sophistication in every phase of their operation, which placed their departments in an unparalleled position when the advocate process was played out.

Perhaps the major defect in the systems effort for the budget practitioners was the enormous array of data prepared. The controller and the city administrator were both hard pressed to find a point of entry into this information, and were compelled to look first and with closest attention at the line-item budget presentation. The fault, I fear, was with the PPB staff which failed, in their desire to produce a complete picture, to highlight, simplify, and order elements and programs of greatest importance and interest. Also, the process of reconciling two budgets, line-item and program, produced a tangle of data that was difficult to sort out under pressure of finalizing the line-item document.

Was this an appropriate way to begin the implementation of a PPB system? Was this exercise worthwhile? Did it provide a more accurate understanding of city programs for department personnel, management and the council? I have already indicated my belief, along with my reasons, that this was probably not the ideal time to introduce this system. However, I do believe, that in spite of the difficulties and frustrations, the process had value. It provided an education by immersion for the department personnel. It insured the completion of program descriptors and projections which provided the base against which the department would be able to measure their own progress during the next fiscal year, that is, the basis for the 1969-1970 management information reporting system. It also built up an understanding and, I now believe, a commitment among several department people to the principles and objectives of an overall PPB system. The psychic costs, however, were often high.

MANAGEMENT INFORMATION PROCESS

With the completion of the 1969-1970 budget, the PPB staff was prepared to consider the application of additional elements of the formal PPBS program, principally the refinement of departmental program structures and the application of management or analytical tools to the department activities. Funds were provided in the budget for the extension of this effort to other line departments; however, it was generally recognized that the city bureaucracy was not yet prepared (motivated) to expend efforts in the development of an overall program structure for the city, nor, I should add, did we have adequate information to do so.

A second decision made at this point was to concentrate on operations programming and planning, emphasizing a single year time horizon rather than a long-range time frame of five or seven years. Considering the several problems uncovered during the budget process, the staff felt it could provide better control of program outputs; therefore, concentration of the long-range elements seemed to be of a lower priority. This position is certainly debatable if PPBS is to gain relevance as a planning as well as a management process.

Using the new program budget as a guide, the staff began at once to develop reporting forms and procedures which would keep the budget effort relevant throughout the fiscal year. Traditionally, budgets are prepared and adopted, staff changes made where funds are provided, and then the budget document is set aside to be used primarily by the controller who periodically checks to see that expenditures are staying within the budgeted constraints. The line-item budget is rarely used, nor does it have the information to act as a program guide for the fiscal year. With the program budget, however, a guide for action was produced, that is, a contract commitment to provide a defined amount of service. It was important that this contract not be relegated to the back shelf since it provided a visible description of projected activity. Whereas previously only financial control was aided by the budget, now operational control was also possible.

Not only did the department heads have to make their program commitments a guide for the fiscal year's activities, they also had to develop relevant procedures through which they could monitor and control their department's activities. Creation of this monitoring system became a high priority consideration. It followed the budget preparation procedure, with the PPB staff reviewing with all levels of department personnel the information needed and the method by which it could be generated and reported.

The reasons for undertaking the effort were carefully reviewed in order to avoid the generation of extra work with no visible benefit to the department staffs. This process was exhausting and, at this time, it is not totally completed. Forms and procedures were developed and information produced on a daily basis relating services completed to the inputs expended, for example, personnel time, materials, and equipment. This process is currently going through a shakedown period.

Phasing in a primary and all-pervasive data collection and information system on top of existing procedures, at least for the short run, requires tenacious supervision, handholding, understanding, and commitment. In the case of Ann Arbor, this has been provided by a staff member from the Community Systems Foundation. Much of the work involves detailed checking and rechecking of the nature and accuracy of the data required and the data actually being recorded. Since the process is time consuming, individual staff members must be regularly supported and assured that, in fact, benefits will result both to the organization and to themselves.

While new as well as formerly recorded information is being generated through this process, there remains a reluctance to abandon former data collection procedures for fear that this new report may be transient. As a result, there is duplication of effort. An obvious question to be considered is whether this is the most effective way of introducing changes in data collection. Rather than terminate procedures with which the personnel felt comfortable we decided to phase in the changed program and let the department staffs make the decision to eliminate previous reports.

A major consideration in the introduction of the reporting system was the requirement that a rapid feedback system of information be instituted. Interest might well be lost without it. The large volume of data produced points toward a mechanized processing system. In the short run, however, it was necessary to hire personnel to do nothing but process daily information and array it for monthly reports. The reports now being produced provide, for the first time, the status of each program in the four operating departments. They relate actual and projected service levels with dollars budgeted and expended for the month and include the department's projections for the following month. They also include several statistical calculations which point up degrees of deviation from projections, both in terms of service levels and cost.

The value of this data to first-level supervisors, department heads, and the city administrator are immense. Knowing how to use the information, however, is dependent to a large degree on the management sophistication of the particular department personnel, and on the ability of the PPB staff to demonstrate programming implications.

Phased in with this activity, the city has developed a series of management training programs which, in the short run, have focused on personal interaction and basic techniques of management motivation. These programs have been developed in conjunction with personnel from the University of Michigan and provide an essential complement to the programming activities.

At this writing, the city is just beginning the application of the most significant element of the system, namely, the analytical techniques, that is, the modification and refinement of existing practices and procedures based on the data generated from the new information system. The four departments have a respectable model of their existing operations through PPBS, and are compelled to evaluate this model in terms of modifying and improving performance in order to achieve or better achieve their stated objectives.

Even at this early stage, several of the department heads are considering organizational changes and the application of specialized systems techniques to further modify and improve certain practices. Our objective at this stage is to concentrate the efforts of the PPB staff in assisting department personnel in this refining and program modification task.

Ann Arbor still has a long way to go before a PPB system is operational in all city departments. Over the next three years, the program for introducing the system will include further refinement of data and procedures for the four operating departments and, within one year, the articulation of operating objectives for five-year planning periods. Additional departments will be included in the system, working first with line operations and ultimately including staff services. In the process, however, certain staff functions will necessarily have to be modified (e.g., elements of the accounting function) to provide essential information for refinement of the operating programs and information.

At a point yet to be defined, a comprehensive program structure will be developed, articulating not only specific program objectives, but also overall city goals which in turn will necessitate the defining of programs along functional rather than departmental lines.

It is all too clear that some of the original time objectives for introducing additional departments and working through long-range operating plans have to be modified. This is due not so much to the limitation of resources (which is a factor) but rather to the time required in building understanding and support for the PPB system by the operating personnel.

Perhaps one short observation is due at this time: specifically the issue of interfacing the changes in operating programs with the political process. I have intentionally focused on the administrative considerations and elements of the PPBS because it is at this level that initial changes must begin. Unless these changes can be translated into services and an ability to describe and evaluate services by the body politic, its relevance is nullified. At each step or phase the city council is potentially involved; they require information and are constantly evaluating the performance of the administration. However, at the budget stage, that is, the point where alternative courses of action must be selected, their involvement, understanding, and commitment should take on a primary status.

CONCLUSION

PPBS provides a means of rationalizing the activities of an institution by formally considering the long-range implications of capital and operating programs, systematically developing an annual program, and creating mechanisms for measuring actual services (outputs) against the predefined objectives. The system can be reviewed and interpreted from several different perspectives and as previously stated, implemented by cutting into the operations at any of several points. No matter what the perspective or point of entry, the system requires both control and change—two factors which can be unsettling.

Any new approach, whether it be a comprehensive program innovation or the modification of current practices, has the potential of generating apprehension and fear among those who perceive a potential for change in their status or position. To accommodate this type of reaction, a major focus of the PPB system must be one of a continuous reselling of the objectives and benefits of the system.

Prior to its introduction, therefore, time should be allocated to discussing and reviewing the reasons and need for *considering* the application of a PPB approach to the particular institution. During this period, two factors should be clearly articulated: (1) a desire to increase the objective rationality of the total operation of the organization, ultimately to increase and improve the services that are provided, and (2) a desire to *assist* employees in carrying out their responsibilities. From the outset, it is imperative that employees feel, perceive and truly believe that the introduction of PPBS will be supportive of both their job responsibilities and their personal status.

The time consuming process of discussions, debates, and philosophizing are essential to assure and reassure that a prime objective is not to undercut individuals, but rather to assist them in improving performance. A simple statement to this effect is useless. People are apprehensive to the prospect of change and, therefore, statements or memos to the contrary are of limited value. To reiterate this point, I would argue that the first efforts at introducing change into the organization must provide some immediate payoff or gratification to the participants, that is, some process which in action demonstrates clearly and honestly that the system will benefit the individual as well as the institution as a whole. One model is to begin the implementation phase with the development of information systems which affect a large number of individuals, both generators and/or users of internally collected data.

That this point of entry had merit in Ann Arbor is not to suggest that it is a universally relevant opening point. In other organizations this point of critical relevance may be in defining the interrelationships of activities to program structures or in applying one of many systems tools to a highly visible and important organizational activity.

The strength of the PPB system, then, is its applicability at any one of an infinite number of points within an operation—points which, in the particular case, have the highest visibility and the greatest potential of rapid feedback to those whose support and participation is basic to the success of the total effort.

PPB brings together a multiplicity of tools and techniques most of which are familiar to organizational participants. However, it pulls them together into a total system, one which is concerned with long-range planning, the relationship of long-range planning to current activity, and the consideration of alternative courses of action. To grasp all of that, particularly if you are an individual who is required also to perform urgent daily tasks which you have been performing over many, many years, is difficult. This is not said in any paternal way but rather from the perspective of an administrator who lives under the daily pressures of crises generated both by the public and by institution members themselves. Therefore, to introduce a system, it is important to provide stability.

It is important to provide a sense of trust and not a situation in which apprehension of change generates subversion and resistance. Even in the best situations, people will be apprehensive and, as a result of these apprehensions, they may resort to the most bizarre behavior in order to protect themselves. By being aware and emphasizing the generation of internal support, this can be minimized.

The thrust of this program should be to implement change to insure its perpetuation and to avoid the reoccurrence of past practices when new systems are abandoned once those introducing them leave. Therefore, it is imperative that those who are ultimately going to be responsible for carrying out new procedures and approaches be involved both in system development and in the initiation of ideas that go into its implementation.

This, as stated above, requires a great deal of talk and talk often is seen as nonproductive. Unless this training and support-building is permitted to flourish, I would contend that powerful and long-lasting change is nearly impossible. In fact, it will generate the seeds of its own failure, namely, apprehension on the part of the participants.

A working atmosphere supportive of procedural changes and modification of previous practices is imperative for a PPB system to be effective. Such an environment is one in which the top administrators recognize that certain new approaches will be unsuccessful but remain committed to the long haul. This notion of commitment is easily articulated, but more easily forgotten when the day-to-day pressures of producing services in response to regular demands come in conflict with the expenditure of efforts and resources necessary to modify and change regular practices.

Unless the inevitable resistance generated at such times can be absorbed, and unless those responsible for implementation are willing to reward the effort to modify practices when not altogether successful, PPBS will become a costly exercise in administrative futility.

12

Leave City Budgeting Alone!: A Survey, Case Study, and Recommendations for Reform

ARNOLD J. MELTSNER
AARON WILDAVSKY

☐ THERE IS LOOSE TALK to the effect that budgeting involves resource allocation. So far as the few American cities we know about are concerned, we believe this rumor to be unfounded. The literature on municipal budgeting likewise tells us little about the allocation of resources. One approach takes as its problem (dependent variable) the determination of municipal expenditures and seeks to explain statistical variation among cities mostly through the use of multiple regression analysis. The explanatory factors (independent variables) used to account for variations in expenditure are data such as property valuation and income per capita. These aggregate measures of wealth may explain a great deal of the variance in total per capita city expenditures but they do not tell us how resources are allocated.

Another approach examines the behavior of city officials and attempts to explain budgetary decision-making by the ways in which these individuals try to solve their problems. Since cities are in a financial straitjacket and officials can make only small changes

AUTHORS' NOTE: *Financial support for the research on which this paper is based came from the National Aeronautics and Space Administration and the Urban Institute. Helpful comments were made by Otto Davis, Michael Dempster, Harry Hatry, Allen Schick, and John P. Crecine. We are grateful to these organizations and individuals but we stubbornly insist that the ideas are our own and take full responsibility for resisting much of the good advice urged on us. This paper is also part of the Oakland Project, an action-research project in which service to the city is combined with policy analysis, teaching, and training of graduate students.*

in their budgets, the rationale for resource allocation is not entirely evident. Our detailed analysis of the budgetary process in Oakland shows that choices among alternative allocations of resources are not at the center of anyone's concern. This does not mean, however, that budgeting is useless. Budgetary activity does serve to control bureaucratic behavior and facilitate tax policy. In view of what the budget does (rather than what it is supposed to do), what can be done to improve the allocation of scarce resources for our cities? We set the stage for this inquiry by reviewing the aggregative and problem-solving approaches to municipal budgeting. Then, in context of the detailed study of a single city, Oakland, we shall suggest a modest reform proposal which cities can reasonably expect to implement without excessive expense.

THE AGGREGATE APPROACH

The objective of the aggregative approach is to develop an empirical theory of the determinants of municipal expenditures using multiple regression analysis. The theory as such is therefore contained in the choice of independent variables which, when their effects are added together, purports to explain much of the variance in governmental expenditures from one city to another. The aggregative approach is not concerned with how the independent variables, such as per capita income or urbanization, get translated into the dependent variable expenditures. Most of these studies share the implicit assumption that something called the political process acts as translator. With few exceptions, the aggregative approach assumes that local government and its officials are a passive entity through which inputs like wealth are converted to outputs like police or recreation expenditures. A further assumption is that (a) per capita wealth is converted into demands for governmental services, (b) the service demands filter through a sieve known as politics, (c) the expenditures incurred satisfy these demands. While these circular models purport to describe resource allocation, they have nothing to say about how demands are generated, how the political system processes these demands, or whether citizens feel satisfied with the outcome.

In a major study of state expenditure, Fabricant (1952) identified key variables and established a model for future aggregative research in cities. He found that income, urbanization, and density of population accounted for over seventy percent of the variance in interstate per capita operating expenditures. Personal income per capita was the major factor. In comparing the change between 1900 and 1942, Fabricant (1952: 136) concluded that "the chief cause of rising per capita expenditures would be rising income." He thought he had found some support for Wagner's law in which increased governmental activity is related to increased economic growth and development (Musgrave and Peacock, 1967: 7-8). With higher income, people would want more services and would be able to pay for them. He established that the wealth of a community was an important determinant of its expenditures. When Fabricant examined expenditures for functions such as fire and sanitation, however, he found that as an explanatory variable, urbanization was more important than income. Here we have one of the difficulties of the aggregative approach: the explanatory power of the independent variables changes with the unit of analysis and with the particular category of expenditures. The same set of independent variables is sometimes statistically useful and sometimes not when the analysis is conducted individually for states, cities, metropolitan areas, and towns or for fire, police, education, sanitation, and street expenditures.

Campbell and Sachs (1967: 5, 172) hypothesize that the same variables can be helpful in explaining fiscal differences between governmental jurisdictions such as central cities and suburbs. Income is one of these variables. But they also add variables to improve the explanation for particular jurisdictions; home ownership, for example, works for central cities since it has a significant negative association with noneducation expenditures and taxes.

Thus, we see that once the analysis focuses on smaller units of government and on particular functions, a variety of variables can be introduced which improve the "fit" between expenditures and their particular determinants but which lose generality. Despite Fabricant's elegant start, there is no general theory of the determinants of expenditures.

Since Fabricant, the research has involved introducing new units of analysis and adding new variables to the trinity of income, urbanization, and density. Some investigators, for example, consider federal and state aid to be major determinants of expenditure (Osman, 1966; 1968; Sachs and Harris, 1964; Kurnow, 1963). Others have questioned the independence and utility of these variables (Fisher, 1964; Morss, 1966; Pogue and Sgontz, 1968). How, they ask, does federal aid differ from the expenditures it is supposed to explain? Campbell and Sachs (1967: 43-67) find that state and federal aid have a positive relation to state-local and local per capita taxes and expenditures. They report that aid acts to stimulate state and local fiscal activity. Brazer (1959) found that intergovernmental revenue per capita was positively related to per capita expenditures in cities.

We believe that aid is not very different from income. Both are measures of potential fiscal capability. It seems reasonable to expect that a community with a high income level and federal and state aid can spend more than a community which is deprived of such affluence. The more money a city raises, the more it can spend. Only if the richer cities spent less and those with lower income spent more would the phenomenon be sufficiently surprising to deserve serious explanation. Certainly the particular mix of aid and local resources can affect the preferences of citizens and officials. But we question whether such aggregative analyses can be sensitive to likely substitution effects.

The partisans of aggregative analysis have extended their tools to a variety of sites:

(1) regions (Sharkansky, 1968a);

(2) the nation's counties (Schmandt and Stephens, 1963; Shapiro, 1963; Adams, 1967);

(3) Iowa's counties (Wright, 1965);

(4) Standard Metropolitan Areas (Campbell and Sachs, 1967; Woo, 1965);

(5) the St. Louis area (Bollens et al., 1961);

(6) the Philadelphia area (Williams et al., 1965);

(7) the Cleveland area (Sachs and Hellmuth, 1961: 68-132);

(8) a national sample of cities (Brazer, 1959; Lineberry and Fowler, 1967);

(9) cities in Ohio and Michigan (Booms, 1966);

(10) California cities (Scott and Feder, 1957).

The list of reported research is actually longer, but since we must confine ourselves to a brief review, we will comment on a few studies which are connected with municipal expenditures.

Brazer (1959) in his examination of 462 cities, found wide variation in total and functional operating expenditures. City expenditures differed by location and type of city. His major conclusion, which seems born of frustration, is that "there is no facile means of explaining the tremendous range of differences in the levels of city expenditures" (Brazer, 1959: 68). He found no relation of population size to city expenditures (with the exception of police protection). However, some students have detected a U-shaped expenditure curve for counties where the very large and the very small in population incur diseconomies of scale (Schmandt and Stephens, 1963; Shapiro, 1963). Brazer also found a positive relation between density of population in cities and per capita expenditures (with the exception of roads and recreation). Per capita expenditures for all functional categories (particularly for recreation and sanitation) increased as median family income increased.

One of Brazer's findings raises the issue of suburban responsibility for the financial dilemmas of the central city. Brazer discovered that central-city expenditures are related to the city's share of the metropolitan area's population. The smaller the central city's proportion of the metropolitan population area, the greater is the central city's per capita expenditure. Campbell and Sachs (1967: 155-166) also examine the fiscal differences between central cities and their outlying areas. They find income and state aid (in addition to the population variable) useful in explaining these fiscal differences.

Scott and Feder (1957) analyzed the variation in per capita expenditures for 196 California cities. They report that per capita property valuation and an index of retail sales are the most significant independent variables. In other words, measures of ability to pay were important while demographic variables of population size and density were not. The St. Louis study also supports the importance of fiscal capability. For the combined expenditure function of fire, police, refuse collection, and street service, Bollens and his colleagues (1961) found that per capita assessed valuation was the most important determinant of per capita expenditures. The more aggregative the expenditure category, it appears, the more frequently measures of economic resources such as the level of income or property valuation become primary determinants. Detailed analysis of local expenditure functions, however, raises some question about the applicability of these measures of wealth.

We suspect that in order to provide better explanations of specific activities, it will be necessary to employ measures which are sensitive to the distribution of income and the kind of property or the source of wealth. A recent study of the municipalities in the Philadelphia area downgrades the importance of ability-to-pay measures. Other factors such as the social rank of the community, public attitudes toward taxes and government, and particularly land use (industrial or residential) were considered more significant than financial resources (Williams et al., 1965: 91-138, 292-293). The importance of other variables is apparent when particular local functions, such as fire or recreation, are examined. Adams' work on counties is instructive in this regard.

In an analysis of the operating expenditures of a national sample of counties, Adams (1967) explored the utility of eleven independent variables (e.g., density, region, level and distribution of income, percentage of nonwhite, and number of jurisdictions) to explain the variation in seven functions (e.g., police, fire, sanitation, recreation). Unlike many other studies, the income variable was not particularly significant in his multiple regression analysis for each of the seven functions. Distribution of income was important in explaining deviations. Per capita police expenditures, according to Adams (1967: 21), "are higher in those county-areas in which a large percentage of families have incomes of $10,000 and over and are lower in those county-areas in which a large percentage of families have incomes of $3,000 and under." Fire expenditures increase with the age of housing, the number of multiple dwellings, and the percentage of the population which is foreign-born. Furthermore, fire expenditures vary by region; the West has the highest deviation from the mean. Sanitation expenditures increase with population density, urbanization, tourism, and the percentage of the population which is nonwhite. Recreation varies by region; it increases with density, urbanization, and the number of multiple dwellings. Adams (1967: 32) also found that there was an income distribution effect for recreation expenditures: "Regardless of the size of the $3,000 and under group, an increase in the percentage of families earning $10,000 and over is associated with an increase in per capita expenditures." Not merely wealth, then, but the way it is distributed, not merely numbers of people, but their cultural habits, determine the level of expenditures. It is important, therefore, not only to discover the independent variables related to expenditures but also to ask whether some general classes of variables—economic, social, or political—are more useful than others.

The question of whether economic or political variables are more important is explicitly raised in Dye's (1966) pathbreaking book. It turns out that political variables (such as party competition, partisanship, and apportionment) are much less powerful than the usual economic factors in determining expenditures within states. By separating state and local expenditures, however, Sharkansky (1968b: 60-64) shows that economic variables like per capita income drastically lose power to explain variance among states in total expenditures or in changes in these amounts.[1] While a state's total wealth does determine its gross expenditures, wealth does not fix the distribution between state and local governments. Some investigators have found political theory (if not political variables) useful for purposes of explanation. Davis and his colleagues, for example, have attempted to test a simple model of voter utility maximization as an important determinant of expenditure and taxation decisions of politicians. In these studies (Davis and Harris, 1966; Barr and Davis, 1966), assessed valuation and property ownership carry political as well as economic meaning.

One of the problems in explaining variation in local fiscal activity is the difference in functional responsibility among the governmental units. Campbell and Sachs (1967: 44-45) handle this problem by introducing an assignment variable which expresses the governmental unit's share of expenditures or taxes. Another approach, taken by Booms (1966) in analyzing a sample of Ohio and Michigan cities, is to restrict the analysis to common functions. He shows statistically that cities with mayors differ significantly from municipalities with city managers. The form of government does have an effect: average per capita expenditures for manager-cities are less than for mayor-cities. Lineberry and Fowler's (1967) analysis of data from a national sample of two hundred cities also found that manager-cities spend and tax less than mayor-cities. What is not clear from these studies is why the structural form of government is an important determinant of expenditures. They present findings but not explanations. They contain no evidence to suggest that manager-cities are more efficient or that mayor-cities contain people who want and get a higher level of services. No one except Davis (whose work is outside the traditional theory of public expenditures) has suggested a general Downsian explanation in which the elected executive, the mayor, is more interested in (or susceptible to) giving rewards to community groups for electoral purposes.

We conclude that most of the research on aggregative determinants of municipal expenditure lacks theoretical statement. There are few attempts to explain the general significance of the statistical manipulations. If the trend does not change in the future, much of this work will continue to exhibit a mindless empiricism with relations established not on grounds of explanatory relevance but simply by the availability of census data.

While the aggregative approach may prove useful for some purposes, from our point of view another basic defect lies in its inability to guide public policy. One learns neither about what public officials have been doing to achieve present results nor what they might do to improve the unhappy situation. Aggregates like per capita income and level of taxation must be connected to individual behavior if the research is to be instructive. For wisdom the advocate of the aggregative approach can offer only, "Mr. Mayor, if your city were richer, it would spend more." If cities were not poor, they might be rich. This is hardly enlightening and not much can be done about it in the short run. To know how and why the participants make decisions and to discern the objectives explicit or implicit in their actions is to place oneself in a better position to make policy recommendations. One may suggest different decision rules in order to produce better outcomes or to attain more worthwhile objectives. Knowledge of the process of decision should also enable the analyst to determine points of leverage for accomplishing change. Where the aggregative approach may determine boundaries beyond which the system cannot go, conscious changes in desired directions depend upon changes in individual behavior. Therefore, we now consider the behavior of individuals engaged in making municipal expenditures.

THE INDIVIDUAL APPROACH

The individual, problem-solving approach belongs under the general rubric of rational choice theories. This approach assumes that actions are ultimately the results of individual behavior; individual human goals are thus central to rational choice theories. These theories are attempts to explain, first, why individuals have the goals they do and, second, why they adopt certain behaviors in order to secure their goals. Explanation proceeds by analysis of the goals of individuals and the incentives that the environment provides for them to adopt one move rather than another in pursuit of their objectives. Local government officials want to stay in office. In order to do so, they must follow the constitutional requirements for balanced budgets. Their desires to hold down costs are reinforced by the necessity of raising taxes if expenditures go up too far. Since expenditures are pushing tightly against the limits of constitutionally permissible (or politically feasible) taxation, strategic choices available to municipal officials are few.

Municipal budgetary behavior is simple. Officials do not review their total budget but make small changes every year around the margins of their base amount within an overall resource constraint. They use simplifying procedures not just to reduce the budget but to cut the decision problem to size. The difference that budget officials make is in the small changes, and it is the accumulation of these changes which the aggregate approach usually analyzes. While the aggregate statistical analysis looks at the total (base plus increment), the individual approach usually looks only at the increments, unless it is considering a new agency or program that lacks a historically determined base.

By examining individual budgetary behavior at a point in time, one can derive suitable explanations of process and procedure, but one cannot understand the total pattern and substance of resource allocation. Individual studies tell us how and why two firemen were added to the fire department for a particular fiscal year; they do not tell us, however, about the total allocation of the community's resources for fire protection. The regression equation of the aggregate approach, on the other hand, does express the total allocation pattern. If nothing else, the constant term signifies a statistical norm from which variations can be analyzed. Part of the problem of the individual approach lies in method. Students of resource allocation who focus on individual decision makers have to find a way of looking at the base as well as the increments. If the agency base (its normal and expected expenditure) has been determined over many years, then it is necessary to go back in time to understand the pattern of growth.

Davis, Dempster, and Wildavsky (1966; forthcoming) in their study of federal appropriations, are making an effort to determine how budgets change over time through the actions of organizational decision makers. Basing their work on knowledge of strategies and calculations used in budgeting (Wildavsky, 1964), they seek to explain and predict agency requests to Congress (through the Budget Bureau) and congressional appropriations for those agencies (Bollens et al., 1961). By observing periods of instability in the time series of bureau-congressional relationships, they are able to determine conditions for change such as replacement of presidents, new congressional majorities, war, recession, and so on. Analysis of time-series, then, offers one way of penetrating the historical base. These authors are also using cross-sectional analyses of growth rates to distinguish the conditions that enable some agencies to grow faster than others. The variables used include congressional estimates of clientele support and confidence in specific agencies, the impact of departments and congressional appropriations subcommittees, and more. Although our analysis of Oakland thus far does not include investigation of the historical base for agency appropriations, there is no barrier in principle to accomplishing that end.

There are several descriptions of municipal expenditures that tell us how budgeting is done, which is certainly appropriate for those who wish to consider how budgeting should be done. In his work on boards of finance in Connecticut, Barber (1966: 33-46) discovered rules by which decision makers simplify the problem of making resource decisions.[2] Board members eliminate items that cannot be touched because they are mandatory in law or politics; they look at salient, large dollar items, largely increases and decreases from the previous year; they consider the budget horizontally (that is, they view the budget across years, rather than between departments); they deal with the concrete, nuts-and-bolts aspects of a problem, restrict attention to the near term by use of past solutions for present problems, and make incremental decisions to take advantage of future feedback. The dollars-and-cents aspect of their calculations also simplifies the decision process (with the possible problem of a divorce from substance) because they need not consider the basic merits of a program every year.

In his study of budgeting in three Illinois cities, Anton (1964) indicates some of the variation in municipal budgetary behavior. One city does not prepare a budget; instead the city clerk adds up expense estimates which are submitted to the council as a proposed appropriation ordinance. In another municipality, the city manager reviews the line-item budget by concentrating on changes from last year's budget. Both these cities allocate their resource incrementally. Bargaining takes place when their councils attempt to balance the budget. Anton cites an example where a police department increased its budget through political pull. In a third city, the manager uses some version of performance budgeting with activity cost accounting to control city operations. As a strong city manager, he influences the council to approve his budget as it is.

In an original empirical study of resource allocation, Crecine (1969) developed a computer simulation model of the municipal operating budgetary process. He re-created the central steps and calculations in the process in order to generate the outcomes. His research in Pittsburgh, Detroit, and Cleveland should be applicable to a number of central cities. Crecine divides budgetary behavior into three submodels: (a) Departmental Request, (b) Mayor's Budget Recommendation, and (c) Council Appropriation. The model begins with the departments' receipt of the mayor's budgetary guidance letter.

Departments do not compete with each other for funds. The budget request is based on last year's appropriation plus an increment whose size depends on the department head's calculations of "what will go" and the tone of the mayor's budgetary letter. In the Crecine model, departments follow the mayor's instructions.

The key feature of municipal budgeting is that the budget is balanced to meet an independent revenue constraint. The mayor's submodel is crucial because the budget must be balanced at this stage in the process. The mayor, like department heads, solves his problems by simple decision rules. Increases should be no greater than a certain percentage of last year's appropriation. He makes minor adjustments to last year's budget or "governs by precedent."

After public hearings, the council rubber stamps the mayor's budget. Crecine feels that the council's minor role is due to the lack of staff and to the balanced budget requirement. In addition, because they are subjected to demands for increased expenditures and decreased taxes, councilmen do little.

Crecine's model of incremental line-item budgeting emphasizes *continuity* of existing policies. It does not focus on programs or policies. City decision makers are insulated from short-term environmental forces. The environment may influence what a department head worries about, but it does not influence the department's total budgetary level. Crecine reports some pressure to keep property tax rates constant, but he finds no direct connection between political pressure and departmental budget totals. Expenditure levels tend to "drift" over time until the decision system responds to some change in the revenue situation. The environment does not correct budgetary drift by directly changing specific expenditures. Budgetary change is more likely to be "filtered through the revenue constraint" (Crecine, 1969: 167-168).[3]

Crecine notes several important limitations to his model. The model lacks a realistic priority scheme for treating departmental deficits and surpluses: departments are treated by account number and not in terms of individual characteristics. The model excludes capital and nongeneral fund expenditures. Political scientists may object to the lack of attention Crecine pays to environmental forces or constituency pressures. But we feel that Crecine's work adequately reflects political reality. The *operating* budget is insulated from the environment. There are undoubtedly occasions when political pressure results in specific expenditures, but these are likely to be infrequent. If social scientists want to ascertain how, if at all, environmental demands are met, they will have to examine detailed allocations within city departments; whose street is repaired and what neighborhood gets the traffic signal are appropriate questions.

Most of the literature we have discussed treats only the operating side of the budget. In the aggregate studies, capital expenditures are usually sifted out to avoid distortions in the analysis. In the individual studies, capital decisions at the local level are artifically separated from operating budgeting decisions. We want, therefore, to consider a fine work that treats capital budgeting directly.

In their book on municipal investment in Philadelphia, Brown and Gilbert (1961) found that political and administrative factors there have produced instability in capital programming. Projects in Philadelphia's six-year program are not necessarily budgeted and those in the fiscal year capital budget may not be executed. The council, which is sensitive to community values, is active in scheduling projects, so that items are shifted backward and forward in time. Related to this instability are the differing time perspectives of the participants; planners have the longest time perspective and councilmen the shortest. City officials have few formal criteria for evaluating projects and allocating resources. Economic analysis is not used and standards are lacking. The result tends to be a choice of small projects over large ones in order to facilitate awards to interest groups, to distribute projects by region as an attempt at parity, and to take advantage of free funds (i.e., federal and state funds).

The conclusions derived from the few previous studies of municipal budgeting are not likely to inspire joy in the hearts of city officials. The revenue constraint is critical. The little money available is divided according to simple decision rules that hold down the burden of calculation and/or ease the most blatant political pressures. Though there are some attempts to introduce efficiency studies, there is virtually no substantive analysis of policies.

Having discussed the work of others, we are now in a position to describe the budgetary behavior of the city of Oakland. The reader should not be surprised when he learns that budgeting in Oakland is not much different from budgeting elsewhere.

OAKLAND: THE ALL-AMERICAN CITY

The local chamber of commerce usually describes Oakland as the "All-American" city, and in one sense this is not an exaggeration since, indeed, it exemplifies the many dilemmas which central cities face. The elements of the urban crisis are all there: racial conflict, a high unemployment rate, substandard housing, poor educational achievement, and low family income.

Behind the demographic picture of Oakland lies the misery of the poor. According to Oakland's Model Cities application (Oakland, City of, 1967: 1-3, 26-37), the city is a "port of entry" for low-income households. Although the total population has tended to stabilize around 370,000, the mix within this total has changed quite dramatically. In 1950, the Black population in Oakland was 47,562. By the 1960 census, this population was 83,618, an increase of seventy-six percent, and it is estimated that the 1966 Black population was 110,000. Similarly, there has been an increase in the number of Spanish-speaking (Chicano) people in the city from 16,500 in 1950 to 23,700 in 1960. But between 1950 and 1960, there was a decrease of eighteen percent in the white population. With this change in population and the reduction in defense and other industries within the city, the unemployment rate in 1966 was about six percent for males and ten percent for females. The Model Cities application also reports that in the poverty target areas the percentage is much higher—around eleven percent for men and sixteen percent for women. Among Black teen-agers, unemployment runs as high as twenty to thirty percent. Fifteen percent of all housing units in Oakland are considered substandard. In addition, educational achievement is lower than in other central cities in California: seventy percent of Oakland's nonwhites over twenty-five years of age have not completed high school. Considering the lack of education and job opportunity, it is not difficult to understand why Oakland's family income is lower than that in the rest of the East Bay metropolitan area. In 1960, family income for one-fourth of the city's families was under $4,000. In a single year, from 1964 to 1965, the number of welfare cases increased by thirteen percent.

City officials, with a few exceptions, have not enlisted in the War on Poverty. They make contingency plans to defend against possible riots, but the "war" takes place on other sites, in a maelstrom of quasi-private and independent public agencies. Welfare is run by the county. In one year, the federal government spends about $95 million in grants for social programs. But most of the money goes to separate agencies such as the Redevelopment Agency and the School District. Only about one to two percent of the city's annual budget ($53 million in 1967-1968) is conceivably supported by federal funds (Oakland, City of, 1968).

THE RECURRING CRISIS: THE REVENUE-EXPENDITURE GAP

The main problem city officials try to solve every year is simply to find sufficient revenue to maintain their current payroll. Their problem is that the city's budget increases at a faster rate than the tax base. Fiscal atrophy is the city's chronic malady. The city manager put it this way to the city council in a report on the fiscal potential of Oakland: "Simply stated, the disquieting fact is that under existing policies and within the constraints of the financial structure, the City of Oakland's expenditure requirements will soon exceed its income." The manager based his projection on the fact that in the 1956/1966 decade, the city budget had risen eighty-six percent while the property tax assessed valuation had gone up only thirty-six percent. The atrophy of Oakland's tax resources has resulted in the city's making efforts to maintain itself rather than trying to cope with the larger social problems it confronts.

Oakland's tax problems are compounded by its land usage and by the business and population that have migrated to surrounding communities. The city has no more land to annex; it is boxed in within its fifty-three square miles of land. One cause of tax base erosion is that not all property within Oakland's fifty-three square miles is included on the tax rolls. Much of the excluded real property is occupied by governmental utilities, public institutions, streets and freeways or is merely vacant. The city's finance director has estimated that over forty percent of Oakland's real property is tax exempt. City officials lament not only the loss of taxable property but also the flight of heavy industry and citizens who pay taxes instead of consume services.

Previous studies of California finance have indicated that the rise in city budgets has been due to increases in population and prices (Shelton, 1960).[4] We know that Oakland's population has not been growing, but one might conjecture that Oakland's budget increase is due to changes in the mix of population. We call this conjecture the blame-the-poor-Blacks hypothesis. In our estimation, however, Oakland could have remained all white and still would face the same fiscal problems today. The city budget has expanded to meet the increased costs of operation, not to meet the needs of Oakland's deprived population. The budget of the city is an operating budget: seventy percent of the total budget is for personnel-related costs. Oakland is not a case of expanding bureaucracy; it is a thrifty, almost frugal city. In 1960, there were 3,014 budgeted full-time personnel positions for the city. By 1968, this number had increased to 3,254, an increase of roughly one percent per year. While the number of personnel has not increased appreciably, the costs of personnel have gone up every year, with built-in merit raises and rising expenditures for retirement plans. Thus, it is easy to see that with the possibility of a two to three percent price increase for nonpersonnel costs, with standard salary increases of four to five percent, with slight increases in number of personnel, and with no increase in productivity, Oakland could have an eight percent growth rate in city expenditures each year without a marked increase in the level of service.

A specious sign of financial stability is the lack of new or ambitious projects, although there are a couple of highly visible ones—the museum and the coliseum complex. Only $5 to $6 million is annually spent for capital improvements, and seventy-five percent of this small sum is from earmarked funds, such as street construction under the state gasoline tax program. Without an aggressive bonding program, city officials defer capital improvements and incur increased maintenance and operating costs. Repair rather than replace has been the prevailing motto. This deferral syndrome exacerbates the general fiscal problem, because many of the facilities which were built twenty or thirty or even eighty years ago are now wearing out. All the city has to show for its pay-as-you-go policy is an excellent bond rating; the low total indebtedness, $18.9 million, is only seventeen percent of the legal limit. The less Oakland does to meet its problems, the more bonding capacity it has left. Because school district and city improvement issues have been turned down in the past, city officials generally are afraid to turn to the voters for approval of bonds. At the same time, population needs are changing and no resources currently exist within the system to meet these demands. Although the system may appear to be stable, it is in a precarious position.

The local property tax, the main source of revenue for cities, is exhausted in a political if not an economic sense. The common perception among officials is that "we cannot raise the property tax rate." Taxpayers complain that the rate is too high. There is some justification for this, since Oakland has one of the highest municipal tax rates in the state of California. In fact, it was *the* highest during twelve of the past seventeen years. City officials believe that raising the property tax would be an act of political suicide. (For additional comments on this problem, see Meltsner, this volume.)

Oakland is living with a fiscal crisis. Officials believe the property tax cannot be touched, assume that voters will be hostile to all bond proposals, and find it increasingly difficult to acquire revenue sources which will be politically feasible and have a worthwhile yield. As its costs go up, the city finds minor sources of revenue, augmenting the property tax each year by adopting a stopgap measure such as a tax on cigarettes or hotel rooms. Since it is difficult to discover painless methods of raising money, top officials have every incentive to hold down expenditures. Their motivation is clear: the more they can cut, the less they have to worry about finding new sources of revenue.

For the past decade, the annual budget guidance to department heads has included some phrase like "hold the line"; every year the financial constraint on the city is apparent. The city manager has often said, "I don't know why we spend so much time on budgeting. After all, it should't make any difference—we won't have any more money." Yet the city manager and his small finance staff spend days and weeks worrying about the budget.

To what extent, we ask, is this activity a meaningless ritual that might be of more interest to an anthropologist than to a student of public policy? How may the budget process in Oakland be considered important? Is Oakland an example of a city where revenue constraints drive out policy considerations so that the emphasis is on cutting the budget rather than meeting community needs for services? A short answer is "no," budgeting is not meaningless, "yes," it has important uses, but "no," it does not serve to allocate resources according to any well defined notion of public policy other than keeping down the property tax.

CONTROL VERSUS STEERING

Although budgeting can have a variety of functions, control and steering are two uses of central importance to students of public policy. Steering (or strategic planning, or policy-making) determines the goals and direction of a public organization. Control connects means and ends by ensuring that long-term goals are approached and, perhaps, reached. Since Oakland's budgeting reflects the acceptance of the premises of the past as goals for the future, public officials slight steering and emphasize control.

The concept of control has many meanings when applied to budgetary behavior. It can mean cut the estimates because there is no money, reduce estimated expenditures in the process of budget execution, or implement the policy preferences of the person who hands out the salary raises. The city manager and his finance staff have a single motivation, however, which underlies these objectives of budgetary control: they wish to balance the budget by controlling uncertainty. Once officials used budgetary control to keep people honest; now it is used mainly as a short-term hedge against unanticipated events. Balancing the budget means that the city always has at least a slight surplus on hand.

We begin our discussion of budgeting for control, therefore, by pointing out that the operational objective of the official is to create a surplus. We go on to show that the problem of control stems from the conflict between budget-spenders and budget-cutters and is exacerbated by the city's overall revenue constraint.

Budgeting has become, in effect, a form of revenue behavior. We next analyze the tools of control by detailing the decision rules which the Budget and Finance Department and the city manager use in balancing the budget. Finally, we suggest some limitations of such efficiency devices as program-oriented budgeting and analytical studies, for control and steering purposes. We will then be ready to raise our basic normative concerns: what advice can be given to city officials (operating under the constraints we describe) to make their budgetary activities responsive to citizen needs? how can municipal budgeting be made relevant to public policy?

THE OBJECTIVE: CREATE A SURPLUS

The operational goal of city officials is to have a little more to spend than they need. This creation of slack (excess resources) essentially provides officials some discretion for adapting to unforeseen events. The first decision rule officials adopt, therefore, is always to estimate revenue lower than they actually expect it to be. The conservative bias in revenue-estimating is cumulative. At every level, supervisors reviewing the estimates tend to lower them. A case in point occurred when the city was installing an urban transit system in the downtown area. There was considerable uncertainty as to what effect this construction would have on retail sales. The revenue analyst had taken this into account in lowering his sales tax estimate, but as his estimates went up the line they were continually reduced. The actual sales tax revenue was much higher than had been anticipated, and the city had the unusual task of disposing of a surplus.

The second rule for increasing slack is to estimate expenditures higher than one expects them to be. In this way, the finance director tries to reduce the departments' surpluses and to create a financial surplus for the city manager to use. Each level of government tries to increase the money available to it, guard it against those above, and squeeze it out for its own use from those below. An estimate that expenditures will be higher and revenues lower than they really will be will certainly create a surplus for someone which can be used to finance future programs.

Allen Schick has observed that budget people in Oakland could not properly be called "revenue maximizers" but rather "surplus protectors." It is true that they do not seek to raise all the money they conceivably could if they viewed the world differently. They attempt to maximize revenue within the constraints of the system as they understand them. Under those constraints they might first be described as "surplus producers" and then, in their budget behavior, as "surplus protectors."

In order to make certain that his office controls as much slack as possible, the city manager seeks to protect the general fund. This fund is his, subject to his ability to convince the city council. Therefore, he is constantly looking for ways to charge special funds for costs that are currently being incurred under the general fund or to move items from the general fund to special funds.

THE PROBLEM: REVENUE AND EXPENDITURE CONFLICT

For the city manager and his finance staff, the purpose of budgeting is not to allocate resources but to meet the revenue constraint. Indeed, the link of revenue to budgetary behavior is obscured if one views budgeting only as a resource allocation device. For the most part, Oakland's resources were allocated years ago when the present governmental structure was established. Previous decisions determine present allocations. The decision to have a manager-council form of government, the decision to have a free library, and the decision on the fire insurance rating of the city were all made by other participants, but they govern today's actions. If one allows for slight increases in the cost of Oakland's employees and expenses, then last year's budget is this year's budget. Although the participants will argue for marginal increases, no one expects that there is enough money to make a dramatic change in a particular budget. The acceptance of the "no-money" premise by most participants reinforces Oakland's incremental budgetary behavior.

The City Charter requires that Oakland balance its budget. Custom dictates that money be on hand to pay all bills. Every year the city manager, his finance director, and the budget officer face the problem of cutting the budget to match the city's revenue constraint; for fiscal year 1967/1968, the total amount requested was about ten percent more than estimated revenues. Cutting is an intrinsic part of any budget process. But what is unique about smaller governments like Oakland, is that the people who do the cutting are the city's revenue maximizers. Cutting the budget is a form of revenue behavior.

When one examines the influence of revenue on the budgetary process, the city bureaucracy appears to be dichotomized into budget-spenders (the departments) and budget-cutters (the manager and his finance staff). The latent conflict between these two sets of officials is related to functional differences: throughout the budgetary process the revenue- or resource-maintenance function of the cutters conflicts with the service-performance function of the spending departments. This conflict is particularly evident when one examines the guidance part of the process. In the fall, the Budget and Finance Department issues instructions to the departments for budget preparation. One of these instructions is the manager's guidance letter which states his assessment of the city's probable fiscal condition for the future budget year. How do budget-spenders respond to the guidance of the budget-cutters?

One might expect from Crecine's (1969) work that the city's revenue considerations would be linked to the department's budget preparation by the city manager's guidance letter. The manager would state his no-money premise and operating officials would lower their requests accordingly. Yet we must ask why most officials in Oakland do not pay attention to budgetary guidance.

The manager's letter contains nonoperational guidance. Because everybody knows there is no money, the letter does not express anything new. A city official complains that for twenty years he has listened to the same old story from a variety of managers and budget officers: there is not enough revenue and the budget must be cut. City officials have cried wolf too often.

An examination of the guidance letters since 1957 reveals two themes. The first states that the city has financial problems, but that the answer is to work together to cut costs and be efficient. This "efficiency" theme appeals to the conscience of the department head to give the taxpayer the most for his money. If the departments can be efficient, then the city can avoid revenue increases. The second theme is more dismal; it stresses the revenue-expenditure gap. It states the no-money premise and emphasizes that services cannot be expanded. The cutters expect the departments to submit expenditure requests similar to their current funding levels. Taken together the two themes state: (a) you must be efficient because (b) there is no money. In Oakland, it is not a question of pulling in the belt one year and then next year we will do better. Department heads soon get cynical and adopt the stance that they have nothing to lose by asking.

In a situation of continuous fiscal deprivation, to be told to cut costs and not to ask for more loses all significance. Furthermore, the guidance is ambiguous:

> *Fiscal Year 1968/69.* As you begin the preparation of your departmental budget request, I would ask that you bear in mind that the City of Oakland continues to face a difficult and debilitating financial problem. The economic condition of the community is such that we as professional administrators must be constantly aware of the fiscal limitations within which we must function.

The department head who received this guidance can say, "Yes, I am aware, but so what?" Lacking specific instructions, department heads are free to interpret the guidance according to their own spending inclinations. Oakland does not make use of ceilings or acceptable percentage levels of deviation. Since the cutters do not quantify the instruction to hold the line, the spenders see the guidance more as a request than an order. Ambiguity provides room for maneuver.

Another interesting feature of the hold-the-line guidance is that budget-cutters do not make much effort in developing it. When they write the letter, they do not try to estimate revenue for the coming fiscal year. Detailed revenue estimates are not computed until late in March and April when the cutters must balance the budget. Considering the importance of revenue to the municipal budget, one might expect that guidance would be based on some investigation of future revenue conditions. Instead the budget-cutters assume that they are not going to have any significant amount of money and write the letter on the basis of their general feeling of financial pessimism.

Thus, both sides in the budgetary arena, cutters and spenders, are well aware of the city's financial problems. Budget-cutters do not have to estimate revenue to put out the guidance, and budget-spenders do not need to have the guidance because they already know. The guidance letter simply initiates the budgetary process; it does not structure or delimit the actor's decision problem.

One could think of guidance as structuring the department head's decision problem if, when the manager said to hold the line, the department head did so. If department heads were trying to hold the line, one might expect a slight increase in the requests as a result of price changes, but not a thirty percent increase. In many cases, however, the difference between a department's 1967 budget request and its 1966 budget appropriation is substantial. Although departments vary in their responsiveness to the manager's guidance, it is reasonable to conclude from Table 1 that most of the large departments do not hold the line.

TABLE 1

MOST DEPARTMENT HEADS DO NOT HOLD THE LINE

Major Departments	1967 Request Minus 1966 Budget Appropriation	
	1966 Budget Appropriation	(percentage)
Museum	-12.7	
Police	+ 5.9	
Library	+ 6.0	
Fire	+ 9.5	
Building and Housing	+11.8	
Municipal buildings	+13.8	
Street and Engineering	+17.3	
Park	+32.3	
Traffic Engineering	+34.1	
Recreation	+38.7	

SOURCE: City of Oakland, Budget and Finance Department.

The hope of getting additional funds is not the only explanation for the failure of most departments to follow guidance. The typical department head explains his response to the guidance as, I am a professional; no one, not even the manager, can understand this job unless he sits here; I know our needs, and I would not be doing my job unless I expressed them. These department heads use the budgetary process for communication: it puts the problem where the money is. Often the department head is like a doctor ordering medicine for a miserly patient; his professional responsibility stops when he writes the prescription. Similarly, the city engineer states that a path needs a railing or someone might be hurt, the fire chief feels the level of protection is in jeopardy because the city should have a new pumper, and the recreation superintendent believes the golf course should be reseeded as the city is competing with other courses. Department heads feel they have done their jobs and relieve their anxiety when they express their needs. Not only do they satisfy their own professional norms, but, by putting the monkey on the manager and council's backs, they abate pressure from their employee and community constituents.

Since noncompliance is widespread, it is accepted as part of life and it carries no penalties. Departmental estimates are insulated from the city's revenue constraint and cease being fiscally realistic. The spenders ignore the revenue constraint and concentrate on their service-performance function. Budget-cutters would also like to ignore the revenue constraint, but they have no choice. They know that it is not easy to get new revenue sources, so they must control the budget by cutting, an activity facilitated by the cutters' minimal service and program orientation. The revenue constraint thus creates two disjunctions in the budgetary process: (a) the spenders ignore revenue limitations in formulating their programs, and (b) the cutters ignore program and service considerations in balancing the budget.

TOOLS OF CONTROL: BUDGET REVIEW BY THE BUDGET AND FINANCE DEPARTMENT

Most of budgetary control is a matter of paring a department's request down to the previous year's base—getting rid of extra requests and refusing to make major reallocations of resources. Oakland's line-item budget facilitates cutting. One does not have to worry about the effects on service when cutting out small items.

Budget officers usually worry about their analysts becoming advocates for the department to which they are assigned, but this is not a problem in Oakland because the analysts do not stay long enough. The Budget and Finance Department is relatively small. Each analyst is assigned three or four departments and, given the workload during the rest of the year, it is unlikely that he will have sufficient opportunity to learn much about any one of them. The few experienced analysts in the office have left for other jobs, and the considerable turnover has reinforced this problem of lack of experience.

The analyst does not have to reckon with the impact of his acts on service or on policy and can be quite ruthless in identifying issues on which the city manager must make the "final" decisions. But the more experienced he becomes, the more knowledgeable he is about the impact of cutting on his departments, and the more reluctant he will be to make rash suggestions.

The very existence of a revenue constraint makes it easier to cut. One does not have to be convinced to cut certain items when there is no choice. The amounts requested by the departments for expenditure are not consonant with the revenue constraints and must therefore be reduced.

How does an inexperienced budget analyst who knows little about the department he is reviewing cut that department's budget? Inexperience itself forces the use of certain rules of thumb common among people engaged in budgeting at all levels. The budget analyst looks at last year's appropriation to see if there has been any marked change. If not, he approves the item; if so, he is likely to reject it. Personnel needs comprise the main costs of the budget. Normally, departments will come in with requests that total over one hundred additional people. Given the belief that for every person added, a fixed cost is incurred that cannot be eliminated (because once somebody is hired they cannot be fired), the standing rule is to cut requests for additional personnel.

When there are expensive capital items these will probably be cut out. A lot depends on who is doing the submitting. If a department head has a good reputation with the Budget Office, then there will be the tendency to investigate further. On the other hand, if the department head has a reputation of always submitting certain items and never getting them, these items are automatically cut out.

Up to this point the procedures are standard (see Chart 1). The analyst approves estimates that are the same as last year's and allows two or three percent for cost of living. He cuts out personnel increases and most capital items. But there are always a few borderline cases.

The budget analyst will generally look to see whether it is possible to trade off machines for people because of the high personnel costs in the city, but this type of review usually results in a deferral of the requests for personnel or equipment because of the need for "further study." He tries to postpone any cost which possibly can be deferred. Equipment and buildings are repaired rather than replaced. No thought is given to the increase in maintenance costs as far as trade-offs would be concerned. The only rule that tends to interfere with the deferral policy is one of the analyst's notion of safety or health. When a basement in the Fire Department needed shoring up because heavy fire trucks driving over the basement ceiling caused the foundation to shake, the

CHART 1

OBSERVED CRITERIA USED BY BUDGET ANALYST
IN REVIEWING A DEPARTMENT[a]

---

(1) Compare request with actual spending—adherence to base.

(2) Allow three percent for inflation.

(3) A capital outlay item is not urgently needed. It would be better to delay a decision on expenditure until more study can be done.

(4) There was an arithmetic error in the budget request.

(5) A facility is being used more and so should get more money.

(6) Recommend repair rather than replacement, unless replacement is obviously preferable.

(7) Don't put money into something when you are uncertain about its future.

(8) The expenditure would produce income.

(9) Cut out department contingency reserves.

(10) High cost, but low returns. A facility cost the City $33,000 and netted $300 revenue. Therefore, the City should consider disposing of the property.

(11) Compliance with respected advice—estimation of needs by an administrator.

(12) We were hard on them in this last category last year. Allow a little increase.

---

[a] Observation by Ken Platt, graduate student in Public Administration.

analyst went to the scene, observed the shaking, and decided that this was indeed a health and safety hazard. On the other hand, when a department requested that new restrooms be built in the park, the analyst examined the old restrooms and decided that they could be renovated.

Because of their inexperience, budget analysts tend to cut from the bottom of a departmental priority listing. Department heads anticipate this and often will rearrange their listings so that important items appear near the bottom. Through this device, they hope to confound the inexperienced analyst and thus influence him toward making no cuts at all. More experienced analysts are naturally suspicious about such rankings. This is particularly true of the city manager, who does not use the submitted lists but rather looks at the items themselves and determines what he thinks the priority should be.

The more experienced analyst (defined operationally as a person who has been in the job for more than two years) develops a conception of his role. One analyst expresses his role as that of "a voice in the court," a buffer between the department and the manager. He feels that his relations depend very much on establishing confidence with the department. Another analyst feels that his actions are mainly attempts to bring realism into the budget process, to give the department some idea as to what the traffic will bear. The less experienced analysts, however, do not have a well defined conception of their role, and they operate primarily on the basis of the decision rules in Chart 2.

CHART 2
RULES OF THUMB FOR CUTTING THE BUDGET

---

(1) Cut all increases in personnel.

(2) Cut all equipment items which appear to be luxuries.

(3) Use precedent; cut items which have been cut before.

(4) Recommend repairing and renovation rather than replacing of facilities.

(5) Recommend study as a means of deferring major costs.

(6) Cut all nonitem operating costs by a fixed percentage (such as ten percent).

(7) Do not cut when safety or health of staff or public is obviously involved.

(8) Cut departments with "bad" reputations.

(9) When in doubt, ask another analyst what to do.

(10) Identify dubious items for the manager's action.

When the budget analyst cuts out an item, it does not necessarily mean that it is forever lost from consideration. Cutting, at this level, is merely the identification of a decision that has to be made by some higher authority. Oakland's budgetary process contains numerous levels of review and appeal. In theory, if a department head had a point of disagreement with an analyst, he could appeal to the budget officer, then to the finance director, the manager, and the council. With so many levels of appeal, one might expect that if a department head were persistent enough, he should be able to eke out a slightly more favorable compromise at each level. The result would be that the analyst's original recommendations might bear little resemblance to the final outcome. But this does not happen. An important thing to understand about the process of appeals is that there is a strong built-in bias dictating that superiors will hesitate to overrule the decisions of their subordinates. They know that the review procedure would be undermined if decisions at the lower echelons were not respected most of the time. The system would be swamped by appeals. The finance director and the manager still reserve the right to make further cuts and, in a few cases, to restore part of the requests. The outcome is that a great many of the analysts' decisions prove to be final but the departments cannot be certain that an appeal is hopeless.

TOOLS OF CONTROL: BUDGET REVIEW BY THE CITY MANAGER

During our period of investigation in Oakland, we were fortunate to be working in a city which was undergoing bureaucratic transition caused by the recent appointment of a new city manager. One would think that in an old city like Oakland rules would be well defined, role expectations would be fixed, and the system of behavior with respect to budgeting would be established. Considerable evidence, however, indicates that the locus of budget decision shifts with the occupant of the city manager role.

In a movement toward reform, Oakland installed the city manager type of government in 1931. At the same time, it voted a charter which establishes independent boards, an independent auditor, and generally creates an environment of fragmented government. With a mayor who enjoys little formal authority, the city manager is relatively free to define his own role.

For something like sixty years or more, students of public administration have been arguing about the merits of the politics-administration dichotomy. It is interesting to see in Oakland, a living example of how the ideology of administration can take hold in role definition. The former city manager defined his role as political. One of his assistants said that the budget bored him. He was not a detail man; rather, he was interested in mobilizing the community to adopt new programs, although most of these programs were never developed enough to be implemented. The new city manager, on the other hand, veers more toward the administrative side of the continuum. He is interested in pulling together a fragmented government by developing centralized control over the departments. For the former city manager the budget was dull; for the new city manager it is the most significant tool he has to control the fragmented government.

This shift in role definition from expressing policy to enforcing administrative control led to other changes. The former city manager's finance director encouraged the departments to express their needs in presenting their budgets. Under the new administration, the budget-cutters would like the departments to be realistic. In the old days, budget analysts would consult with departments and come to some agreement as to what ought to be cut. In the new budget procedures, analysts are discouraged from contacting departments or even from letting them know what actions are going to be taken. In the current procedures, there is still negotiations and compromise, but the site has shifted to the city manager's office.

Accompanying this redefinition of roles has been a shift in the other sites for budget decision-making. Under the former administration, the budget was mainly handled by the finance director. He was Mr. Budget. He presented the budget to the City Council; he was the one who negotiated with the departments, and a great many decisions were made at his level. When the former manager's budget was presented to the council, his formal recommendations were often overturned. The council was a source of appeal by commissions and departments, and it did make changes in budgets. The new city manager, however, works quite closely with the budget subcommittee of the council to develop support for his budget.[5] When the budget is discussed by the full council, no actions are taken to countermand the manager's recommendations.

The new manager makes budgeting decisions in the context of his definition of the total job. Since he is trying to centralize administration in the city, he uses the budget to find out whether department heads are fulfilling programs which they previously informed him about. It provides him a great deal of information, not so much because of the direct form of its submission, but in its function as a stimulus to further inquiry into the status of various projects. One department, for example, requested a number of mechanized streetsweepers. In focusing on this issue, the Budget and Finance Department suggested that perhaps the department should buy several mechanized streetsweepers to replace six men. The department head was not in favor of terminating these men. The need to make a decision, however, revealed the fact that the department head had not outlined an overall streetsweeping program as the manager had requested of him previously. The manager's budget actions, therefore, are not only concerned with money; they are really an integral part of his overall attempt to supervise.

As the city manager tried to gain control and began to institute his ideas, he caused conflicts within his budget staff. In the clash of attitudes between those favoring departmental expression of needs and those advocating greater control, the manager came to believe that his budget analysts were advocates of the departments rather than of his office. The clash between the "need" and "control" orientations led to considerable turnover in his staff. The new budget team appointed by the city manager has adopted his control orientation and is looking for ways to extend his discretion. In the budget for 1968/1969, for example, money which would have been allocated to the auditor-controller for the use of his ADP equipment was diverted to facilitate the transfer of automatic data processing to the manager.

Another element of the control orientation is indicated by the manager's desire for almost complete secrecy with respect to budget actions. He does not want the budget analysts to tell the departments what he is considering. In review sessions with department heads, he asks questions about their operations but does not tell them where he is going to cut. In this way, he makes sure that when he submits the budget to the council, it is indeed his budget.

The city manager does not rely on the budget submissions alone. In addition, he asks the budget analysts to prepare a set of ancillary data which includes detailed analyses of personnel requests, overtime, and travel, for council consideration. He has a simple three-stage process of examining the budget. First, he sees whether the revenue estimate is as pessimistic as his own expectations. Then he asks that those fixed costs which he can do nothing about (such as the built-in pay raises under a city formula) be subtracted from the budget. And finally, he gets down to looking, as he says, "at the nitty gritty details."

Although the manager claims he does not want to get involved in details, he continually does so because ability to use the budget for control lies in small things. When submissions reach him, the issues for decision have already been identified. The original budget submitted by the department is there, however, and he will often rifle through page by page to pick up an item and consider it. His review does not involve any type of meat-axe-cutting—no ten percent here or five percent there. It is strictly a line-item deletion governed by his own set of preferences.

No item is too small for the manager's attention. It is not true that he just pays attention to the lumpy or big dollar items. While an item must become important to him because it requires some kind of action, amounts are only one index of saliency. For example, the city manager made considerable effort to reduce the mayor-council appropriation in order to cut down the expenditure for business cards and some minor operating costs. He did this not primarily because of the money involved but to symbolize the need for economy. Similarly, he restored to the budget a request for a typewriter because the manager of that department had made a considerable case for need of the typewriter, and he wanted to reward good work.

A general rule the manager follows is to be less generous to a department which had gotten more than the other departments the previous year. In one case, the budget analyst had cut back a submitted budget request to the previous year's level. When it got to the manager he felt that the total budget had not been cut back enough, because he remembered that the previous year the department had won a much greater share than usual of the available resources.

An ancillary decision rule followed by the manager is that he should not give resources to departments which have had insufficient time to absorb what they have already been allocated. It is the manager's prerogative to determine how much the department can actually use. An incremental approach to budgeting is usually described by a short time perspective, and essentially as a maintenance of the basic budget. But holding the line does not always mean keeping expenditures at the same level as last year, especially when that level may have been high.

No doubt the manager's review exhibits a greater concern with program than the somewhat blind cutting of the Budget and Finance Department. But because of the lack of information, the time constraint, and the volume of minor cuts which are made, it is not usually possible to trace the effect of a particular cut on a department's output. The manager does not like to perceive the budget job merely as cutting. He wants to be able to reallocate but often feels that he does not have sufficient information. So he continually emphasizes the need for detailed study of department operations. Because of the lack of skilled staff, such studies are seldom carried out. Hence, most of the manager's activity is involved simply in cutting with insufficient information on program implications.

Because the city manager is only one man, a great deal slips by in the budget review process. In no sense can we say that the basic budget is intensively reviewed. But it is true that many items within the basic budget and incremental to it are reviewed if they are given sufficient visibility. The budget officer provides visibility in focusing certain issues for decision. The dollar amounts focus others. But a more important source of visibility is the manager's own experience in operating city government. His expectations about what things ought to cost or what departments ought to do appear when he is looking at the budget. Similarly, his preferences about what constitutes a reasonable program are revealed. When he was reviewing the Police Department budget, for instance, his predispositions about civil defense resulted in a reduction of uniformed personnel for that activity. The language of social psychology—selective perception, the provision of cues—seems to describe the kind of decision process which operates during the city manager review.

Experience has taught the manager that the form of budget can be as important as its substance. It is not always clear, for example, who will be allowed to present a budget. The submission of a budget is also a representation of jurisdiction. The Off-Street Parking Commission's budget is normally included under the Traffic Engineering Department. The city manager would not agree to the Parking Commission's request for a separate budget, because this would mean increasing the fragmentation of government. Sometimes departments want to disassociate themselves from a particular activity. A good example is the Police Department's repeated attempts to get rid of the Pound activity and the policemen's insistence on having the Pound's budget separated from theirs, even though they manage the animal control appropriation. The city architect, who works for the Municipal Building Department, has never liked to consider himself under the jurisdiction of that department, and so he repeatedly tries to submit a separate budget, which is then sent back to him to be incorporated with the normal municipal building budget. Thus the right to submit a budget is important to the participants in the process.

From the city manager's perspective, the department's budget can be used to commit him the following year; he is concerned about what is said in the narrative, so that the department head cannot come back and say, "See, you said I was going to get this next year." If he had his way, he would have no narrative statements in the budget at all.

The budget can also be used to avoid jurisdictional problems in the future by judicious placing of funds in accounts like "Contractual Services," as in the case of the appropriation for the city's Central Yard. Since the city manager did not want anyone to feel a proprietary interest in the Yard, he placed the money in a neutral category that did not belong to any operating agency.

In his desire to gain control, the city manager has insisted that the departmental requests do not show in his submission to the council. He wants the budget to appear to be his own and not to create any cause for controversy by showing the requests of the departments. For the city manager, the budget's form is almost as important as its substance.

Much of the manager's activity is devoted to the preparation of the budget for presentation to the City Council, especially to its budget subcommittee. The manager operates essentially as a gatekeeper for fiscal policy issues with respect to this subcommittee, describing what he considers to be the key issues. Meetings with the subcommittee have certain mutual advantages to both participants. The city manager feels out what the council possibly might accept, and at the same time the council gains more confidence that the manager is taking care of running the city efficiently, so that they do not have to worry about it. If the budget subcommittee goes along with the manager, this operates as a fairly strong lever with the council.

The manager knows that councilmen do not have the time to pay attention to the whole budget, that they have "pet projects," and that he must pay attention to these. He learned in his previous assignment that, as a former councilman told him, "You've got to give the man a bone to play with, and then the rest of it goes by without review." For the Oakland City Council, the bones are currently the community improvement projects in which the city provides subsidies for civic-benefit organizations such as the symphony, flower societies, and so on. In this case the manager simply recommends a total for such activity and allows the budget subcommittee to allocate within this total.

Councilmen generally are reluctant to reduce the number of city employees and, in order to counteract this hesitation, the manager goes to some length to prepare an analysis indicating that a majority of the city employees do not live in the city and therefore are not voters. Another device that he uses with the council budget subcommittee is to make comparisons with other cities, such as San Diego, Sacramento, and San Jose, to show that Oakland is really in bad shape, either in terms of a low level of service or a high level of cost.

"It's your budget, Mr. Manager," said the vice-mayor to the city manager. The actions of the City Council in 1967 document this: after a four-hour presentation of the budget by the manager, it was passed without any change.

The vice-mayor feels that the budget is pretty well locked in when it comes from the city manager; councilmen really cannot do much by then, even if they wanted to. Oakland councilmen are part-time members and they do not have the staff or resources to understand the consequences of the decisions they are called upon to make. In the past they have had to cut the budget when they did not trust the submission from the manager, but they had no means of procedure, so they had to resort to across-the-board cuts. They are not adverse to cuts, but they would like to have substantiation for them. And this is one clue to the city manager's success with the council: he tries to document his cuts and provide a rationale. The manager takes the time to get to know councilmen and their interests. As a result, his preferences are often adopted.

Once the budget is passed, the budget process is by no means over. While one year's budget is being executed, the next year's is being influenced. Policemen's associations, interest groups, and various employee associations are trying to improve the status of their clients throughout the year. Decisions such as the extension of vacation time for senior policemen have to be included in the budget. Often the manager is presented with these decisions as completed actions which he cannot change. At other times, he is given the opportunity to make a staff report and influence the council. But, in any event, incidents occur during the year which tend to delimit his discretion and thereby further constrain the amount of financial resources which will be available for the budget.

We can now understand that although the city manager may claim that he should not spend much time on budgeting because there are no extra resources to distribute, he actually spends a great deal of time. He is the key figure in making most of the decisions. After the budget and finance people review the submissions from the departments and focus on some issues for decision, the city manager reviews all the budgets and, for the most part, makes the decisions. He guides the City Council in its consideration. He feels that it is his budget. And he uses it to make his influence felt throughout city government.

CONTROLLING UNCERTAINTY THROUGH EFFICIENCY

There are other ways to apply budgetary control. The finance director thinks an appropriate means is to introduce administrative efficiencies. He estimates that something like $2 million a year could be saved with the introduction of such efficiencies. For any single budgetary year, however, the efficiency approach has limitations. The budget cannot be balanced simply by appealing to officials to be efficient. Determining that one way of doing something is better than another is not easy. Even when officials feel they understand which way to go, implementation is difficult because each administrative efficiency has to be sold to the council on its merits. Administrative improvements can only be approached piecemeal in the context of Oakland politics. Generally, the council cannot and could not accept a mass of these at the same time, nor would many citizens accept them, because it is hard to distinguish a so-called "efficiency" from a simple cut in service.

While it is politically possible to introduce administrative improvements singly, a major limitation in the operations of the city is the inadequacy of the resources of officials who are seeking change. Oakland's budget-cutters lack information, time, and staff. Inexperienced analysts and line-item budgeting facilitate balancing the budget, but unfortunately inhibit the determination of administrative improvements. During the budget season, officials are so busy doing arithmetic that they can identify only a few of the soft spots in the city's operation. Their basic tool of analysis, comparison with other cities, shows that Oakland may be out of line but not why. The city may have too many building inspectors or the cost of circulating a book may be excessive, but what to do about it requires study. The budgetary process acts as a catalyst suggesting avenues for exploration and resulting in the "if we have time we should investigate" syndrome. Off-season, officials are continually fighting small fires, taking care of minor transfers of funds or getting themselves involved in detail. Consequently, they have limited opportunity to do in-depth studies of department operations and problems.

The studies they have done are neither dramatic nor would they reshape Oakland's basic configuration. When contemplating big changes such as the reorganization of public works activities, they call in consultants. The typical study is more mundane. Four man months of study effort went into the recommendation for consolidaton of the library's janitors into a unified maintenance force under another department. By consolidating its janitors, Oakland was able to eliminate six personnel positions. Budget and finance officials feel that they saved the city over $40,000 a year. The savings, however, were just on paper. Nobody was fired. The janitors were absorbed into the city's increased requirement for building maintenance brought about by the opening of Oakland's new museum.

The type of studies which budget officials do—minor reallocations of personnel, control of overtime, cost control on construction work orders, centralization of equipment maintenance activities, avoidance of duplication in forms and procedures—are congruent with their analytical capabilities. Recognizing the inadequacies of their study capabilities, city officials are seeking ways to improve. Besides adding analyst positions (not readily filled), they have taken the trade journals seriously and have attempted to orient the budget more toward program and activity rather than to line-item and organization. Although they have been using what they call activity budgeting for several years, the line-item approach is still apparent underneath the new formats. Why this is so is worth exploring.

There are two reasons why a new format does not necessarily change past incremental budgetary behavior. First, officials rely on their experience as a short cut to decisions. They know that the old procedures work. Second, the new format and procedures are not much different from the old. They do not provide officials with the type of information that will cause a shift in decision rules. It is all too easy to claim bureaucratic inertia when the problem may lie in the inadequacy of the new procedures. Oakland officials found that they could not avoid making decisions by line items. Indeed, they spent a great deal of time juggling numbers to calculate a comparable previous year so as to identify significant changes in items. Comparisons were made across years for items, and activities were ignored.

Officials ignored the activities because they reflected organizational designations which are artificial and useless. The Police Department delineated its activities by referring to organizational components such as the Criminalistics Section, the Inspectors Division, and the Juvenile Division. Over forty percent of its budget was in one activity, the Patrol Division. The Library saw its program in spatial terms and classified most of its budget into two activities: main library and branch department. Other departments took the easy way out and put their budget in as few categories as possible; the city's parks appeared under one activity called "field services" and the Street and Engineering Department divided its budget into "new facilities" and "maintenance of existing facilities." The Municipal Buildings Department just listed the buildings it maintains (City Hall, Hall of Justice, etc.). The net result was the submission not of programs, but of a set of intra-departmental budgets that were mainly organizational components of a department. The city's need for a responsible budget executor and the budget officials' willingness to argue with departments encouraged the organizational bias against the activity structure.

Activities which involved the support of more than one department were troublesome; the city's small civil defense activity was separately identified but most of its personnel costs appeared in the Fire and Police Departments' budgets. Even activities within a department which involved an allocation of joint costs proved difficult. These problems were resolved by maintaining the organizational identification of the activity. Thus, the Fire Department had a separate activity for its Training Division, and the Police Department identified its Transportation Section as an activity involving its vehicular fleet.

Workload statistics were supplied for many of the activities, but the cost information in the budget was not linked to these statistics. The Fire Department, for example, showed a workload statistic of 217 fires investigated for arson, but this number was not connected to the costs for its Fire Prevention Bureau activity. Instead, the department maintained line-item identifications of capital, equipment, and personnel to show costs. In short, allocations of costs to activities were inconsistent, "activities" were not always activities, and there was too much paper work, especially since the decision process was not changed by the new format.

At this time, we cannot tell whether the costs of Oakland's effort at a new approach to budgeting are worth the benefits. The city manager does not like the added paper work, but is delighted that the new format shows more costs in an operating department's budget instead of leaving these costs in overhead departments or special accounts. When all the costs are out in the open, so to speak, it is easier for him to gauge the efficiency of a department head. If everybody were "one big family" and agreed about goals and procedures, he believes that program or activity budgeting would not be necessary because he views these innovations as another control mechanism and not as a steering or strategic planning device. The city's study effort is similarly conducted within the framework of established goals. Fundamental questions such as whether the city should be in the recreation business are never seriously addressed. Without sufficient pressure to change the basic configuration of city services, there is little incentive to worry about steering. Policy-making for most of Oakland's budget-cutters is a matter of efficiency for control.

AGENDA FOR REFORM

Oakland is a poor city. It has far more than its share of poor people who need services they do not get. Budgeting is not a matter of allocating resources, but of keeping costs down so that expenditures will not exceed income and necessitate new taxes. Nor is there a concerted effort to deal with policy problems outside of the budgetary context. Oakland is also poverty-stricken in the human resources available to work on its problems. While the city has many intelligent and devoted men on its payroll, few have been trained to evaluate the effectiveness of existing policies or to devise new alternatives. The city has no policy analysts and few economists who could serve in that capacity. It lacks the analytical capability to make effective use of outside consultants.

With little talent and less opportunity to analyze policies (in view of the lack of money), top city officials are confirmed in their view of the budget as a servant of lower taxation and administrative control. The budget does keep expenditures down and provide information useful in keeping tabs on departmental activities.

Proposals for reform of the budgeting process and for increased attention to policy analysis must pay heed to the realities of city government that we have discussed. No reform proposal that makes the budget appreciably less useful for control and cutting purposes is likely to be adopted. No proposal that calls for a large increase in personnel, thus adding to the city's major cost, is really feasible. No proposal that requires a high level of analytic talent to make it work will get far. No proposal that depends on the existence of, or the likelihood of getting, good data relevant to actual decisions has much chance in the next few years.

There is, of course, one change that should be radical and swift if only anyone knew how to accomplish it. We refer obviously to the substantial increasing of available city resources. Cities like Oakland need help from the state and federal governments in the form not merely of soft advice but of hard cash. They need leaders who can mobilize citizen support for bond issues and other forms of revenue. Interest in resource allocation would rise appreciably if only there were resources to allocate. In the meantime, city officials need advice on how to do a little better within existing constraints.

Our cities do not need to undertake, individually, the creation of a comprehensive policy-making system such as PPB (see the Mushkin and Borut chapters, elsewhere in this volume) where the costs of the system will undoubtedly outstrip the conceivable implementable benefits. Most planning-programming-budgeting systems theoretically have three elements: (a) an analytical activity designed to evaluate existing policies and suggest new policy alternatives, (b) a multiyear financial plan containing extensive lists of output-oriented categories (the program structure) and costs and output measures for these categories, and (c) explicit links (program memoranda, "crosswalks," etc.) to the annual budgetary process with its own appropriation codes and customs. We wish to argue that cities do not have to swallow the whole system to improve their decision-making and enhance resource allocation.

At the federal level, PPB, even with presidential support, has run into difficulties. Marvin and Rouse (1969) have reported on the numerous difficulties of transferring a Department of Defense model across-the-board to the domestic agencies. Duplication of effort with budgeting, confusion as to whether the Bureau of Budget was a client of PPB or was to provide help and guidance, agency head indifference, congressional problems, lack of qualified analysts, program structures which resembled agency organizational structure and did not facilitate analysis, and the lack of valid cost and output data were among the central factors which they identified.

The experience of the State-Local Finances Project of the George Washington University is also instructive. The project's staff, charged with a technical assistance mission, worked for one year with a group of state and local governments to install PPB. This demonstration effort was designated the 5-5-5 project because the group of governments included five states, five counties, and five cities. Recently the staff discussed the lessons which they learned from their efforts to install PPB. They found these fifteen governmental units lacking in some of the important requisites for a successful PPB system. There was a lack of significant study capability and, more importantly, a lack of strong chief executive support. The program structure was not used "significantly," operational outputs were not defined, and only three governments actually completed a study or a program analysis (Mushkin et al., 1969: 12-26). We make no criticism of the project staff's efforts because we recognize the difficulties associated with the installation of PPB. We also think cities should be aware of these difficulties. As the project staff said, "Governments wishing to implement PPB should expect to invest considerable time and effort before significantly useful products can be achieved" (Mushkin et al., 1969: 23).

But can cities afford to make the investment? In the Department of Defense, an analytical effort of, say, $200,000 (about five or six analysts for a year) might bring a payoff in millions of dollars. In cities like Oakland, there is not enough resources to put that much effort into a study, and the payoff is not likely to be as great. Planners in local government have to do double duty. Physical planners have to spend their time on zoning variances and budget-cutters have to look for revenue and cost reductions. Short-term needs drive out long-term considerations. Even if a city were rich enough to afford PPB, where would it get the people to put theory into practice? There is a shortage of trained policy analysts as well as a dearth of money.

The creation of multiyear financial plans usually results in the production of dreadful data with which no one knows what to do (Wildavsky, 1969). The categories represent organizational compromises and confuse rather than clarify programs. The cost data are artifacts of the imagination and are not derived from existing information systems. Indeed, most municipal accounting systems are currently incompatible with the data requirements for an extensive program orientation. Credible output data is virtually nonexistent.

Putting aside the data question, what does the local official do with his multiyear financial plan? The plan will show a simple-minded projection of current activities. It looks into the future by translating the resource requirements of the status quo. Although there is much talk about the inspirational nature of financial plans, the plan itself will seldom generate alternatives. Imagine the city staff exhausted by the creation of one plan and then the manager or mayor (just back from an indoctrination session in systems analysis) asks for ten additional plans because he has been told that he should consider alternatives. Consequently, the mayor/ manager will find out that (a) that *the* plan is not susceptible to change, or (b) that he requires a group of analysts to consider alternatives.

If the plan is really comprehensible and specifies objectives in terms of who gets what, then we wonder about the political wisdom of such an undertaking. As the multiyear plan moves to the status of a locked-in budget with its approved programs, we would expect the level of community conflict to increase. The plan will reflect the preferences of the city's bureaucrats. In all likelihood it will not reflect the preferences of the neighborhoods or the elected representatives of the city. Councilmen will be caught in a crossfire between neglected neighborhoods and unhappy bureaucrats.

Given the fiscal condition of our cities, most councilmen have been content to let the budget drift. Occasionally they make changes in line items. A program-oriented budget will probably not change their behavior. First, there will still be the no-money problem; without fresh funds, there will not be much to allocate. Second, reallocations will be difficult because of the increased visibility of expenditures.

Neighborhoods will want more, not less. And there is no guarantee that increased demands will result in a commensurate tax policy. Finally, given the embryonic state of financial planning, it will not be any easier to make allocation decisions. Showing projected expenditures by program categories will not inform councilmen about the possible service impact of their decisions. Measures of service output tied to expenditures would be required, and these measures are hard to come by. How does one measure crime, fire, and disease prevention?

With enough resources like time and intelligent analysts, we expect that such measures could be developed. But why should each city attempt to measure everything at once? The problem, as we see it, is for city officials to be selective in what they do so as to husband their own limited resources. Selma Mushkin (1969: 167) reports that fifty or sixty cities have entered into the PPB business. Before other cities take the plunge, they should reflect on their own resources.

There is no need for a city to develop everything on its own. If performance or work-measurement information has proved of use in another city, why not use it? Performance measures are different from those which might be used for PPB. They are designed for efficiency and control and not for steering purposes.[6] Since some of the problems of our cities are efficiency problems, a selective borrowing from other cities seems reasonable. The same logic applies to cities that insist on going all the way with program structures and multiyear plans. We believe that every city does not require a unique structure. A model structure (or at most a few) could be developed as a product of research and application. If nothing else, neighboring cities could get together and divide up the work. Each could work out the mechanics for a particular functional area (such as recreation) and share its experience with others. But our preferences differ.

We recommend that:

(1) Most budgeting should maintain a line-item orientation because of its ease of administration and its advantages as a mechanism for control of expenditures. The line-item budget has traditionally been criticized because its overspecificity and focus on inputs inhibits policy analysis. The fact is, however, that men who could do policy analysis could work around line-item budgets and men who cannot must have them if they are to make decisions at all. In any event, top executives are unlikely to give up the known

advantages of line-item budgeting in favor of program categories that result neither in control nor analysis.

(2) Performance budgeting should be implemented on a selective basis where there is a known and predictable relationship between costs and the unit of work measured. This is more likely to occur in a public works department than in a "paper-mill" operation like accounting. One of the difficulties with reform is that we are prone to act as if the last generation of proposals had been successfully adopted and the time had come to get on with the new. Actually, there are few places that have successfully adopted performance budgeting. Indeed, it may be that the accounting orientation of city officials has been too narrow and too rigid to derive performance measures that demonstrate full understanding of the policy problems involved. One result of the trend toward policy analysis may conceivably be bringing in people who, though they do not know how to make PPBS work, are actually able to develop measures of output.

(3) A group of policy analysts, separate from the budget staff, should be established who would work directly for the manager or mayor. This group would work full-time on short-term studies within their own analytical capabilities. Unless a city has a top-notch policy analyst, it must go out and hire one to head its policy analysis unit at a salary of something like $25,000 a year. This sum is the minimum contribution the city must make to get analytical capability. The rest of the people can come from internal transfers. The engineering department can certainly provide a man with technical skills who could perform quantitative studies under the supervision of the analyst. The budget department can provide an analyst who is sensitive to the cost implications of policy problems. The rest of the small group, anywhere from three to six people, should be transfers from city planning, which is likely to be the only reservoir of semianalytical types in a city. Although this is not a paper about the irrelevance of city planning, a few brief words of explanation are in order.

"Planning" is another term for "the authoritative allocation of values" without authority, and it is not surprising that neither city officials nor citizens wish to let planners dictate important decisions now, let alone years in advance. At the same time, planners do not know how to make worthwhile comprehensive plans even if by a quirk of fate some city would be prepared to act on their recommendations, so they are stymied. We do not refer, of course, to men engaged in zoning or physical planning, which are well defined activites. The rest of the planners may be considered itinerant social scientists looking for a role and a client. In some cities they undermine the mayor and go about drumming up business from any interest group they can find. They have invented a new ideology called advocacy planning to explain why, in the name of participatory democracy and the maximum feasible participation of the poor, professional planners must be employed by local citizens to do their thinking. In still other cities these planners become de facto aides to the mayor or city manager and perform as policy analysts though they are not called that. Our proposal would simply legitimate this marriage. It would finally give planners an honorable client and a role they could at last understand, so they would cease wandering around cities like Banquo's ghost, believing that someone, somewhere, someday, should listen to them, but unable to say "who" or "why."

Thus, with the beginning of an analysis group, top city officials would at last have the ability to undertake short-term studies of major policies. If they should wish to exert leadership through policy change, they would have a group that belonged to them. If they merely preferred to concentrate on efficiency studies, they would have people who could really help them.

Working for the manager or mayor, the policy analysis unit no doubt will have trouble in securing the full cooperation of the departments. There must be some incentive for departments to become sensitive to policy questions. One incentive would be to allow the departments to keep any surplus money which might be created as a product of analysis. Analysis should not be directed merely to the cutting of costs. It should be directed at improving the output of services to meet community demands and needs. If departments feel that analysis will enhance their *own* programs, then they will cooperate. Moreover, if they see that the mayor or manager takes these analyses seriously during budget review, departments will certainly want to participate in them rather than be subject to somebody else's formulation of their problems.

(4) For more elaborate studies that involve large-scale change and more sophisticated techniques, officials should resort to established contractors or to the relevant experience of other cities. The policy analysis group could monitor the output of various research corporations and public agencies to see if they had come up with anything worth adopting. It could perform the function of adapt-

ing ideas initiated elsewhere for local purposes. It could also prepare proposals for grants. But these activities should not be overemphasized. In view of the fact that the analytical group works for the highest city officials, the danger exists that they would become coordinators of other people's work. The main purpose of the group must be to do policy studies.

We certainly agree with partisans of PPB that an emphasis on program or output can have several advantages. A program orientation can encourage public support by helping a citizen understand the connection between taxes he pays and the services he receives. Moreover, PPB can encourage civil servants to be aware of fundamental rather than trivial questions. Asking what his goals are or why he is in this business can be unsettling for a conservative bureaucrat. On the other hand, partisans of PPB place too much emphasis on requiring a multiyear financial budget with its program structure and supportive data. They are convinced that without a program structure in which the whole city gets involved, policy analysis would go down the drain. PPB, as a system, becomes more important than the men who must implement and live with it. We do not agree with this view. We see no reason to accept costly parts of a system that have no short-term payoff on the vague grounds of possible institutionalization or coordination effects. If we want to encourage the responsiveness of our local institutions to the demands of its citizens, then we have to question whether total acceptance of PPB is the best way to proceed.

If we were writing a book on city government rather than a narrowly focused paper on city budgeting, our recommendations would be much broader. The way to change patterns of resource allocation in a city is not merely to tinker with its budget but to make fundamental alterations in political structure. As Arnold Meltsner suggests in his companion article in this volume, the revenue problem is prior to and more important than the problem of resource allocation; unless there is a great deal more money to distribute, only small changes in allocation between large purposes will be feasible. In order to develop new sources of income, vast amounts of political support will have to be generated.

Failing to effect fundamental alterations in the structure of politics, there might still be room for substantial increases in the use of existing political resources by city officials and citizens. Without some such change, the politics of budgeting becomes a narrow subject. A demand model of the kind implicit in *The*

Politics of the Budgetary Process (Wildavsky, 1964) is inappropriate because the revenue constraint and lack of citizen mobilization mean that few demands are generated and fewer still can be met. A bureaucratic model which focuses on small adjustments such as the one we have described for Oakland is clearly more appropriate. We do not argue here that to change budgeting you must change the city politics, although in the largest sense this is profoundly true. Rather, our recommendations accept for the moment the existing political structure and ask how bureaucratic decision-making can improve within these constraints. We would prefer that the world were different, but we do not think that city officials will benefit by statements of regret. They need help now, and we are trying to provide it in terms they can use.

Strong political leadership and an intelligent use of policy analysis for steering may move us closer to addressing our cities' dilemmas than would a program structure. In any event, our reform proposals are aimed at cities that do not have money to spend on nonexistent analysts working with bad data.

We believe that our recommendations take advantage of the analytical aspects of program budgeting and, at the same time, free city officials from the onerous burden of creating nonoperational financial plans and meaningless program structures. At a relatively small cost, these proposals enable a city to expand its analytical capability. The weakness of our proposals is, of course, that they depend on the presence of city officials who are interested in analysis and willing to nurture and reward it. If the will exists, however, we have suggested a modest way for city officials to gain experience in using analysis and, therefore, to learn whether they want more.

NOTES

1. An advantage of Sharkansky's work, from our perspective, is his use of previous expenditures as one determinant of current expenditures (see also Sharkansky, 1967). The independence of the previous expenditure variable has been questioned by Robert L. Harlow in an exchange in the *National Tax Journal*, XXI (June, 1968: 215-219). Unlike most of the independent variables that are used in aggregative studies, the previous expenditure variable approximates the budgetary decision rules of public officials.

2. There are striking similarities between Barber's (1966) experimental findings at the local levels and the results of Wildavsky's (1964) interviews at the federal level.

3. The revenue constraint is not nearly so important for federal agencies because their expenditures (except for a few of the largest) are too small a proportion of the total to make that much difference. There is, in addition, no requirement for a balanced budget and the prospects for raising revenue are much brighter.

4. For the fact that the per capita local government burden has increased, see State of California, Legislature, Assembly, Interim Committee on Revenue and Taxation, *Financing Local Government in California,* prepared by Wilma Mayers, December 1964, pp. 13-14.

5. The budget officer and finance director were so conditioned to the former city manager's lack of interest in the budget that when the new manager came into the office and the budget was being prepared, they kept it up in their office wondering what to do with it. Should they take it down to the new manager for his review? After a month of worrying, they finally asked him what he would like to do with the budget, and he said that he certainly wanted to review it.

6. A few years ago, the prevalent reform fad among governmental officials was the performance budget. Practitioners may remember the nights they stayed up creating activities, functions, work or performance measures, and cost factors almost out of thin air. Undoubtedly performance budgeting was a secret plot to solve the unemployment problems of cost accountants. The objective of the performance approach, as Schick (1966: 250) indicates, is to "help administrators to assess the work-efficiency of operating units." The approach has three ingredients: (1) the projection of activities (what is to be done) by work or performance measures (number of books circulated, tons of garbage collected, miles of street resurfaced); (2) the application of cost relationships (X dollars per mile of street resurfaced) to the performance measure to create a budget estimate for the particular activity; and (3) the evaluation of costs against previous experience or established standards. Combined with a program of progress reporting during the fiscal year, performance budgeting, on a selective basis, can be useful in controlling costs by increasing the efficiency with which well-accepted and recurring activities are carried out.

REFERENCES

ADAMS, R. F. (1967) "On the variation in the consumption of public services." Pp. 9-45 in H. E. Brazer (ed.) Essays in State and Local Finance. Ann Arbor: Institute of Public Administration, University of Michigan.

ANTON, T. J. (1964) Budgeting in Three Illinois Cities. Urbana: Institute of Government and Public Affairs, University of Illinois.

BARBER, J. D. (1966) Power in Committees: An Experiment in the Governmental Process. Chicago: Rand McNally.

BARR, J. L. and O. A. DAVIS (1966) "An elementary political and economic theory of the expenditures of local governments." Southern Economic Journal, XXXIII (October): 149-165.

BOLLENS, J. C. et al. (1961) Exploring the Metropolitan Community. Berkeley: University of California Press.

BOOMS, B. H. (1966) "City governmental form and public expenditure levels." National Tax Journal, XIX (June): 187-199.

BRAZER, H. (1959) City Expenditures in the United States. New York: National Bureau of Economic Research.

BROWN, W. H., Jr. and C. E. GILBERT (1961) Planning Municipal Investment: A Case Study of Philadelphia. Philadelphia: University of Pennsylvania.

CAMPBELL, A. K. and S. SACHS (1967) Metropolitan America: Fiscal Patterns and Governmental Systems. New York: Free Press.

CRECINE, J. P. (1969) Governmental Problem-Solving: A Computer Simulation of Municipal Budgeting. Chicago: Rand McNally.

DAVIS, O. A. and G. H. HARRIS, Jr. (1966) "A political approach to a theory of public expenditure: the case of municipalities." National Tax Journal, XIX (September): 259-275.

DAVIS, O. A., M. DEMPSTER and A. WILDAVSKY (forthcoming) "A predictive theory of the budgetary process."

––– (1966) "A theory of the budgetary process." American Political Science Review, LX (September): 529-547.

DYE, T. R. (1966) Politics, Economics, and the Public: Policy Outcomes in the American States. Chicago: Rand McNally.

FABRICANT, S. (1952) The Trend of Government Activity in the United States Since 1900. New York: National Bureau of Economic Research, Inc.

FISHER, G. W. (1964) "Interstate variation in state and local government expenditure." National Tax Journal, XVII (March): 57-74.

KURNOW, E. (1963) "Determinants of state and local expenditure re-examined." National Tax Journal, XVI (September): 252-255.

LINEBERRY, R. L. and E. P. FOWLER (1967) "Reformism and public policies in American cities." American Political Science Review, LXI (September): 701-716.

MARVIN, K. E. and A. M. ROUSE (1969) "The status of PPB in federal agencies: a comparative perspective." Pp. 801-814 in U.S. Congress, Joint Economic Committee. The Analysis and Evaluation of Public Expenditures: The PPB System. Ninety-first Congress, First Session, Volume 3.

MORSS, E. R. (1966) "Some thoughts on the determinants of state and local expenditures." National Tax Journal, XIX (March): 95-103.

MUSGRAVE, R. A. and A. T. PEACOCK [eds.] (1967) Classics in the Theory of Public Finance. New York: St. Martin's Press.

MUSHKIN, S. J. (1969) "PPB in cities." Public Administration Review, XXIX (March/April): 167-178.

――― et al. (1969) Implementing PPB in State, City, and County: A Report on the 5-5-5 Project. Washington, D.C.: State-Local Finances Project, George Washington University.

Oakland, City of (1968) Digest of Current Federal Programs in the City of Oakland (October).

――― (1967) Application for Planning Grant Model Cities Program (April 3).

OSMAN, J. (1968) "On the use of intergovernmental aid as an expenditure determinant." National Tax Journal, XXI (December): 437-447.

――― (1966) "The dual impact of federal aid on state and local government expenditure." National Tax Journal, XIX (December): 362-372.

POGUE, T. F. and L. G. SGONTZ (1968) "The effect of grants-in-aid on state-local spending." National Tax Journal, XXI (June): 190-199.

SACHS, S. and R. HARRIS (1964) "The determinants of state and local government expenditure and intergovernmental flows of funds." National Tax Journal, XVII (March): 75-85.

SACHS, S. and W. F. HELLMUTH, Jr. (1961) Financing Government in a Metropolitan Area: The Cleveland Experience. New York: Free Press.

SCHICK, A. (1966) "The road to PPB: the stages of budget reform." Public Administration Review, 26 (December).

SCHMANDT, H. F. and R. G. STEPHENS (1963) "Local government expenditure patterns in the United States." Land Economics, XXXIX (November): 397-406.

SCOTT, S. and E. L. FEDER (1957) Factors Associated with Variations in Municipal Levels: A Statistical Study of California Cities. Berkeley: Bureau of Public Administration, University of California.

SHAPIRO, H. (1963) "Economics of scale and local government finance." Land Economics, XXXIX (May): 175-186.

SHARKANSKY, I. (1968a) "Regionalism, economic status and the public policies of American states." Southwestern Social Science Quarterly (June): 9-26.

――― (1968b) Spending in the American States. Chicago: Rand McNally.

――― (1967) "Some more thoughts about the determinants of governmental expenditures." National Tax Journal, 20 (June): 171-179.

SHELTON, J. P. (1960) "How to keep local expenditure under control." P. 92 in J. A. Vieg et al. California Local Finance. Stanford: Stanford University Press.

WILDAVSKY, A. (1969) "Rescuing policy analysis from PPBS." Public Administration Review, XXIX (March/April): 194-195.

――― (1964) The Politics of the Budgetary Process. Boston: Little, Brown.

WILLIAMS, O. P. et al. (1965) Suburban Differences and Metropolitan Politics: A Philadelphia Story. Philadelphia: University of Pennsylvania Press.

WOO, S. K. (1965) "Central city expenditure and metropolitan areas." National Tax Journal, XVIII (December): 337-353.

WRIGHT, D. S. (1965) Trends and Variations in Local Finances: The Case of Iowa Counties. Iowa City: Institute of Public Affairs, University of Iowa.

Part III

SUBECONOMIES IN THE URBAN SETTING

Part III

SUBECONOMIES IN THE URBAN SETTING

THE ECONOMY OF THE POOR IN THE URBAN SETTING

HOW THE "SYSTEM" AFFECTS THE POOR

Introduction

This volume is about governmental interventions in various aspects of urban society. Our premise is that in order to talk sensibly about, for example, government housing policy, it is necessary to talk about the urban housing and property markets and to understand how they operate in the absence of a proposed government intervention. Once the underlying structure and characteristics of the subeconomy under investigation are understood, it becomes possible to anticipate the effects of proposed government policy. Implicit in the structure and content of this volume is the notion that government, particularly city government, does not control many of the levers on the mechanisms that produce outcomes in urban areas, that it is useless and sometimes harmful for government to push levers without knowing something about the mechanisms that intervene between government policy and environmental outcomes.

This section is offered in the spirit of the above. We have chosen to concentrate on the urban poor as an object of public policy. There are many conditions of the poor that society would like to change. We are becoming increasingly aware that public policy in this area has not always been productive and, indeed, has often been counterproductive. Consider, for example, the increasing suspicion of proposals to put highways through urban ghettos because property is relatively cheap there and because the proposal would also eliminate blight; it also destroys neighborhood societies and decreases the stock of (desired) low-cost housing. Or the distrust of an aid-to-dependent-children welfare policy that tends to break up Black families (see Moynihan, 1965).

Because a major portion of the urban poor are Black and because of the pattern of racial segregation in housing, it is important to understand the economic structure of the largely Black ghetto found in nearly every large city in the United States. In the context of the larger urban economy described by Thompson (this volume), Fusfeld examines the characteristics of the ghetto economy. He focuses on the linkages between the ghetto and its outside world and concludes that much of the activity of the ghetto can be understood in terms of a residual subsystem; individuals and institutions that cannot survive elsewhere in urban society gravitate to the ghetto. The picture he paints is not a pleasant one, with the more successful individuals and institutions escaping to other segments of society at the earliest feasible moment, leaving behind those implicitly or explicitly rejected by the larger urban system.

Pascal examines the phenomena of residential segregation and explores in considerable detail a model of how residential segregation comes about. He concludes that a self-selection process, where each racial (or income) grouping has a positive preference for its own kind as well as a possible distaste for other groupings, seems consistent with the available data on segregation. The implications for public policy are important. Open-housing legislation alone, under a self-selection process, is unlikely to have much of an impact on overall residential segregation, for example.

Muth, in the context of the citywide housing market, examines some of the problems faced by government housing policy. He argues that lower transport costs and population increases have created much of the urban sprawl characteristic of large cities today. He also argues that the concurrent increase in slums is due to an increased demand for low-cost housing. The cheapest way (currently) to supply this demand is through conversion of older units, perhaps vacated by new suburbanites; increased densities in existing low-cost units; and reduction of maintenance and repair expenses. Muth makes a strong case for the notion that much of what is thought to be discrimination in housing is actually an income effect or the result of the cumulative effects of discrimination in all sectors of society, which in turn influences income distributions among racial groups. He argues for attacking the demand for low-cost housing through a system of income subsidies as a means for dealing with many of the urban housing problems.

In examining how the overall state and local tax-expenditure structures affect the poor, Netzer points out that although state-local government expenditures decidedly favor the poor, state-local tax systems work in the opposite direction in powerful ways. The property tax is the most regressive (relatively disadvantageous to the poor) state-local tax, but, in practice, overall regressivity is usually partially offset by the characteristics of remaining portions of the tax structure. Sales taxes without credits or food and housing exemptions are also disadvantageous for the poor. Flat-rate, no-exemption income taxes do not help the poor either. Netzer concludes his article with a plea for increased emphasis on appropriately designed state-level, rather than locally raised, taxation.

Many proposals have been put forth for expensive mass transit systems to get poor, untrained central-city workers to jobs, thought to be increasingly more available in the suburbs than central city. Noll examines the relationships between population and employment locational decisions and the employment characteristics of the urban poor. Like Fusfeld, Noll sees the poor moving out of the central city (ghetto) as soon as they become successful. As employed people move out, a major source of information about employment opportunities for the ghetto unemployed moves out as well. The result is a relative and absolute decline in central-city growth. As skilled labor (successful employees) moves out, partly because they desire and can afford to consume more land, the long-run effects are shifts of higher-skilled employees to suburbs and lower-skilled to the central city. Noll argues that the suburbanization of employment opportunities (commercial and industrial establishments moving out of the central city) is as much a response to this shift in labor location as it is to other factors.

In examining how the "system" affects the poor, we could also have included the Aberbach and Walker paper found in Part IV, in particular, their discussion on how Blacks see city services as affecting them.

—J. P. C.

REFERENCE

MOYNIHAN, D. P. (1965) The Negro Family: The Case for National Action. Washington, D.C.: Department of Labor.

THE ECONOMY
OF THE POOR
IN THE URBAN SETTING

13

The Economy of the Urban Ghetto

DANIEL R. FUSFELD

☐ THE GHETTO ECONOMY is in many ways a world apart. It differs markedly from that of the rest of the country in many ways. Perhaps its most important distinguishing characteristic is its backwardness, its lack of the dynamic, progressive changes that bring advancement to the rest of the economy. At the same time, there are points of contact through which the ghetto is influenced by economic activity in the rest of the nation.

SOURCES OF INCOME

The income of the urban poverty area comes from four major sources, only one of which represents a viable and continuing link with the forces of progress.

(1) Some residents work in the high-wage, progressive sectors of the economy. In Detroit, for example, jobs in the automobile industry are held by a racially integrated work force, a number of whose members live in the urban ghettos. The industry is highly capital-intensive and oligopolistic, and has a strong union; labor productivity is high and wages correspond. Detroit, however, is something of an exception. Most large cities have a smaller portion of their central city work force in such industries.

(2) The economic base of the urban poverty area is the more backward sector of the economy, characterized by low wages, relatively wide cyclical variations, and exposure to all the debilitating forces of competition. A large portion of the workers who are employed full time in these industries earn wages around or below the poverty level.

(3) This low-wage economy is supplemented by an irregular economy, partly legal and partly illegal, that provides further income for the residents of urban poverty areas, largely through provision of services to other residents.

(4) Finally, income supplements from outside the urban poverty areas, some public and some private, provide the transfer payments without which the population could not survive. Welfare payments are probably the largest and certainly the most controversial of these transfers.

Although these four aspects of the economy of urban poverty areas can be relatively easily identified, the studies that would document their extent and significance have not been made. We do not know, for example, the proportion of the residents' income generated by these four sources. Nor do we know which are increasing and which are decreasing. Nor do we have much notion of how the four sectors have changed in recent years. Nevertheless, it is possible to look more closely at several of them.

THE LOW-WAGE INDUSTRIES AND THE WORKING POOR

A large number of ghetto residents work in low-wage industries (Bluestone, 1968a: 1-13; Sheppard, 1969: 2-11). The jobs may be in manufacturing, service industries, or retail and wholesale trade. Their common characteristic is that many full-time employees who work steadily in these industries earn less than a poverty-level income. Table 1 gives some examples of industries in which a large portion of all employees falls in the category of the working poor.

The low-wage industries are not highly visible, and we do not look to them for examples of progress. Typically, the individual enterprise is small, requiring relatively little capital investment. The technology is labor-intensive. Both labor productivity and profits are low. Oligopolistic market control is generally absent, which means that employers are unprotected from the rigors of competition. Unions are largely absent, so that workers are subject to the wage-squeeze characteristic of labor-intensive, highly competitive industries. Sales in these industries are generally quite sensitive to prices, which means that even if competition were reduced, firms would have little chance of improving their revenues through price increases. Since profits are low, little is done in research or product development, which reinforces the technological backwardness of these industries.

The low-wage industries may well be subject to greater fluctuations than other sectors of the economy when aggregate demand falls. We know that labor turnover rates are high, which means that the incidence of unemployment in the labor force is also high. We also know that the bulk of the work force is relatively poorly educated and has relatively low levels of work skills. A large proportion of the jobs are dead ends; there are not many higher paying jobs which a worker can qualify for by his daily work on the job. There are relatively few ladders to better positions. Finally, workers who start out in the low-wage sector of the economy apparently tend to stay there rather than move into similar jobs in the high-wage sector (although more studies of this phenomenon are needed to determine how strong a tendency it is).

TABLE 1

SELECTED LOW-WAGE INDUSTRIES EMPLOYING SUBSTANTIAL NUMBERS OF THE URBAN POOR

Industry	Year	Total Employment	Average Hourly Earnings	% Earning Less Than $1.60/hr.
Nursing Homes and Related Facilities	1965	172,637	1.19	86.3
Laundries and Cleaning Services	1966	397,715	1.44	72.5
Hospitals (excl. Federal)	1966	1,781,300	1.86	41.2
Work Clothing	1964	57,669	1.43	72.8
Men's and Boys' Shirts	1964	96,935	1.45	70.4
Candy and Other Confectionery	1965	49,736	1.87	34.2
Limited-Price Variety Stores	1965	277,100	1.31	87.9
Eating and Drinking Places	1963	1,286,708	1.14	79.4
Hotels and Motels	1963	416,289	1.17	76.1
Department Stores	1965	1,019,300	1.75	59.6
Miscellaneous Retail Stores	1965	968,200	1.75	58.0
Retail Food Stores	1965	1,366,800	1.91	47.6

The low-wage industries are part of the unprotected economy. Workers are not unionized for the most part. Employers are not protected by the oligopolistic industrial structure that shields such firms as General Motors Corporation and the other industrial giants. Most protective legislation, such as minimum-wage laws, is only now being extended to low-wage jobs, and many are still not covered. Many federal programs that might have provided greater protection, such as the loan operations of the Small Business Administration, are only now beginning to provide services. The state employment services have left the area largely unserved. These industries are fully exposed to the "satanic mills" of supply and demand, which grind both workers and business firms exceedingly fine. The higher-income sectors of society have been protected while those needing it the most continue to be exposed.

THE IRREGULAR ECONOMY

The urban poverty area supports an occupational structure and service economy that is quite unconventional and partly illegitimate. The need for it arises from the inability of residents to pay for the usual organized and commercially provided services used by higher income areas, and by the lack of the business enterprises that normally provide them. To compensate for this lack, the ghetto economy has developed an irregular economy that involves: (a) informal work patterns that are often invisible to outside observers, (b) a network of occupational skills unique to ghetto life but which have little significance for jobs outside the ghetto, and (c) acquisition of skills and competences by workers in nontraditional ways, making their use in the larger society difficulty if not impossible.[1]

Louis Ferman has identified several occupational types in the irregular economy:

(1) The artist. Entertainers, humorists, painters and craftsmen.

(2) The hustler. The supersalesman who often operates on both sides of the law; for example, the casket salesman who retrieves coffins from the local cemetery, refurbishes them, and offers them for sale.

(3) The fixer. The individual who receives cash income in exchange for information. Sometimes the information con-

cerns the availability of stolen merchandise, sometimes job opportunities, sometimes the details of the welfare system.

(4) The product developer. Products such as rum-raisin ice cream, sweet potato pie and barbecued spareribs enjoy a large sale in some ghettos. They are also produced there by ghetto residents.

Some of these irregular occupations are practiced full time, some part time, some almost as hobbies. They all fill needs not served through regular economic channels. Most are not illegal, although there is a substantial amount of illegal activity as well, carried out by fences, thieves, bookies, narcotics pushers, pimps and prostitutes.

The irregular economy's work has certain advantages over work in the regular economy. The worker is not accountable to any authority for his earnings, no records are kept, and taxes can be avoided. The work is individualistic in nature, and can give the worker a sense of competence and control over his existence that a regular job may not provide. Entrepreneurship and risk give the activity some of the aspects of a game, yet the risks are usually not high. Finally, people who work either part or full time in the regular economy can supplement their incomes in the irregular economy, and vice versa.

The irregular economy has one major disadvantage, however. It encourages patterns of behavior and attitudes toward work which makes it difficult for the worker accustomed to the irregularity, lax work standards, and high rates of labor turnover in the irregular economy to move easily into jobs in the regular economy, where work rules are more rigid, lost time and absenteeism is not tolerated to the same extent, and supervision is more rigorous. In some respects the work habits of the irregular economy are similar to those of the preindustrial work force that economic historians are familiar with, and some of the same difficulties are found in adapting workers in the irregular economy to jobs in the mainstream of a modern industrial society.

One point more: the irregular economy enables people to develop production skills, entrepreneurship, and sales ability that could be developed and put to use in more systematic ways for the economic development of the area. Its very existence indicates an unfortunate waste of ability and intelligence. Although the specific skills of the irregular economy may not be highly applicable in the regular economy, they indicate the presence of a high degree of initiative and entrepreneurship.

THE WELFARE SYSTEM

When David Caplovitz (1963) surveyed a sample of residents of four low-income housing projects in 1960 in New York City for his book, he found that twenty-eight percent of the sample received incomes wholly or partly from welfare or pensions, and only seventy-two percent obtained their income exclusively from earnings. The proportion would certainly be higher today. In 1968, almost one million persons in New York City were receiving public assistance payments alone, or about twelve percent of the population of the city as a whole. In 1960 the number receiving public assistance (not pensions) was around 375,000.

A November 1966 survey of slum areas in major U.S. cities showed that forty-seven percent of the families in the survey received transfer payments, including unemployment insurance (5.1 percent), welfare or aid for dependent children (18.1 percent), and other nonemployment income (24.6 percent). In the slums of Los Angeles, thirty percent of the population received welfare payments, and in Cleveland, twenty-one percent of the families received aid for dependent children (AFDC) payments. In all of the areas surveyed a very large proportion of the city's welfare cases were slum residents.

A substantial number of people received public assistance through the present federal-state system. In January 1968 the total was 9,057,000, divided as follows:

Persons in families receiving aid for dependent children	5,436,000
Old-age assistance	2,070,000
Aid to the blind	83,000
Aid to the disabled	651,000
General assistance	817,000
Total	9,057,000

The figure for 1968 was twenty-three percent greater than the 1961 total, in spite of economic growth and reduced unemployment over the intervening years.

Although the number of recipients is large and the amount disbursed is now running at almost $5.7 billions annually, there is very little good information on the economic impact of the welfare economy. We do not know its effect on the mobility of labor, its effect on the tax base and financial structure of cities, its relationship to the rest of the ghetto economy, its impact on the structure of the community, its effect on property ownership and real estate values, or its relationship to the incidence and nature of crime. Very few aspects of the welfare economy have been studied, including its place in the economy of the urban ghetto.

In one respect, the welfare economy is notorious. Administration of the welfare system has been blamed for a serious deterioration in the family structure. By withholding payments from families with an able-bodied male worker, the system encourages the unemployed to desert his family in order for them to receive assistance. The welfare system has undoubtedly had that effect, but other aspects of the ghetto economy work in the same direction, including low wages, erratic employment, and general conditions of poverty. Family stability is discouraged by the difficulties that poverty brings. Desertion has always been the poor man's divorce (Clark, 1967; U.S. Department of Labor, 1965).

Another major question about the welfare economy concerns its effect on the supply of labor. It has been argued that many ghetto residents would rather be on welfare than keep a job. There may be something to this argument, but the extent of this effect is undetermined. It is true that welfare payments come regularly twice a month, while employment may be irregular. In such cases, welfare is less risky than employment. It is also true that many low-wage jobs open to persons on welfare are boring, dead-end jobs that don't provide adequate income anyway. Why bother? The result appears to be that welfare payments reduce labor-force participation by a significant amount.[2]

The welfare economy often does discourage people on welfare from working. Any earnings are usually deducted in full from the welfare payment. Since working also involves expenses (travel, clothes, and so forth), the welfare recipient who works often ends up worse off than if he didn't work. Some efforts are now being made to allow for work expenses, but they have not made much headway.

A family on welfare can avoid these effects only by breaking the law. Many do that, without avoiding family breakup and discouragement in the labor market. The result is a further disintegration of social controls and decreased respect for the institutions of organized society. Efforts to beat the system are generated by the system itself.

The welfare economy has a perverse effect on the ghetto economy. It encourages family disintegration, discourages work, and promotes a breakdown of social controls. These long-run effects may well be more costly than the short-run gains obtained by seeing that needy people receive the nominally necessary food, clothing, and shelter.

BUSINESS ENTERPRISE IN THE
URBAN GHETTO

The retail business districts of central-city ghettos have an outward appearance much like any retail district, with the exception that they are usually a little shabbier and obviously cater primarily to a low-income clientele. There are important differences that are not obvious to the casual observer, however. Business practices are exploitive, and the customers often have few, if any, alternative sources of supply. In addition, the racial composition of the ghetto has tended to create a partially segregated, Black retail market.[3]

The underworld of shady and often unethical business enterprise which flourishes in the urban ghetto has been studied and described in depth by David Caplovitz (1963), whose book illustrates the extent to which a distinct business way of life flourishes in the relatively isolated markets of central-city poverty areas.[4]

Retail firms, particularly those selling consumer durables, flourish through provision of easy credit, selling high-priced and often shoddy merchandise by use of high pressure tactics, and personal methods of attracting customers (peddling) and getting payment. This system of business enterprise is highly exploitive; high prices, high cost of credit, and shabby merchandise effectively strip the poor of whatever assets they may have and ensure their continuation in poverty. These practices are often supplemented by use of the law courts to get payment; repossession and garnishment of wages are devices used by furniture, appliance, and jewelry stores in particular.

Business practices of the ghetto perform an important function in the economy, however; they permit the poor to buy major durable goods that normal channels of trade do not provide. The poor pay more, but they obtain goods they would otherwise be unable to buy.

Caplovitz points out that one reason for the high prices and credit costs of ghetto retailing is the higher risks associated with selling to the poor. These risks cause businessmen in other retail areas to turn down such customers. Ghetto business establishments respond to this gap in the market by devising the business practices that enable them to sell to the poor and still make a profit.

A second feature of ghetto retailing is the fact that a very large portion of the customers are Black. This has led to a racially segregated market in some areas of retailing which has enabled Black business enterprise to develop.

The segregated market exists in part because of geographical considerations. Some types of retail stores stick close to a neighborhood simply because people don't want to travel far to buy food, personal care products, and similar items. At the same time, business capital requirements and the size of the market make possible dispersal of many small retail units of neighborhood size.

A second reason for the segregated market is the historical unwillingness of enterprises catering to whites to serve Blacks. Until World War II this included many substantial retail outlets such as department and furniture stores and, until very recently, restaurants, hotels, and similar establishments.

For example, in a segregated city like Washington, D.C., before the 1950s, Negroes who shopped in the downtown retail district could expect rude treatment designed to discourage their patronage, while inability to use eating places (and rest rooms) made shopping even more difficult (Green, 1967: 3, 298).

Closely related pressures from the white community restricted the practice of most Black lawyers, doctors, dentists, accountants, and other professionals to the urban ghetto and Black clients.

Finally, certain types of personal services such as hair cutting and cosmetics have become Black business specialties because of the special needs and fashions of Black customers.

Black-owned business enterprise reflects the quasi-segregated nature of the ghetto market. Firms are centered in personal services and retail trade, with a scattering in other services and construction. Although very little precise information is available, the

Small Business Administration has recently estimated that some 158,000 business enterprises were owned in 1969 by nonwhites, out of a national total of some 5,420,000, or 2.9 percent. The great bulk of the nonwhite owners were, of course, Black people. The nonwhite-owned businesses were distributed among industries as shown in Table 2.

TABLE 2

NONWHITE-OWNED BUSINESS ENTERPRISES, 1969

Industry	Number of Firms (000)	Percent of Nonwhite Business Enterprises
Retail Trade	71.0	45.0
Personal Services	33.0	20.9
Other Services	20.7	13.1
Construction	8.1	5.1
Wholesale Trade	5.8	3.7
Manufacturing	2.6	1.6
Other	16.8	10.6
Total	158.0	100.0

SOURCE: U.S. Small Business Administration (1969) Quarterly Economic Digest 2 (Winter): 26.

Most Black-owned business enterprises are located in central-city areas, and especially in urban poverty areas, in contrast to white-owned businesses, which are not located in slum areas to any significant extent. The contrast in location is shown in Table 3.

Although Black-owned business enterprises are heavily concentrated in central-city areas and in slums, they do not comprise a majority of business enterprises, even in the urban ghettos. Even in New York's Harlem, a study of street-level businesses showed that only some fifty-three percent were Black-owned. Studies of similar areas in other cities show nonwhite ownership varying from twenty to forty percent. The Small Business Administration (1969: 26-27) estimates that in central-city ghettos in the nation as a whole, nonwhites own only twenty-six percent of business enterprises.

Black-owned enterprises are small operations, often run by an individual or a small family, and have few employees. In most size dimensions, they are smaller than white-owned enterprises. Some relevant comparisons are shown in Table 4.

Although very heavily concentrated in small retail trade, food services, and personal services, Black business enterprise has also developed a rudimentary network of manufacturing and wholesale trade to back up the Black retail and service enterprises. For example, Black-owned enterprises manufacture cosmetics for Black-owned beauty parlors, and some of the distribution is in the hands of Black-owned wholesale enterprises. Since the Negro markets are relatively small, however, the supporting production and distribution network is also relatively small.

Finally, there is an infrastructure of Black-owned newspapers, banks, and insurance companies that is also relatively small, although highly significant for Black business and the Black community. Thus, in 1967 there were seventeen Black-owned banks (out of a national total of some 12,000) with total assets of about $147 millions (0.039 percent of the national total). These banks were small, weak, and not highly profitable, although their growth record has been comparatively good in recent years (Brimmer, 1968). Life insurance companies owned by Blacks have a similar record: twenty companies in 1962, which held almost all the assets of some fifty Black-owned insurance companies, had 0.75 percent of industry sales while holding only about 0.25 percent of industry assets. This reflects both the growing demand for insurance by Blacks and the tendency of smaller insurance companies to grow faster than the industry average. Nevertheless, their proportion of the business is very small, they tend to follow conservative investment policies, and have lagged in entering some of the newer areas of the industry (Brimmer, 1966: 309-322).

The segregated nature of the market for the products and services of Black business enterprise, together with its small size and concentration in retail and service industry, create some significant problems of economic development. Two conflicting trends can be observed.

The long-term trend until the mid-1960s was one of relative decline. Black-owned business enterprise appeared as a decreasingly significant factor in the national economy, even within urban ghetto areas. Several developments lay behind this trend:

(1) Small business enterprise in general, while growing in numbers, has been declining in relative importance. This is particularly true of retail trade, where large chain-store distribution has become the dominant pattern.

TABLE 3

LOCATION OF BLACK-OWNED AND WHITE-OWNED
BUSINESS ENTERPRISES, 1969
(percent)

Location	Black-Owned [a]		White-Owned	
Urban, 50,000 and over	71.9		45.6	
Slum		33.3		3.4
Other Central City		35.1		30.3
Suburban		3.5		11.9
Urban, under 50,000 and rural	28.1		54.4	
	100.0		100.0	

SOURCE: U.S. Small Business Administration (1969) Fact Sheet (May 19, mimeo.).

[a] Does not include firms owned by other nonwhite minorities.

TABLE 4

SIZE COMPARISONS, WHITE-OWNED AND
MINORITY-OWNED BUSINESS ENTERPRISES

Number of Paid Employees	Minority-Owned (percent)	Other (percent)
0	32.5	26.0
1-9	61.0	55.0
10-49	5.0	13.2
50-99	1.2	2.7
100 and over	––	3.1
Gross Receipts ($000)		
0-9.9	33.4	18.7
10-19.9	14.8	12.5
20-49.9	18.5	19.4
50-99.9	11.1	14.7
100-999.9	19.8	25.8
1000-4999.9	1.2	6.2
5000 and over	1.2	2.7

SOURCE: U.S. Small Business Administration (1969) Fact Sheet (May 19, mimeo.).

(2) Urban renewal and low-cost housing projects have taken a heavy toll of the small enterprises located in urban ghetto areas.

(3) As Black incomes have risen, a growing proportion of Black spending has gone into big ticket items such as automobiles, household appliances, furniture, and related purchases. These sectors of the retail market are not segregated, and are served almost exclusively by white-owned enterprises.

(4) As the urban Black population has increased, and the size of the Negro market has grown, it has been penetrated by white-owned enterprise taking advantage of larger opportunities for profits.

The situation is indeed paradoxical. A once highly segregated Negro market grew in both numbers of customers and in average income. The result was not a corresponding growth of the Black business enterprise that grew up to serve this market, but a penetration of white enterprise. At the same time, the segregation of the Black market tended to break down as Blacks began to use the ordinary white channels of trade to a greater extent.

The second trend has been more recent. Following the uprisings of 1966 and 1967, white-owned business enterprises have tended to move out of ghetto areas or, if destroyed, to remain closed. This has opened the way for expansion of Black-owned enterprise to replace them. How extensive this shift has been is a matter of conjecture, for hard data is lacking and the evidence is largely that of casual observation. The source of the shift has been dual. On one hand, whites have moved out of ghetto businesses in large part as a result of uncertainty and fear. On the other hand, a growing awareness of Black business by ghetto residents themselves—both a cause and a result of the uprisings—has stimulated Black buying from Black businessmen.[5] This shift in consumer and community attitudes in the urban ghettos stimulated the movement to promote Black business enterprise, community economic development, and the expansion of Black banks and other financial institutions.[6]

It is too early to tell whether these recent developments will have a significant impact on the ghetto economy. The fact remains that most ghetto residents spend their incomes with business enterprises whose capital comes from outside the ghetto, whose ownership is outside the ghetto, whose sales and administrative personnel live outside the ghetto, and which sell products produced outside the ghetto. The flow of purchases, wage payments, profits, and other expenditures sets up a flow from the ghetto to the larger economy outside the ghetto. By contrast, the circular flow of spending within the ghetto, which runs from customer to enterprise to employment and payments back to customer, is meager. It is limited to a narrow and fragile segment of small retail and service enterprises which, until very recently, was declining in importance rather than growing.

THE DRAIN OF RESOURCES

One of the most striking characteristics of the urban poverty area is the continual drain of resources out of the area and into other sectors of the economy. Although largely unmeasured, and perhaps unmeasurable, the drain includes savings, physical capital, human resources, and incomes. As a result, urban poverty areas are left without the most important resources needed for development and improvement, and the economic infrastructure is seriously deficient.

The drain of resources can be seen most clearly in the process of transition as an area becomes part of the spreading urban ghetto. As migration and population growth spread the boundaries of the ghetto into neighboring parts of the city over the last two decades, middle-class whites moved out. With them went most of the professional personnel who provide personal and business services. Doctors, dentists, lawyers, accountants, insurance agents, and related professionals left and were not replaced.

Other human resources leave by way of the educational system and the high-wage economy. Drawn by opportunities outside the urban poverty area, many of the most intelligent, capable, and imaginative young people move into the progressive sectors where rewards are greater and opportunities wider. This drain of human resources leaves the economy of the ghetto—whose chief resource is manpower to begin with—without many of its best products.

The drain of capital is equally striking. A substantial portion of the savings of the urban ghetto goes into financial institutions such as banks and savings banks whose investment policies draw the funds out of the area and into business loans, mortgages, and other investments elsewhere. Little comes back to support the ghetto economy or promote its development.[7] Even though the owner-ship of the original savings or thrift accounts remains with ghetto residents, the funds are generally used elsewhere.

Probably the largest flow of capital out of the urban poverty area, however, takes place in housing. Failure to maintain housing facilities enables the owner to withdraw his capital while at the same time maintaining his income. Ultimately the property will be worthless simply because of wear and tear, but while it is being used up, the owner has been getting his capital back and has been deriving a nice current income. Housing authorities in most major cities are quite aware of this process, and have found no way to stop it. Its basic causes lie in overcrowding and very high rates of deterioration through overuse, together with the failure of most cities to develop effective methods of preventing neighborhood decay (Schorr, 1963; Frieden, 1967: 148-191).

Two aspects of the drain of capital out of housing should be noted. If one or two property owners take their capital out by refusing to replace depreciation, all of the surrounding owners have incentives to do likewise. One deteriorated building draws down the value of surrounding property. One house broken up into small apartments and crowded with numerous families makes it difficult to sell or rent to single families next door. These neighborhood effects cause the drain of capital to accumulate and accelerate once they begin, and are almost impossible to stop.

In addition, families owning their own homes may find them-selves locked into the ghetto because of income, age, or race. Their investment in property either deteriorates or its value rises much more slowly than that of families outside the ghetto. A white, suburban family discovers that economic growth creates a windfall gain in the form of rising property values. The effect on the ghetto family may be just the opposite; its house, located in a deterio-rating urban ghetto, may well decline in value as the neighborhood deteriorates. At the very least, the ghetto resident discovers that over his lifetime the windfall gains from growth in property values is considerably below that of his white suburban counterpart.

Capital also flows out of urban poverty areas through public facilities. Local governments have allowed their capital investments to fall by not replacing depreciation of buildings and other invest-ments. Schools and libraries have been allowed to deteriorate, parks to run down, streets and curbs to go unrepaired, fire and police stations and medical facilities to deteriorate. In part, this process has been due to the added strains that a denser population has placed on public facilities. In part it has been the result of the financial problems that have prevented cities from increasing their expenditures for public services to meet expanding needs. In part it has been due to the traditional tendency of city governments to maintain facilities in the middle- and upper-income areas and to put the priorities of the slums last. But whatever the reasons, the process represents another drain of capital out of the urban poverty areas.

The transition to an urban ghetto also features the loss of many organized institutions. Hospitals have moved out. So have churches and other organized community groups. One of the striking features of today's urban poverty area is a deficiency of those groups and associations that have traditionally provided a community with stability and order and with a sense of continuity and participation. This gap in social needs has tended to be replaced by the development of those informal community groups present in every neighborhood. In the urban ghetto, however, much of this informal organization has been outside the law—juvenile gangs, for example. The lack of community groups, together with the strains inherent in poverty and the breakdown of family structures brought on by the system of welfare payments, contribute to and intensify the isolation and anomie inherent in urban life. The social and psychological problems of the urban ghetto add a further dimension to the urban and racial crisis, but it is important to realize that they are related to the fundamental economic problems of the area.

The net result of the drain of resources from the ghettoized slum is to create a backward, underdeveloped area in the midst of an otherwise progressive and expanding economy. Capital moves out, human resources move out, community structure is weakened, public services are inadequate, and the professional skills needed for improvement are largely lacking. The economic infrastructure required for development is primitive. The major resource of the area is its manpower, which is used by the rest of the economy as a pool of low-wage and relatively unskilled labor and, as the Vietnam war has shown, by the armed forces for front-line infantry service. Although the charge of colonial exploitation may be an exaggeration, there are some striking similarities to the old-fashioned colonial economy.

THE DRAIN OF INCOME

Income flows out of the urban poverty area in much the same way as do capital and other resources. Earnings of residents are spent in retail markets through which purchasing power flows to the outside economy, where its multiplied growth-stimulating effects are dispersed without affecting the ghetto significantly. Products sold in the ghetto are imported from outside. The stores, particularly those selling higher-priced items, are owned by people who live outside the ghetto. Sales personnel and other employees live outside the ghetto, for the most part. Business services are obtained from banks and other firms outside the ghetto. A steady stream of payments—wholesale purchases, profits, wages, and so forth—moves outward from the ghetto residents to others. Instead of nurturing expansion of incomes and greater economic welfare within the ghetto, the money spent there promotes growth elsewhere. The ghetto is drained of the expansion-stimulating effect of its own purchasing power.

No community is self-sufficient. The goods purchased in any community are imported, except for a very small proportion of local products. In this respect the urban poverty area is like any other. But in other communities a significant portion of the retail and wholesale trades is owned locally and most of the employees are local. The profits and wages earned by those people are spent locally, and serve to help support the local community. A chain of spending and respending is set up which adds strength and variety to the local economy.

These internal income flows are of growing importance in contemporary urban areas. Cities used to be noted for their export industries, such as steel in Pittsburgh, automobiles in Detroit, meat products in Chicago, and so on. Those industries formed the economic base around which retail and service industries developed. Although the economic base industries are important, students of urban economics are coming increasingly to recognize that a very large portion of the economy of any metropolitan area is self-sustaining. Each sector of the city's economy strengthens and give support to the other sectors by means of the income flows that each generates. Cities still specialize in certain types of export products, but they are becoming increasingly general in their economic activities (Thompson, 1968, 1965).

The urban poverty area lacks the highly developed internal income flows that might lead to a viable economic pattern. Aside from the irregular economy and relatively small enterprises requiring little capital, business enterprises are not owned by ghetto residents. Labor is the chief export, and the incomes earned are spent in chain supermarkets, outsider-owned furniture and appliance stores, and other enterprises with ownership, management, and many employees, almost always of nonghetto origin.[8] The profits and wages received by the outsiders do not come back into the ghetto to support other enterprises or employees. They flow, instead, into the economy of the progressive sector located elsewhere.

These patterns are exaggerated in urban poverty areas by the low incomes which prevail. Compared with the rest of the economy, relatively small amounts of ghetto incomes are spent on services. With a larger than usual amount spent on goods, the income drain which purchases of goods create in any community is proportionately larger for urban poverty areas than for others.

Low incomes also mean that housing costs are a larger proportion of family budgets than elsewhere. For many ghetto families the cost of housing ranges upward to 35 percent of family incomes. This means that a significant portion of ghetto income is transformed into withdrawals of capital by owners of rental property.

These patterns of income flow help to explain why the welfare economy is needed to stabilize the ghetto economy. Flows of income in the private sector are generally outward, requiring a compensating inward flow via the public economy. The outward flow of income also helps to explain why increased welfare payments may help the individuals or families who receive them, but have little or no impact on the ghetto economy as a whole. The bulk of the increased payments leaks out rapidly.

A word of caution about the drain of income is in order, however. Aside from general descriptive accounts, mostly of a nonscholarly nature, there are few firm data on which to estimate the income flows out of the ghettos. It is equally difficult to determine the extent to which urban poverty areas differ from other urban areas in this respect. At this stage the best we can do is sketch a noticeable, qualitative difference without being able to quantify it.

PERMANENT DEPRESSION

One of the most striking economic characteristics of urban poverty areas is the condition of permanent depression that prevails. Even though the rest of the economy may be prosperous, even booming, and may feel the inflationary pressures of rising aggregate demand, unemployment in the urban ghettos will remain high. Ghetto unemployment rates stay at levels that would signal a serious depression if they prevailed in the economy as a whole.

Documentation of the permanent depression of urban poverty areas has come only recently, although the condition has been well known to observers for a long time. In 1968 the Department of Labor reported on a study of employment and unemployment conditions in America's urban slums. Based on the March 1966 Current Population Survey, the study showed an overall unemployment rate of about 7.5 percent in the poorest 25 percent of all census tracts in U.S. cities with over 250,000 population (selected on the basis of the 1960 Census). The United States as a whole has not seen a national unemployment rate that high since the 1930s. The highest it has been since the end of World War II was 6.8 percent (in 1958, calculated on an annual basis). The national unemployment rate in March 1966 was just half that of the slums, 3.8 percent.

The nonwhite unemployment rate was even higher. Forty-two percent of the residents of these poor census tracts were nonwhite. Their unemployment rate was 9.4 percent—about two and one-half times the national average and approaching the national rates of the more depressed years of the 1930s. Among nonwhite teen-agers (fourteen to nineteen years of age), the unemployment rates were thirty-one percent for boys and forty-six percent for girls, although the sample was quite small, and there is substantial possibility of error in these last figures.

This preliminary survey was followed by a more intensive study of urban slum areas in twelve major cities in November 1966. At a time when the national unemployment rate was 3.7 percent, the average unemployment rate in the slums was about ten percent, or almost three times the national average. Conditions varied from one city' slum to another, but all showed serious unemployment problems.

TABLE 5

UNEMPLOYMENT RATES IN U.S. URBAN POVERTY AREAS, 1966

Area	Unemployment as a Percent of the Labor Force
Boston (Roxbury)	6.9
Cleveland (Hough and surrounding neighborhood)	15.6
Detroit (central Woodward area)	10.1
Los Angeles (south Los Angeles)	12.0
New Orleans (several contiguous areas)	10.0
New York (Harlem)	8.1
(East Harlem)	9.0
(Bedford-Stuyvesant)	6.2
Oakland (Bayside)	13.0
Philadelphia (north Philadelphia)	11.0
Phoenix (Salt River Bed area)	13.2
St. Louis (north side)	12.9
San Antonio (east and west sides)	8.1
San Francisco (Mission-Fillmore area)	11.1

These unemployment figures include only those persons who were actively looking for work. They do not include those who should be in the labor force but are not because they feel, rightly or wrongly, that they can't find a job (or for other reasons). This nonparticipation rate in urban poverty areas was eleven percent among men in the twenty-to-sixty-four-year age group, as compared with a seven percent rate for men of that age in the economy as a whole.

The unemployment figures also exclude a substantial number of adult men that other statistical sources indicate should be part of the slum area population. The November 1966 survey failed to find between a fifth and a third of the adult men of the slum areas. This parallels the Census experience with the undercount problem.[9]

Finally, the unemployment figures do not include persons who are working part time but would like to have full-time jobs. The November 1966 survey showed that 6.9 percent of all employed persons in urban poverty areas fall in this category. The national figure was 2.3 percent.

These latter considerations indicate that the situation is worse than the figures on unemployment reveal. In particular, the contrast with the rest of the economy is sharp enough so that national unemployment rates become meaningless in describing conditions in the slums. This does not mean that when unemployment rates decline in the national economy the slums feel no impact. They do. The manpower resources of urban poverty areas are used by the larger economy, but they form a pool of workers, many of whom are unemployed or employed part time, which is always underutilized. Even when markets everywhere else are tight, and inflationary pressures in the progressive sectors of the economy are pushing up prices, wages, profits, and interest rates, unemployment in the urban slums is pulled down only to the levels that in the economy as a whole are characteristic of serious recessions. Charles Dickens' famous phrase, "It was the best of times, it was the worst of times," applies to this situation with something of an ironic twist; when the first part applies to the rest of the economy, the second part would be a good description of the urban poverty area.

Even the presence of inflationary pressures in the economy as a whole does not bring prosperity to the ghetto. In early 1969 the new Urban Unemployment Survey of the U.S. Department of Labor showed unemployment rates two and one-half times the national average in six slum areas, teen-age unemployment rates of thirty percent, and earnings of less than $65 per week for one-sixth of all employed workers living in slums.

CIRCULAR CAUSATION AND CUMULATIVE EFFECTS

The condition of the ghetto economy is a classic example of circular causation in social processes (Myrdal, 1968: Volume III, Appendix 2). The ghetto economy perpetuates its own poverty. Low incomes mean low levels of living. This style of life has obvious deficiencies; poor food, bad housing, poor health, and bad sanitation. These conditions lead back to low labor productivity and a perpetuation of low incomes.

The drain of resources out of urban poverty areas—manpower, capital, income—serves to reinforce the poverty. Social overhead capital is inadequate. The public services that might overcome part of the deficiencies in private incomes are insufficient. In particular, deficiencies in the educational system lead to inadequate training, low skill levels, and low productivity.

Employment patterns, especially in the low-wage industries and the irregular economy, reinforce the pattern of poverty and create barriers to movement of workers into the high-wage sectors outside the urban poverty area. At the same time, those ghetto residents who do move up and out take with them much of the entrepreneurship that development of the ghetto economy requires.

Economic development is further retarded by ineffective instruments for local control; the destinies of urban poverty areas have been largely in the hands of outsiders. Weak political representation and control of local governments by an establishment power structure have kept the poor out of power. The result is a weak infrastructure of voluntary organizations and a low level of popular participation in the decision-making process. This, in turn, retards the development of decision-making and entrepreneural abilities. This dual lack of entrepreneurship and effective power means that decisions which affect the ghetto economy will be made largely by the outsiders who dominate the decision-making process.

One result has been that many policies and programs have hurt, rather than helped, the ghetto. Welfare payments have tended to weaken attachment to the labor market. Urban renewal has increased the overcrowding of housing rather than diminished it. Highway construction has had the same effect. Educational programs have been unable to prevent a serious deterioration of the schools. Even the benefits of low-cost housing have been relatively small compared with the incomes generated for nonghetto residents.

In this context, the racial attitudes of whites and the long heritage of Black repression take on key significance. Together they have kept the great majority of Blacks in the ghetto, unable to move out of the vicious circle of self-generating poverty that prevails there.

The pattern of the ghetto economy, then, presents a series of self-reinforcing influences:

(1) Poverty breeds a style of life which reinforces the conditions that lead to poverty.

(2) Resources that might lead to betterment and development are drained out.

(3) Lack of political power has brought public programs that are often harmful to the ghetto economy.

(4) White attitudes toward race have kept most of the ghetto residents from moving out.

A social system in which a pattern of circular causation operates will generally reach an equilibrium in which causative factors balance each other. It may be a moving equilibrium if growth processes operate. If outide forces impinge on the equilibrium, and if they set up secondary effects moving in the same direction, a self-sustaining process of growth can be established, particularly if the social system can move beyond the position, or threshold, from which the old equilibrium can no longer be reestablished. These basic propositions from the theory of growth and economic development embody the concept of cumulative effects; circular causation can lead to maintenance of the existing equilibrium, but it can also lead to cumulative movements toward either growth or retrogression.

Economic growth is particularly difficult for the ghetto economy. Its weak infrastructure, the lack of local initiative and entrepreneurship, and the shortage of capital make it difficult to get a growth process started. This is part of the self-reinforcing process by which poverty creates the conditions that preserve poverty. More important, the tendency for resources and income to drain out of the ghetto economy means that even if the forces of development were to appear, much of their strength would be dissipated before they had a significant impact on the ghetto itself. Any program (or programs) that seeks to improve the economy of urban poverty areas must reverse the drain of skilled manpower, capital, and income if a cumulative process of growth is to be established.

Rather than growth, a cumulative process of retrogression has been the fate of urban poverty areas over the last twenty years. The key outside influence was the migration of the 1950s and the ensuing population explosion. These demographic changes both expanded the size of problems and worsened the condition of a social and economic system already suffering from permanent depression and a self-reinforcing pattern of underdevelopment. Retrogressive forces were set in motion which worsened the poverty, speeded up the drain of resources, and further weakened the social and economic infrastructure. Only massive increases in transfer payments and large increases in public expenditure programs have been able to stem the tide and re-create (perhaps) a new equilibrium.

Conditions may still be getting worse rather than better. The incomes of Blacks appear to be rising more slowly than those of whites, creating a widening income gap between races (for example, see Batchelder, 1964: 525-548). There appears to be a tendency for the gap between the high and low wage industries to widen (Bluestone, 1968b). Recent studies of the interim and special censuses indicate that housing segregation is growing rather than decreasing (Farley and Taeuber, 1968: 953-956). As the second wave of the urban ghetto population explosion starts—the children who comprised the first wave in the early 1950s are now entering young adulthood—a second lost generation is about to appear in the cities. The current picture is one of continuation of ghetto conditions and the circular causation of poverty and deprivation.

THE URBAN GHETTO AS A
RESIDUAL SUBSYSTEM

Although it is a world apart, the urban ghetto also serves an essential need of our society as a repository for people for whom the larger social and economic system has little or no use.

Every social system rejects individuals who do not meet the standards established for membership in the various subsystems that make up the larger social order. For example, the requirements for acceptance into the northeastern suburbs of Detroit—one of the subsystems that make up the whole—include the income required to buy or rent an expensive home. Those families unable to meet that requirement are rejected, and find residences in less expensive Detroit suburbs—one of the other subsystems. Other family characteristics, such as race, religion, and national background have also been important for acceptance or rejection, and at one time were formalized into a point system operated by realtors to determine a family's eligibility for the wealthy northeastern suburbs.

Rejection mechanisms are many and varied. Test scores and grades are important for admission to college. Certain requirements must be met to obtain a civil service job. Formal or informal restrictions keep some country clubs and fraternal organizations free of Jews or Blacks. A middle-class pattern of behavior is necessary for acceptance in the typical suburban community.

Examples can be multiplied, but the principle is clear; each of the subsystems into which the social system divides itself accepts or rejects individuals according to its own formal or informal criteria. One of the most unfortunate aspects of our society is that race is one of the major criteria for acceptance or rejection.

Rejection from one subsystem does not mean rejection from all, in most cases. The individual or family unable to find a place in one will usually find another. For example, the scholar whose work is unacceptable for employment at Harvard or Yale takes a job at a university a little down the ladder of academic prestige, or at a small liberal arts college, a former teachers college in the southwest, a community college. There may be frustrations and disappointments, but a place is usually found. Most people are not rejected from the social system as a whole.

But some are. Dangerous or particularly bizarre (irrational) behavior takes some to mental hospitals. Crimes draw others to prisons. Others gravitate to the urban ghetto, often because of a syndrome of characteristics such as race, poor education, low skill, and lack of adaptability to the behavior patterns of the more successful sectors of the social order. Mental hospitals, jails and slums—these are the chief depositories our society has created for those who cannot fit, or be fitted into, the dominant way of life. They make up the residual subsystems into which society's rejects congregate.

An example will show how the mechanism works. The Black sharecropper, poor, almost uneducated, lacking in skills, who found employment in agriculture in the Mississippi Delta region in 1950 found that he was redundant there in 1955. He was rejected from the economic system in which he had formerly been able to subsist, even though he was at the bottom of the economic and social order. Cast out there, he made his way to the ghetto of a southern city or a northern city. He moved, by a process which we do not understand, to the place where other economic rejects also had moved. Partly by choice, and partly because there was no other place to go, he ended up in the urban ghetto.

People rejected from the social-economic system are not scattered at random. They do not stay in the suburbs, or in southern agriculture, or in the universities, if they are caught up in the rejection mechanisms. The residuals tend to collect or be collected at specific points in the system as a whole. The best analogue is physical refuse—trash dumps, auto graveyards, and so forth—

although cemeteries also qualify. This tendency for residuals to congregate in specific areas is characteristic of humans, too, and we have our jails, mental institutions, and slums.

Residual subsystems are usually separated from the functioning central sectors of the social order by barriers of various types. Mental hospitals and jails are walled or fenced, and entry is by a formal process of legal commitment. The urban ghetto is different. There are no physical barriers between the ghetto and the rest of society, and no formal methods by which individuals are committed to life in the ghetto. The barriers are economic and social rather than physical, and the selection process is informal.

The barriers between the urban ghetto and the rest of the economic and social system do not cut off the ghetto completely. A substantial number of individuals move out and up into other sectors of society. Others move down and into the ghetto. Individuals are always moving from one subsystem to another in our highly mobile society. The process is not that simple, however. The urban ghetto has three characteristics which make it different:

(1) Mobility outward for many people is not based on merit, or even income, which often acts as a proxy for merit. Race is an important criterion for movement out. It is a barrier for Blacks and Puerto Ricans that is not present for whites.

(2) The culture of the ghetto creates and fosters a way of life that makes it difficult for individuals to be accepted in other sectors of the economy and society. This includes such factors as work habits and attitudes toward work developed in the irregular economy, and patterns of behavior fostered by the welfare economy.

(3) The fact that the urban ghetto is a depository for people rejected from society influences the attitudes of the rest of society toward the ghetto. These attitudes are reflected in inadequate provision of public services, such as education and health, which tend to preserve ghettoization and reduce the upward mobility of ghetto residents.

These informal and unseen barriers show up in the economic statistics of inordinately high levels of unemployment, relatively low income levels, relatively low residential mobility, and many other distinctive features of the ghetto. More important, they tend to keep within the ghetto a large number of people whose native abilities and potential for development are largely wasted. This waste of human resources is particularly tragic in the case of young people whose opportunities are restricted and whose futures are bleak simply because of the environment into which they are born and in which they are raised.

The national economy is not a seamless web of relationships that make up a unified, articulated whole. Rather, the urban ghetto is a quasi-enclave. It is part of the national economy and is affected by what goes on in the national economy, but the connecting links are seriously deficient. Barriers exist which prevent, for example, an increase in gross national product from having the same effects in the ghetto that it has in the rest of the economy. Barriers prevent the rising living standards due to economic growth from having the same effects in the urban ghetto that they have elsewhere.

The quasi-enclave of the ghetto has developed its own characteristic system of relationships within which its inhabitants function. We have described some of the economic aspects of this social subsystem: the low-wage sector, the irregular economy, and the welfare economy. We have noted some of the results: poverty, permanent depression, and underdevelopment. We have also identified some of the relationships between the ghetto subsystem and the larger social system: the flow of transfer payments into the ghetto which serves to stabilize and helps to support it, and the outward flow of capital, income, and human skills. And we have emphasized the fact that the ghetto subsystem tends to preserve the conditions that lead to ghettoization, generates the attitudes that keep people ghettoized, and largely prevents a process of economic development from getting started. The ghetto subsystem, in short, tends to preserve itself in a relatively static position, and reproduce itself from generation to generation. It may also grow in size, as it has done in the last fifteen years.

This concept of the urban ghetto as a residual subsystem—a place where society maintains its outcasts and semioutcasts—is essential to the development of policy and programs. In this context the welfare system or other forms of income maintenance are not solutions, but ameliorative devices. Improved education and training programs may help individuals, but they will not transform the ghetto unless they operate on a scale large enough to enable more people to overcome the barriers to outward and upward mobility than are moving inward and downward to the ghetto. Better housing may improve living conditions but will not necessarily make an impact on ghettoization—our experience with public low-cost housing confirms that view. Better health services and other public facilities, while also needed and desirable in themselves, likewise fail to attack the basic causes of ghettoization.

This is not to say that concerted and large-scale programs to improve incomes, education and training, housing and health—and to provide jobs—are doomed to failure. If they are large enough and if they are sustained long enough, they can make an important contribution to a solution of the problem, but, in essence, they are addressed to symptoms rather than basic causes.

The causes of the urban and racial crisis are the rejection mechanisms of our society which turn people into residuals. And the crisis will be with us until those social processes are changed.

(1) The rejection mechanisms must be identified and their operation modified or ended. In particular, the relationship of Blacks to the larger opportunity system must be drastically transformed.

(2) Better feedback mechanisms must be developed to bring society's rejects back into the mainstream. The greatest potential lies in employment and job training, but they have never been used on the massive scale required.

It is in connection with programs to transform the rejection mechanisms and generate better feedbacks that programs for income maintenance, education, housing, health, and all the rest begin to make sense. Standing by themselves, or even used together, they can have little effect, because they do not change the fundamental workings of a social system that continually creates residuals. However, they can be useful when developed in conjunction with policies and programs directed toward significant changes in some of the more dysfunctional aspects of our economic and social system.

One promising development is the movement for community self-determination. It takes such varied forms as community control of schools and programs for community-oriented economic development. This movement is a realistic and practical reaction to the semi-isolation, permanent depression, and exploitation of the ghetto, on the part of people who realize that they themselves are the only ones who can move their little subsection of society toward a more desirable condition. It is an effort to reconstruct the relationships between their world and the larger world outside on more favorable terms. These new relationships are yet to be defined, but the community self-determination movement promises for the first time to give ghetto dwellers a degree of control over their own destiny. If it can be combined with a thoroughgoing effort to reconstruct those elements in our social and economic order which create and preserve ghetto conditions, some real progress may ultimately be made.

NOTES

1. This section is based on Ferman (n.d.). See also Piore (1968).

Descriptions of life styles based on the irregular economy can be found in: Liebow (1967); Brown (1965); Lewis (1965); Thomas (1967); X, Malcolm (1965); and Gans (1962).

2. Preliminary results of a study by Marjorie Hanson of the Labor Workshop, Columbia University, Department of Economics, show that welfare payments under the program of Aid to Families with Dependent Children deter the parent from seeking work. The deterrent is greater for nonwhites than for whites. The study also showed that the availability of jobs (as measured by employment and unemployment rates) is a factor in determining labor-force participation among AFDC parents, and here again the effect was greater among nonwhites than among whites. See Hanson (1968).

3. Very few studies have been made of ghetto business enterprise, although many impressionistic descriptions are available. The two most useful recent studies are: Caplovitz (1963) and Raine (1967). Studies of Black business enterprise include: Brimmer (1966 and 1968). Earlier studies of Black business are: Dubois (1899); Myrdal (1944); Harris (1936); Drake and Cayton (1945); Pierce (1947).

4. Caplovitz' (1963) study empirically verified conditions in ghetto consumer markets that were well known from casual observation and impressionistic accounts. Its chief contribution was its demonstration that a systematic and functional relationship prevails between conditions of ghetto life and the organization and practices of ghetto markets.

5. A parallel and related development has been a decline in white employment in the urban ghettos, which has occurred on a larger scale than the shift in business activity and for similar reasons.

6. A similar development occurred in the first decade of the twentieth century, for much the same reasons as the present upsurge in Black business enterprise. See Franklin (1966). A fuller account can be found in Franklin (1956: 389, 395-397).

7. Very little is known about the details of these money flows, and recently some efforts have been made by some banks and insurance companies to redirect the flow of some of these funds back into the urban ghetto. We should also note that some ghetto savings are invested in rental housing by ghetto residents themselves. Studies are needed of the ownership of property in the urban poverty areas to get a better picture of these patterns.

8. The riots of 1966-1967 and the rise of Black nationalism have changed these patterns somewhat. Many people formerly employed in the urban ghettos have given up their jobs, leaving more jobs for ghetto residents. This appears to be one reason for the unexpectedly large decline in unemployment rates in urban poverty areas in 1967-1968. Tight labor markets elsewhere in the economy have assisted by providing other employment opportunities for whites employed in ghetto areas.

9. One result of the Census undercount of men in the slums is a serious political underrepresentation of slum areas. Election districts are based on Census counts of population. The undercount means that the slum constituency is underrepresented at the local, state, and national levels.

REFERENCES

BATCHELDER, A. (1964) "Decline in the relative income of Negro men." Quarterly Journal of Economics 78 (November).

BLUESTONE, M. (1968a) "Low wage industries and the working poor." Poverty and Human Resources Abstracts III (March-April).

——— (1968b) "The secular deterioration of wage terms among industries in the United States, 1947-1966" (mimeo).

BRIMMER, A. F. (1968) "The banking system and urban economic development." Address delivered to the joint session of the 1968 Annual Meetings of the American Real Estate and Urban Economics Association and the American Finance Association, Chicago, December 28 (mimeo).

——— (1966) "The Negro in the national economy." Pp. 251-336 in J. Davis (ed.) The American Negro Reference Book. Englewood Cliffs, N.J.: Prentice-Hall.

BROWN, C. (1965) Manchild in the Promised Land. New York: Macmillan.

CAPLOVITZ, D. (1963) The Poor Pay More. New York: Free Press.

CLARK, K. B. (1967) "Sex, status and underemployment of the Negro male." In A. M. Ross and H. Hill (eds.) Employment, Race and Poverty. New York: Harcourt, Brace & World.

DRAKE, S. C. and H. R. CAYTON (1945) Black Metropolis. New York: Harcourt, Brace & World.

DuBOIS, W. E. B. [ed.] (1899) The Negro in Business. Atlanta, Ga.: Atlanta University. Reprinted in the Atlanta University Publications, Numbers 1-18. New York: Arno Press and the New York Times, 1968.

FARLEY, R. and K. E. TAEUBER (1968) "Population trends and residential segregation since 1960." Science 159 (March 1).

FERMAN, L. A. (n.d.) "The irregular economy: informal work patterns in the urban ghetto" (mimeo).

FRANKLIN, J. H. (1966) "A brief history of the Negro in the United States." Pp. 53-54 in J. Davis (ed.) The American Negro Reference Book. Englewood Cliffs, N.J.: Prentice-Hall.

——— (1956) From Slavery to Freedom, A History of American Negroes. New York: Alfred A. Knopf.

FRIEDEN, B. J. (1967) "Housing and national urban goals: old policies and new realities." In The Metropolitan Enigma. Washington, D.C.: U.S. Chamber of Commerce.

GANS, H. (1962) The Urban Villagers. New York: Free Press.

GREEN, C. M. (1967) The Secret City: A History of Race Relations in the Nation's Capital. Princeton: Princeton University Press.

HANSON, M. (1968 "The effect of welfare payments on labor force participation: preliminary findings" (mimeo).

HARRIS, A. (1936) The Negro as Capitalist. Philadelphia: American Academy of Political and Social Science.

Joint Economic Committee (1968) "Employment and manpower problems in the cities." Pp. 97-107 in Hearings (May 28-June 6). Washington, D.C.: U.S. Government Printing Office.

LEWIS, O. (1965) La Vida. New York: Random House.

LIEBOW, E. (1967) Tally's Corner, A Study of Street-Corner Men. Boston: Little, Brown.

MYRDAL, G. (1968) Asian Drama. New York: Random House.

——— (1944) "The Negro in business, the professions, public service and other white collar occupations." In An American Dilemma. New York: Harper & Row.

PIERCE, J. A. (1947) Negro Business and Business Education. New York: Harper & Row.

PIORE, M. J. (1968) "Public and private responsibilities in on-the-job training of disadvantaged workers." Massachusetts Institute of Technology. Department of Economics working paper (mimeo).

RAINE, W. J. (1967) "Los Angeles riot study: the ghetto merchant survey." University of California, Los Angeles, Institute of Government and Public Affairs.

SCHORR, A. L. (1963) Slums and Social Insecurity. Washington, D.C.: U.S. Department of Health, Education and Welfare.

SHEPPARD, H. L. (1969) "The nature of the job problem and the role of new public service employment." Kalamazoo: Upjohn Institute for Employment Research.

THOMAS, P. (1967) Down These Mean Streets. New York: Alfred A. Knopf.

THOMPSON, W. R. (1968) "Internal and external factors in the development of urban economies." In H. S. Perloff and L. Wingo, Jr. (eds.) Issues in Urban Economics. Baltimore: Johns Hopkins Press.

——— (1965) A Preface to Urban Economics. Baltimore: Johns Hopkins Press.

U.S. Department of Commerce (1968) "Recent trends in social and economic conditions of Negroes in the United States." Current Population Reports. Washington, D.C.: Bureau of the Census (July): Series P-23, Number 26.

U.S. Department of Labor (1969) "Employment surveyed in slum areas of six large cities." News release (20 February).

——— (1968) "A sharper look at unemployment in U.S. cities and slums."

——— (1965) The Negro Family. (The "Moynihan Report") Office of Policy Planning and Research (March).

U.S. Small Business Administration (1969) Quarterly Economic Digest 2 (Winter): 26.

X, MALCOLM (1965) Autobiography. New York: Grove Press.

14

The Analysis of Residential Segregation

ANTHONY H. PASCAL

□ MY INTENTION in this paper is to assess the state of understanding of the phenomenon called residential segregation. The paper deals mainly with the housing situation of black Americans, but alludes at various points to other groups as well. I begin by asking why segregation is an important topic and then discuss the current magnitude of segregation in American cities and point out some recent trends. A survey of theoretical, empirical, and policy issues follows. In the course of the survey, I try to sketch the rudiments of a new model of housing segregation.

RESIDENTIAL SEGREGATION IN AMERICA

WHY STUDY SEGREGATION?

For several reasons. First, because there seems to be a great deal of it. It is a ubiquitous and obvious characteristic of American cities, as we shall see.

Second, and more important, we investigate segregation because its widespread existence seems to violate a fundamental American ethical tenet. Our society places great value, in rhetoric if not in practice, on equality of opportunity and freedom of choice. That millions live in conditions which are grossly unequal and in which, we suspect, they had little choice, arouses in many a commitment toward change.

Third, we feel that segregation exacerbates other social problems, that it has external effects. Substandard education, shoddy municipal services, obstacles to job opportunities, and barriers to political participation are seen as some of the consequences black Americans derive from segregated living patterns.

Finally and somewhat ironically, given the foregoing, the topic is interesting just because residential integration has lost some of its general acceptability as a social goal. Black power advocates and black nationalists increasingly point the way toward racial separation. A wider group endorses this direction either because its members actively favor this road or because they despair of integration. The reemergence of the debate on the desirability of integration lends urgency to the pursuit of knowledge about it.

THE MAGNITUDE AND EXTENT OF SEGREGATION

Black Americans have vastly different residential patterns than do others with whom they share metropolitan areas. To start at the most aggregative level, they made up seventeen percent of the central-city population of the 216 largest metropolitan areas in the U.S. in 1960, but only five percent of the suburban population. This is by no means wholly a result of the "natural" sorting out by economic characteristics. In fact, a larger proportion of low-income whites (with $3,000 or less) than high-income blacks (with $10,000 or more) lived in these suburbs (Kain, 1969: 90).

Nor was 1960 an unusual year. The Taeubers (1965: 1-8, 28-68), in the most comprehensive treatment of the urban experience of black people, provide ". . . strong and consistent evidence for the conclusion that Negroes are by far the most residentially segregated large minority in recent American history." Using a measure of housing segregation called a dissimilarity index to quantitatively express the neighborhood-by-neighborhood separation of the races, they review trends in a large number of cities going back as far as 1910.[1] They find that as blacks moved North to the large cities in the period up to 1940, they were housed in an increasingly segregated pattern. By 1940, blacks were highly segregated in all cities—segregation indexes averaged about eighty-five—irrespective of region, city size, or relative size of the local Negro population. In the period 1940-1959, segregation indexes remained about constant in the North and West while they increased in the South, especially as newer cities and suburbs grew up in that region. In the fifties, as suburban housing opportunities opened throughout the nation, whites began to desert the central cities. The ghettos grew in both area and population as a result of the continuing Negro migration off the farms. Segregation, however, only very slightly abated in the North, while it intensified somewhat in the South. Besides the North/South differences, the Taeubers also observed a tendency for those cities where the black population increased fastest, relative to white, to exhibit increases in their segregation indexes.

Though in general the Taeubers' findings apply to central cities and not to metropolitan areas, on the basis of some supplementary calculations they were able to surmise that measurements of suburban segregation were correlated with central-city segregation within a given metropolitan area. And, confirming the point, they found that as blacks became less centralized in a metropolitan area, they became no less segregated.

The trend of segregation since 1960 is subject to some debate in the literature. Farley and Taeuber (1968) were able to construct segregation indexes for a nonrandom sample of thirteen cities (with populations between 100 and 800-odd thousand and with at least 9000 Negro residents) for which special enumerations were conducted by the Bureau of the Census during the period 1964-1967. When compared to the same thirteen cities in 1960, the unweighted average percentage of nonwhite rose from seventeen to nineteen while the unweighted average segregation index increased from a value of seventy-seven to seventy-nine.

On the other hand, Sudman et al. (1969: 50), in a recent article based on surveys made in 1966-1967 claim that "integrated neighborhoods are much more common than most Americans think they are." Defining neighborhoods according to local perceptions of boundaries, and integrated as having continued and projected white and black in-migration, they estimate that nineteen percent of all Americans live in integrated neighborhoods.[2] However, the average Negro fraction in these "integrated neighborhoods" is only three percent and many of them appear to be "threatened" by contiguous neighborhoods in which the number of blacks is growing rapidly (Sudman et al., 1969: 55-56, 64).

What kinds of neighborhoods house black people? Sudman et al. investigated this question. They found that the higher the black fraction in a neighborhood, the fewer owner occupiers there were, the older the buildings tended to be and the less the new construction was under way. On the other hand, the value of structures occupied by owners—at least for Negro fractions under ten percent—and the percentage substandard seemed uncorrelated with racial composition. For renters the pattern is a bit more confusing. Lily-white neighborhoods tend to have somewhat lower rents after standardizing in a rough fashion for size and age of units (Sudman et al., 1969: 66-67). The Taeubers (1965: 154-169) explore the socioeconomic status of the Negro population, rather than the characteristics of the housing stock, as a function of the degree of segregation. Their findings are generally consistent though with those for the housing stock—i.e., the lower the fraction of blacks in a neighborhood, the higher the status of the blacks who do live there.

THE FUTURE OF SEGREGATION

What is the likely future trend of racial residence patterns? All authors tend to agree that this will depend on a number of factors such as changing racial attitudes, the strength of the tendency toward convergence in socioeconomic characteristics between black and white, the relative rates of urban population growth for the two races, and certain physical and financial aspects of the housing market.

In stating attitudes toward black neighbors to public opinion poll interviewers, white Americans show a decline over time in prejudicial feelings. Gallup reports that when in May of 1963, whites were asked whether they should move away if a Negro moved in next door, fifty-five percent said no, while in the same month of 1965 the no's had climbed to sixty-five percent (Sudman et al., 1969: 50). Nor is there evidence that black separatist arguments have done much to change Negro preferences for integrated housing. Rather, the preference seems to have grown. Harris polls in 1963 and then in 1966 found sixty-four and sixty-eight percent of blacks respectively expressing a preference for integrated neighborhoods (Kain, 1969: 92-93). A Newsweek poll (1969: 20) of blacks in 1969 indicated that seventy-four percent preferred an integrated neighborhood.

The segregation of most non-Negro ethnic groups has clearly declined over time (Lieberson, 1963; Taeuber and Taeuber, 1965: 67). This might suggest that as the socioeconomic differences between blacks and whites diminish along with gains in black educational status and weakening employment discrimination, we should expect similar shifts in black/white residential patterns. Arguments concerning the role of socioeconomic factors in explaining observed segregation receive substantial attention later on in this paper. Suffice it to say for the present that no student of the subject denies some role to these factors. Socioeconomic differentials have been narrowing, particularly during the sixties when labor markets were continuously buoyant. Miller (1969) pointed to the relative gains made by city-dwelling Negroes in educational attainment, occupational status, and earnings since 1960. A recent federal government report shows similar relative gains in employment, income, and health status in the period 1960-1965/1966 (Bureau of Labor Statistics, 1967).

Increasingly, the growth in black population in metropolitan areas is a result of natural increase and of migration from other metropolitan areas and, decreasingly, of migration off the farm, where fewer and fewer Negroes live (Farley and Taeuber, 1968: 954; Bureau of Labor Statistics, 1967: 9). Also, the migrants are less and less distinguishable socioeconomically from the sitting black population. This suggests that as the source of unskilled, unurbanized Negro migrants dries up, housing integration will tend to be speeded since it can be shown that it is the Negroes of relatively high socioeconomic status who lead the movement out of the ghetto. The "invaders," in fact, are likely to be similar in status to the whites already in the neighborhood (Taeuber and Taeuber, 1965: 6-7, 99-100). In sum, demographic trends will tend to inhibit the total of future segregation both directly, insofar as black house-seekers become more like whites socioeconomically, and indirectly, since, on the basis of past experience, a leveling off in black population growth has been associated with reduced black/white segregation (Taeuber and Taeuber, 1965: 4). The socioeconomic convergence, of course, might also lead to a relative reduction in the Negro birth rate which would reinforce this slowdown in Negro population growth.

Changes in the housing market present a more complicated forecasting problem. If the plans come to fruition for locating publicly assisted housing that tends to appeal to blacks (e.g., public housing and rent supplement projects) outside the ghetto, while white-dominated projects (e.g., urban renewal and rehabilitation) are placed near existing ghettos, the pace of integration will obviously accelerate. To date though, there have been more plans than action.

When we speculate on the effects of changes in new construction rates and costs, the complexity grows. Not only are empirical trends—say in mortgage interest rates—unpredictable, but the direction of expected impact is hard to specify. Further on, I try to demonstrate that a reduction in the cost of new housing units tends to accelerate neighborhood turnover from white to black but may not speed integration.

THEORIES OF RESIDENTIAL SEGREGATION

Using the plural in the subtitle of this section is misleading. If we restrict the use of the word theory to statements about the causal relationships underlying phenomena, statements which are useful in making predictions about the observable world, then there is only one theory: residential segregation is the result of a series of housing site choices, based on both the racial (or ethnic) and the nonracial attributes of various sites, such as price, quality, style, location, and so forth. The spectrum of choices people face may be more or less constrained. Thus, when as in South Africa the law forbids black occupancy of certain districts or, as in the United States before 1948 the courts enforce deed restrictions that prohibit sales to blacks or, as is sometimes alleged, groups of middlemen collude to prevent black entry into a given neighborhood or, as happens, white residents make credible threats of violence against black home-seekers, then severe constraints are present in the market for residential sites.

SOCIOECONOMIC FACTORS IN SEGREGATION

Theoretical treatments of the phenomenon of housing segregation are all based, though sometimes only implicitly, on a model or residential site choice. This applies to the work of demographers, sociologists, planners, and economists. Great debate, on the other hand, rages over the importance of the constraints in explaining segregation, and particularly now in the United States, over constraints of the last two types cited above. Substantial disagreement also continues over the relative weight that should be accorded to racial and nonracial factors in the explanation of observed segregation. I attempted to sort out these factors in an earlier monograph (Pascal, 1967).

Several residential-site-choice models have appeared in the literature in urban theory. Well known examples include Harris (1961), Herbert and Stevens (1960), Hoover (1948), Isard (1956), Meyer et al. (1964), Moses (1958), Muth (1961), Park (1962), and Schnore (1954). Generally, they treat residential sites as though they were chosen through some optimization process in which the attributes of the site—location, amount and quality of housing services, price —are matched against the desires of the household as generated by its individual characteristics—size and composition, income available, pattern of trip destinations. I have reviewed and compared a number of these contributions in Pascal (1967: 38-64).

How relevant are such models to the understanding of segregation? Another way to ask the question is to attempt to isolate from the observed segregation of, say, blacks those components that would be expected to result from variation in the attributes of residential sites and from the differences in household characteristics between the black and white population.[3] In earlier research (Pascal, 1967: 114-161), for example, multiple-regression equations were estimated with data from the Chicago and Detroit metropolitan areas. Variables measuring the required housing outlay, the access to Negro job opportunities, and the type of housing available in a given neighborhood were used to predict the percentage of Negro households in that same neighborhood. Such regressions succeeded in "explaining" from about one-third to about one-half of the variation in neighborhood Negro fractions. Similar regressions were fitted for the low-income white population to test whether the residual or "unexplained" portions of the variance could be attributed to racial factors, i.e., prejudice against Negroes by whites or prejudice in favor of Negroes by Negroes. These findings, together with additional analysis of the housing patterns of the black population, including a procedure in which the Taeubers' segregation index was adjusted to take account of segregation "expected on socio-economic grounds," added up to a strong case for the existence of pervasive and deep-seated racial factors in the explanation of segregation.

The approach just described sought to isolate the minimum amount of segregation which could confidently be attributed to prejudicial attitudes in housing choice. It therefore assumed that black families seek housing locations that are convenient to their work places, among other attributes, in a manner equivalent to the behavior hypothesized for white families (Carroll, 1949). This assumption did not prevent the allocation of a substantial portion of observed segregation to the prejudice explanation. In a series of articles, J. F. Kain (1964a; 1968; 1969) has argued forcefully for an alternative proposition: black workers, given the constrictions of housing segregation, are likely to choose employment close to home, often in firms that deliberately locate near the ghetto in order to tap a low-wage pool of labor. Since the precise nature of the residential-site-choice/work place-choice relation is virtually unidentifiable empirically, each interpretation has merit. The first helps establish a lower bound for the relative weight of racial factors in segregation, while the second highlights an additional important source of employment disadvantage for blacks.

Other studies have sought to dismiss the power of nonracial factors to explain the residential patterns of blacks, though the dismissal has not proved very convincing. The Taeubers (1965: 90-93), for example, present the results of a series of regression estimates for a number of cities in the period 1940-1960. The proportion of nonwhites in a tract was regressed against measures of housing quality, or housing price, or tenure in the same tract. No multiple regressions (i.e., when several housing characteristics are used simultaneously to predict the nonwhite proportion) were calculated, but the simple regressions they report typically explain about fifty percent of the observed variance. Disinclined to stop here, however, they proceed through rather mystifying steps— highly dubious in terms of the theory of statistical inference—to try to isolate a "pure" racial explanation. Their essentially invalid procedure results in the according of only a negligible fraction of explained variance to housing characteristics.[4] Since the authors never use a tract's accessibility to work places as an independent variable in multiple-regression equations where the characteristics of the tract's housing stock is also taken into account, nor examine metropolitan-area (i.e., as opposed to city-limits) patterns, it is difficult to judge whether their findings with respect to the importance of socioeconomic variables are consistent with those of other researchers (e.g., Kain, 1964b; Pascal, 1967).[5]

In a companion attempt to isolate the nonracial factors, the Taeubers (1965: 70-76) utilize a series of "objective" independent variables to predict the change in the segregation index in sixty-nine cities. They include population change among whites and nonwhites, a measure of suburbanization, a measure of new construction, and the change in nonwhite occupational status. For the period 1940-1950, only the new-construction variable was significant (all the variables together explained one-third of the variance), while for 1950-1960 only nonwhite population change and nonwhite occupational status change had coefficients significantly different from zero (in this latter case, all the variables together explained over half of the variance). Their interpretation is unexceptionable as far as it goes: very low levels of new construction essentially prevented any major change in housing patterns in the World War II decade, while, once this constraint was removed, the growth in Negro population expanded the ghetto, and any segregation measure for a given point in time, therefore, caught neighborhoods in the process of population turnover. They ignore their rather interesting finding with respect to the relationship between Negro occupational advancement and the decline of segregation.

ATTITUDINAL FACTORS IN SEGREGATION

Establishing that a useful theory of housing segregation must incorporate socioeconomic as well as attitudinal factors does not resolve the conceptual issues. Recall that questions remain on the handling of the socioeconomic variables and particularly on identifying the direction of causality between these variables and observed segregation (Pascal, 1967: 119-139). Even more difficult problems arise as we attempt to refine our notions concerning the attitudinal factors. Particularly hard to answer are the following two questions: *How important is voluntary segregation in explaining the observed residential separation of minority groups? To what extent are negative attitudes on the part of the majority a reaction to class differences and to what extent are they reactions to race?*

Voluntary Segregation

All minority groups will, to some extent, choose to segregate themselves. The motives are rather obvious. A people who share a culture, a common set of experiences, sometimes a language, and among whom kinship ties are important, will want to cluster residentially. Clustering permits easier communication and encourages the establishment of services and institutions—food stores, clubs, churches, and the like—with specialized appeal for the group. In theory, such reasons for clustering might better be listed under the socioeconomic explanation of segregation; in practice, they will nearly always contribute to the unexplained variance and will be lumped with attitudinal factors. Some aspects of voluntary segregation are truly attitudinal in nature, however. At the two extremes are the minority group whose members feel a disdain or suspicion toward the majority, and the group whose members feel no particular repugnance toward the majority but who anticipate hostility from the majority toward themselves.

The question then is not whether we should expect black Americans to choose some segregation, but rather whether we should expect such a propensity to go very far in explaining their observed segregation. Some empirical tests of the voluntary segregation hypothesis have been attempted, mainly by comparison of the residential patterns of groups who could be expected to have strong preferences for segregation with the residential patterns of blacks. Lieberson's study (1963: 120-132), for example, compares the residential segregation of various nationality groups from native whites and shows that they are invariably less segregated than are blacks (see also Weaver, 1948). Other research (Pascal, 1967: 257, 161-171) has demonstrated that when in the Chicago metropolitan area segregation indexes are constructed for Italo-Americans and compared with those for middle-to-upper-income nonwhites, the index for the latter is over five times as large. The message of these studies is that it would be necessary to posit a truly prodigious desire for segregation among Negroes in comparison to other minorities to generate the drastically different residential patterns they display. When asked, of course, almost three-quarters of blacks express a preference for integrated rather than segregated housing (Pascal, 1967: 5-6).

Race Prejudice and Class Prejudice

When whites attempt to separate themselves from blacks—and all the evidence cited above must convince us that they do indeed attempt it—how much of their antipathy is a result of skin-color differences and how much is a result of presumptions about the class characteristics of nonwhites? The distinction, however difficult it may be to test empirically, is an important one. Class prejudice does seem to affect housing choice. I was able in my earlier monograph (Pascal, 1967: 115, 139-148) to demonstrate that a group differing in class though not in racial terms—i.e., white households with 1959 incomes between two and four thousand dollars in the Chicago metropolitan area—were segregated from the majority more severely than could be expected on the basis of spatial variation in the characteristics of the housing stock alone.

Suppose that people genuinely resent the incursion of neighbors with exotic social standards, e.g., in housing maintenance, street behavior, child-rearing practices, and use racial identification as a predictor of social behavior. Note that the minority-group members actually "threatening the invasion" of a given neighborhood need not themselves be, and often in reality are not, deviant in the sense described. It is enough that the sitting-majority population believes that the entry of "lower-class" people will be eased as the minority fraction in the neighborhood grows. If this description of behavior, which is based on an argument advanced by Glazer and McEntire (1960: 4-10), is accurate, we should expect diminishing efforts to maintain the separation of a particular minority group as that group approaches the majority in socioeconomic characteristics, i.e., as its supply of potential lower-class invaders shrinks. The postwar experience of the West Coast Orientals in attaining a significantly less-segregated residential pattern is often cited as evidence that is consistent with this hypothesis.

Again, we may call on the Taeubers (1965: 67) for information on these points. They present a table that includes calculations of segregation indexes between whites and Negroes and between whites and other races, of whom over ninety percent were Oriental, for San Francisco in the years 1940, 1950, and 1960. The Negro/Caucasian index remained fairly constant—71, 71, 70—while the Oriental/Caucasian declined sharply—76, 68, 51—over this twenty-year period. Observers have argued that the growing tolerance for Orientals was a result not only of the relative socioeconomic advance of those already settled in the United States, but also of Caucasian recognition of the diminishing likelihood of continued large "lower-class" immigration from the Orient. Though the evidence is not nearly so clear (Taeuber and Taeuber, 1965:

65-67; Glazer and McEntire, 1960: 84-109, 144-147), it is supposed by many that the Puerto Ricans in the Northeast and Mexican-Americans in the Southwest have not experienced similar gains in housing integration precisely because of the relatively open immigration from their respective homelands, even though they are often not nearly so identifiable as Orientals.

HOW IS SEGREGATION ENFORCED?

I would conclude that though voluntary segregation plays some role in generating black residential patterns, and though class feeling in the majority contributes as well, race prejudice on the part of whites bears a substantial responsibility for the observed segregation in housing. This conclusion seems not wildly inconsistent with the studies we have reviewed. What these studies fail to do, however, is to treat in any satisfactory way the mechanisms by means of which attitudinal segregation is enforced. Before the 1948 Supreme Court ruling against the judicial enforceability of race-restrictive covenants, the question would have been rather idle. Writers (Kain, 1968; 1969; Taeuber, 1968; Laurenti, 1960) who have tried to frame an analysis of how segregation is enforced in the post-1948 period have tended to fall back, in one way or another, on conspiracy theories.

We economists are probably by training predisposed against conspiratorial explanations except in cases where the conspiracies are protected by force of law. There usually seems to be just too much potential gain awaiting an individual violator of any agreement and this tends to make informal conspiracies unstable and temporary. Yet, the dimensions of housing segregation appear exceedingly stable and remarkably long-lived. Some of the descriptions of enforcement mechanisms which have been advanced in the literature can be examined in terms of their likely stability properties.

We can first dismiss as a conceptually rather uninteresting case the difficulties that blacks have in securing apartments in multiple-unit buildings containing no blacks or relatively few blacks. Here, no conspiracy is required to explain the reluctance of a landlord to accept a black tenant. An explanation requires only the following assumptions: (a) the landlord tries to maximize his profits; (b) he believes his other tenants are racially prejudiced;[6] and (c) he estimates that in the process of a complete racial turnover of his building, his average vacancy rate will climb, at least for some period. Note that we require no assumption about the racial feeling of the landlord. All this is rather obvious and no doubt accounts for a good deal of observed segregation.

The more interesting cases, from a theoretical standpoint, are those in which blacks are underrepresented in single-family housing in a given neighborhood. In the subcases of both owner-occupied and for-rent houses, the existing literature seems to require collusion somewhere in the system. Suppose the existing landlords share a distaste for blacks and are therefore disinclined to deal with them as tenants. What is to prevent a less-prejudiced person from buying one of those houses and, in turn, renting it to a black family? An agreement among existing landlords not to sell to such persons will be stymied by the difficulty of identifying potential agreement-breakers among buyers. Once a landlord with a lesser level of prejudice is in possession of a house, competition among landlords will eventually force the opening of all of the houses to blacks; in the process, the original landlords will swallow their prejudicial feelings in return for lower vacancy rates or higher rents, or they will sell their houses to other less-prejudiced landlords. In other words, the collusive agreement will be unstable.

This is all very well for rental housing, some might argue, where people are faced with the economic costs of their racial attitudes, but what about owner-occupied units? In most neighborhoods of a city we find blacks underrepresented as owner-occupants, and the underrepresentation seems more marked than would be accounted for by lower income levels among blacks. Social pressure—another name for tacit collusion—is typically singled out as the explanation. A white home-seller declines to offend this soon-to-be-former neighbor even in the case where he himself feels no particular color prejudice. His neighbors would take offense at his selling to a black family, the argument goes, because they fear a decline in property values or simply because they dislike living near blacks.

But neither will this line of reasoning stand up. At some point it will pay some whites to specialize in buying houses from other whites and selling them to blacks. Since these intermediaries never need actually move into the neighborhood, they would remain indifferent to the feelings of white neighbors. People do, in fact, assume this intermediary role; when the people are also real estate agents they are called "blockbusters." The interesting question, which we will want to explore more fully, is why blockbusting is not a ubiquitous activity.

Not only, of course, are real estate agents accused of hastening racial turnover in some neighborhoods through blockbusting, but they are also charged with preventing entry in others by refusing to show property to interested black home-seekers. This pair of accusations rests on a latent assumption of a truly massive collusion among realtors to engineer blockbusting in some places while suppressing it in others. That there are bigots among brokers is undeniable, and that on the average they are more prejudiced than other people is certainly possible. But to suppose that through some agreement they can keep every agent in a city from sniffing out profitable transaction opportunities, can make all of them conform to some master plan for racial settlement, seems highly unlikely.

The argument that real estate brokerage firms have stakes in particular neighborhoods and will therefore act to preserve their reputations or their investments in them is undercut by the fact that these ties to a locality never have the force of a monopoly. An outside broker can always come in if the economic stakes are high enough in a transaction opportunity which the neighborhood brokers prefer to ignore. And since it is pretty clear that the primary motivation of real estate brokers is money-making and not the preservation of racially pure neighborhoods, they will come in. A very similar analysis of the role of mortgage lenders would, by the way, serve to absolve them also of major culpability for the preservation of pervasive segregation.

In my earlier work on this topic (Pascal, 1967: 126-127), I expressed a similar dissatisfaction with conspiratorial theories and sought the explanation of segregation mechanisms in other directions. There I reasoned—a la Becker (1967)—that whites as house sellers and landlords demanded compensation for dealing with blacks, for whom they felt distaste, and that this compensation resulted in race differentials in selling prices or rentals charged which, in turn, discouraged blacks from entering white neighborhoods. This idea is echoed by Kain (1969: 98). Reconsideration makes it obvious to me, however, that as long as there exists among whites a distribution of racial tastes (i.e., of prejudice) —rather than a constant level for all whites—it will always pay a white from the low end of the distribution of prejudice to buy a house from a white at a point higher on the distribution in order to turn it over to a black.

EXTENSIONS OF THE THEORY OF RESIDENTIAL SEGREGATION:
THE SELF-SEGREGATION MODEL

How then does attitudinal segregation get enforced? What prevents a Negro family from moving to a neighborhood that is convenient to the husband's work place, where house prices are consistent with its housing budget, and which contains a style of housing that conforms to the family's preferences? The answer, I believe, does lie in white racism, abetted probably both by black racism and by class prejudice. A model of residential choice which properly incorporated a racism variable would go a long way in explaining how segregation comes about. This same model would help, incidentally, in understanding ethnic clusters, spontaneous neighborhood renewal, and other commonly observed phenomena.

What we require, I believe, is a model that specifies how people value alternative residential sites, *including the racial characteristics of sites.* That is, the value a particular household accords to a given housing site should be thought of as a function of the following: the physical and aesthetic features (amenities) of the house and lot at that site; the convenience of the site to important trip destinations of the household—particularly work places; the income and housing tastes of the household; and finally, the value of a term which is the product of the distance a site lies from various concentrations of "bigees"[7] and of the level of bigotry felt by the household in question. Similar functional relationships can be formulated to express the value occupants place on their own residences. Next, an expression that relates the strength of bigotry to household characteristics would be necessary. With the establishment of some reasonable rules about conditions under which change in occupancy of a site will occur as a function of the relationship between values accorded by existing occupants and potential occupants, we will have a general theory of residential-site choice.

Such a theory would permit us to predict the probability that a particular site would be occupied by a household belonging to a certain group (e.g., black, white, rich, poor).[8] We could also then predict the probability that the group affiliation of the occupant will change as the terms in the value function changes. Questions about the likelihood, say, of a site switching from white to black occupancy as a result of a relative increase in the black population, or of a relative enhancement of black socioeconomic status, or of natural aging of the housing stock, or of additions to the stock, would be possible. Extending the analysis to neighborhoods, which could be defined as groups of contiguous sites with similar house/ lot characteristics, would permit statements about expected race or class proportions by sub-area and from there statements about segregation.[9]

In this approach the racial (or class) effects are not treated as residuals but as explanatory variables. Not only do sites vary in racial attributes—here defined, for example, by a potential or gravity index of distance to bigees—but households vary in bigotry. It may be true simultaneously that whites are bigoted with respect to blacks, blacks with respect to whites, and rich with respect to poor. The relationship among the means and variances of these various bigotry-level distributions would have important implications for the resultant housing-site choices of blacks and whites.

Notice that in the model outlined here, a white household with some positive level of bigotry will value a site more highly the more distant it is from concentrations of blacks.[10] This means, in effect, that a white household will offer more for some sites than will a black household and less for others, if each race prefers its own kind, *ceteris paribus.* Then it becomes clear that the reason few blacks are found in some neighborhoods is because they see no particular point in paying the bigotry (against blacks) premium that owners and landlords can demand and receive there. Interestingly, this mechanism permits whites to avoid the tension between ethical precepts and bigoted feelings which Myrdal (1944) points out. Nevertheless, black families are in effect excluded by means of a requirement that they pay for a site feature they do not value, and in fact, may strongly disvalue.

This line of reasoning as to the relationship between site prices and minority occupancy does not accord with the currently conventional wisdom on this theme. It is widely accepted in the literature, e.g., Duncan and Duncan (1957: 168-184, 211-233) and McEntire (1960: 135-145), that blacks must pay more than whites for equivalent housing.

For rental housing, "objective" reasons for charging higher prices to blacks have been advanced. According to Muth (1962: 30), since Negro families will tend to be larger, with a higher proportion of children, and will usually have a less stable income, they may be expected to impose extra maintenance and administrative costs on landlords. These, shifted forward, will result in landlords demanding higher average rents from Negro families. (In the self-segregation model outlined here, this would take the form of a rental housing outlay by black families which was independent of the attributes of the site.) That black families pay more for houses they buy, on the other hand, has not been documented.[11]

A second belief that would tend to undercut the validity of the self-segregation model involves the effect of black entry on neighborhood prices. It is frequently asserted that the impact is nil. The foremost proponent of, and constantly cited authority for, the view that site prices will be unaffected by black entry is Luigi Laurenti (1960). Clearly, the self-segregation model would generate such a result only given a very special combination of values for white and black bigotry levels, other household characteristics, and preexisting racial concentrations.

Unfortunately, the study from which Laurenti draws his conclusions is marred in such a way as to cast substantial doubt on them. He observes changes in residential property transaction prices over time—1943 to 1955 in San Francisco, Oakland, and Philadelphia—in test neighborhoods in which Negro entry had occurred. "Control" neighborhoods were selected for the same three cities to maximize resemblance to the test neighborhoods in every characteristic other than Negro entry. He found no relative difference in property-price changes between test and control neighborhoods. The generalizability of these findings, however, is open to serious question on the basis of the following facts:

(1) Nowhere in the study are data presented which would permit a systematic comparison of the attributes of test and control neighborhoods. Nor are we told how the various criteria were weighted in selecting the controls. He does provide maps of the three cities in which one consistent relationship visibly emerges; almost invariably the test neighborhood lies closer to the center of the city than does its respective control (Laurenti, 1960: 70-71, 124-125, 178-179). This opens the possibility, of course, that urban growth patterns were generating increased accessibility values in the test areas, which masked a decline in the bigotry premiums whites were paying in them.

(2) Both his test and control areas appear to be very uncharacteristic of urban neighborhoods in general for this period. Prices in Laurenti's control neighborhoods grew by an annual average of 1.8 percent; in his test areas, the average yearly price increase was 2.5 percent per annum (Laurenti, 1960: 60-64). Yet during this

same time span, the housing cost component of the national Consumer Price Index grew by more than 3.2 percent per year (U.S. Bureau of Labor Statistics, 1968: 237). And recall that more than nine-tenths of the neighborhoods Laurenti examined were in the fast-growing San Francisco Bay area.

(3) Neighborhoods with relatively heavy nonwhite entry tended to have poorer price performance vis-à-vis their controls than areas with light entry (Laurenti, 1960: 60-64, 69-193).

In sum, then, we can conclude that the evidence for price/race relationships which would be inconsistent with the self-segregation model does not seem to stand up under examination. On the other hand, it must also be admitted that there is available no clear evidence in Laurenti or elsewhere that the less integrated the neighborhood, the higher the site prices. This is not so surprising, however. Standardizing for the physical and amenity features of sites, one would expect that their prices would decrease along with their distance from important destinations. The center of the city still tends to be a dominant work place destination but also usually happens to coincide, more or less, with the ghetto. Thus, when one observes the prices in a suburban neighborhood, there is an off-setting of the effect of high bigotry premiums by low location values, so that in neighborhoods with identical structures the more segregated will not appear as the more expensive if they are also further from the center of town. Sudman et al. (1969: 65, 71-73), report rents and values as a function of the degree of neighborhood integration which are generally consistent with this analysis.

What factors determine the bigotry level of a household? Evidence exists (Allport, 1958; Hyman and Sheatsley, 1964) which seems to suggest that bigotry is correlated negatively with socioeconomic status, particularly education level. The bigotry level relevant to housing-site choice is probably also a function of the presence of children in the household—children being an indicator of the likely amount of neighborhood socializing and the strength of concern about the quality of the public schools and the likelihood of interracial dating. Renters usually occupy a lower socioeconomic status than owners which might incline them to more prejudice but, on the other hand, their households typically have fewer children. Moreover, renters will be free of concern about the effect of race patterns on land values. Some argue also that there is a differential tolerance for black neighbors among various ethnic groups, independent of the factors already specified, but no clearcut evidence for this assertion exists.

The physical, aesthetic, and locational (in respect, say, to work places) attributes of a neighborhood may, of course, outweigh its racial features for a given household. It was, obviously, just this sort of situation that suggested the specification of a site-choice model which explicitly included both racial and nonracial variables. Thus we observe white enclaves in some metropolitan areas (Capitol Hill in Washington, Hyde Park in Chicago), black enclaves in others (Pasadena in the Los Angeles area, Evanston in Chicago), even though, in general, ghettos are extensive and unbroken. Also, the self-segregation model would suggest that the enclave dwellers are typically people with relatively low bigotry levels, e.g., liberals, whites without school children who, therefore, have lesser fear of "pollution" by the lower classes, blacks with school children trying to escape the lower classes, and the like. That nonracial features play an important role also explains why the whites and blacks in mixed neighborhoods tend to be of similar socioeconomic status.

In general, the self-segregation model, though far from completely worked out, can be used to generate a series of interesting predictions. Each of them tends to be consistent with empirical facts.

(1) A relative increase in the black population will cause existing sites to change hands, i.e., shift from white to black occupancy, but may also succeed in pushing whites farther out into new suburban additions to the housing stock and therefore have no important effect on the degree of segregation.

(2) A relative rise in the socioeconomic status of blacks will raise the quality of the housing they occupy but will have no necessary effect on their propensities to seek housing in integrated neighborhoods. When an individual black family enjoys an enhancement in socioeconomic status, on the other hand, it is likely to move into a more integrated neighborhood, since it can thus more easily purchase distance from the poor.

(3) As relative black socioeconomic status rises, whites will pay less to escape from blacks since there is a class component in the bigotry level. Thus, segregation should diminish. Notice, however, that in reality the changes that have been occurring in the socioeconomic position of blacks are quite complicated. In the aggregate, black "normal" families are generally approaching whites, at however disappointingly slow a rate, in occupation and income as a consequence of improved educational opportunities, some lessening of job discrimination, and tight labor markets (U.S. Bureau of

Labor Statistics and Bureau of the Census, 1967: 13-26). On the other hand, in the black slum areas of many cities, conditions have been worsening (U.S. Bureau of Labor Statistics and Bureau of the Census, 1967: 69-91-97; Miller, 1969). The number of nonwhite families headed by females and the number of boys sixteen to nineteen years old in such families have risen dramatically. Both white and black families would have an incentive to flee to the suburbs to escape the taxes required to provide welfare for the former and to remove themselves as potential victims of delinquents among the latter.

(4) A rise in the cost of newly-constructed houses (as, for instance, when interest rates rise) tends to inhibit movement outward. This slows the rate of neighborhood turnover but has no necessary effect on the ultimate pace of desegregation. It does, however, reduce possibilities for blacks to upgrade the quality of their housing.

(5) A declining level of overt hostility on the part of white residents toward black newcomers will not significantly speed the rate of desegregation unless the decline in hostility is matched by a decline in prejudice and thus in bigotry premiums.

The self-segregation model would seem then to be a useful tool for analyzing residential segregation, in making forecasts about its future, and in identifying policy interventions that can alter it. We move now toward a discussion of the data gaps that prevent an early estimation of the parameters of the self-segregation model.

DATA REQUIRMENTS AND DATA GAPS

ESTIMATING THE SELF-SEGREGATION MODEL

We require data to quantitatively estimate the nature and relative importance of the various relationships expressed in the self-segregation model. The data, of course, would also serve to test the appropriateness of the hypothesis embodied in the model, i.e., whether the model is consistent with observable reality. Unfortunately, a goodly portion of the information that would be required for these purposes is not directly available. There also remains questions on data organization and interpretation.

Physical and Amenity Attributes of Neighborhoods

For information on physical features of the housing stock, the decennial census of housing remains the best source. Figures on age, number of rooms, tenure, and condition of dwelling units are available for census tracts, though not for blocks. Nowhere is it possible to ascertain systematically such interesting features of neighborhoods as average land area per dwelling unit, or even gross density. Equally difficult to obtain is information on the variation in amenity levels among neighborhoods. Partly this is a result of the amorphous nature of the amenity concept; people tend to associate with amenity such features as cleanliness (both on the ground and in the air), general attractiveness, micro-climate, terrain, nearness to school, shopping, and recreation, but rarely do they assign weights to the various components described, nor would there be much agreement if they did.

Locational Attributes of Neighborhoods

For locational attributes, the data situation has been even more bleak. The accessibility of areas of a city to the work places existent in that city will for the first time be available for a wide cross-section of metropolitan areas in the 1970 Census of Population, in a form, presumably, which will permit classification by industry, and by age, race, sex, income, and occupation of the job holder. Heretofore, it has been necessary to derive such information from the origin and destination surveys conducted in cross-sectional metropolitan transportation studies, which usually cannot be compared with figures derived from the census.

Measurement of the location of neighborhoods with respect to bigees is simpler in one respect while more difficult in another. The census provides quite detailed information on the race and class of occupants of various neighborhoods, though only at the ten-year intervals noted. How to organize the data is another question. While gravity models have been tested and found useful in measuring accessibility to employment (Pascal, 1967: 101-107, 115-122), I am not aware that anyone has applied this or any other formula to derive an index of distance from "distasteful" groups. In indexing the attractiveness of a site (or neighborhood) to bigots, what relative weights should be assigned to the number of bigees and to their distance from the site? Should distance be measured in physical or economic terms? How should barriers— freeways, aqueducts, city boundaries—enter?[12] Should the index take account only of current patterns or of expected future patterns of bigees as well?

Household Bigotry Levels

Data for testing hypotheses concerning the relationship between bigotry level and other household characteristics, such as age composition, education, ethnicity, and region, are also scant. It is possible that only the quite expensive findings from special survey research will fill this purpose. Also, as time goes on, the ethnic composition of neighborhoods, which many claim has an important effect on its hospitality to dark-skinned minority groups, becomes harder to discover from census materials. A person of Polish descent, both of whose parents were born in the United States would, for example, be indistinguishable from a Mayflower descendent and from a third-generation Italian-American in the census.

The Concept of Neighborhood

To this point I have used "neighborhood" as if it were an unambiguously identifiable entity. Clearly, this is far from the case; it might not be too far from the truth to say that neighborhood is in the eye of the leaseholder. Yet the concept is of crucial importance to the study of segregation, for in these studies we are interested in the distribution of persons among areas that have some intrinsic meaning to people. Perhaps the only sensible solution for establishing boundaries for neighborhoods is to query local residents. This is more or less what was done when census-tract maps were charted. "Tract boundaries were established cooperatively by a local committee and the Bureau of the Census..." but also "they were generally designed to be relatively uniform with respect to population characteristics, economic status and living conditions" (U.S. Bureau of the Census, 1962). The problem, of course, is that color is an important population characteristic and, thus, the census tract, when used as a unit of observation, is not independent of racial residence. The only other unit by which segregation within cities is regularly observable is the block, which has a good deal less economic and social significance. The Taeubers (1965: 223-231) were well aware of these difficulties and, in the context of a very thoughtful and comprehensive discussion of them, computed correlations between block-based and tract-based segregation indexes for a group of sixty cities. For the more useful indexes, the correlation ran about 0.7. There appears to be no generally satisfactory and, at the same time, practical solution to the problem.

KNOWLEDGE ABOUT THE IMPACT OF SEGREGATION

The empirical problems in studies of segregation extend beyond data deficiencies for the estimation of central explanatory models. We are interested in segregation, after all, partly because we think it has a significant impact on other social phenomena. Yet, quantitative understanding of the nature and magnitude of these impacts is surprisingly scant.

Segregation and Job Opportunities

Just as the location of job opportunities will affect residential-site choice, there is no doubt that for bigees the reverse causation is also simultaneously true. The isolation of the ghetto will inhibit the informal spread of job-market information while rapidly suburbanizing manufacturing jobs will increase commuting costs for ghetto residents. A recent theoretical piece by McCall (1968: 7-9, 12) illustrates the relationship between the cost of job search, the likelihood of withdrawal from the labor force, and the expected duration of unemployment. The effect of housing segregation on work opportunity has also been the subject of recent empirical research. Kain (1968), for example, shows that in Chicago and Detroit in the period 1950-1960, the segregated pattern of Negro residences in the inner city combined with an outward movement by manufacturing firms appreciably reduced the net employment opportunities for Negroes, even though at the same time Negroes became overrepresented in their own neighborhoods in service jobs that required customer contact. Although here again Kain's interpretation of the direction of causality is not above question, he no doubt has pointed out an important effect.

A follow-up study by Mooney (1969) which used data for twenty-five cities over the period 1948-1963, concluded that the negative effect of housing segregation on employment was more likely to cause withdrawal from the labor force than unemployment. Mooney found the segregation effect was rather negligible for black female workers and that, in general, the local level of economic activity was considerably more important than were residence patterns in explaining the employment situation of blacks. The improvement in the relative employment situation of inner-city females which the industrial location shifts brought about may have also resulted in even further diminishment in the economic role of the black male as compared to that of his wife. Some observers have pointed to the marginal status of Negro men as bread-winners in the explanation of Negro family instability and in the failure to inculcate work orientation in male youth (U.S. Office of Policy Planning, 1965: 30-37).

The trend toward suburbanization of jobs is not, of course, unrelated to residence patterns. Here again we can identify a feedback. As the centers of cities began to fill up with impoverished, unurbanized, and, for a time, politically apathetic people, the incidence of crime, fire, and sanitation problems probably increased, while the availability of municipal services to cope with such problems was decreasing. The results of this on insurance rates and on employee morale no doubt effected the avidity with which management tended to seek alternative suburban locations for plants.

Segregation and Educational Opportunity

The Coleman Report (1966) makes a strong case for the impact of student-body characteristics on a student's scholastic achievement. For black students it suggests that attending an integrated school has, on the average, a significant and positive effect on achievement over and above that attributable to the student's own family background, to the quality of his teachers, and to the educational facilities he enjoys. Though critics have disputed the relative importance the report accords integration vis-à-vis the other variables (e.g., Bowles and Levin, 1968), none has denied that it appears to have a substantial independent effect. That segregation may result in a decline in the quantity and quality of education resources allocated to minority neighborhoods is explored in the next section.

Segregation and Political Power

The typical view of the relationship between the residential segregation and the political power of a minority group has been negative; that is, isolation and the resultant apathy toward municipal affairs lead, it is argued, to a loss of power by the minority group and a diminishment in the share of local public services it can command. In fact, it is alleged (Logue, 1968) that neighborhoods inhabited by blacks are consistently short-changed in such public services as sanitation, recreation, transportation, education, and fire and police protection. Little is known, however, about the quantitative significance of inequity in the allocation of municipal resources.

On the other hand, there has occurred recently a growing recognition of the political power derivable from minority-group concentrations in certain areas. The Black Power movement seems to have become aware of the potential difficulties in political organization which residential integration in cities would bring about. More than twenty years ago, Drake and Clayton (1945: 114) noted this in observing: "Negro politicians and civic leaders have a vested interest in maintaining a solid and homogeneous Negro community where their clientele is easily accessible." Negro clustering and its political consequences need not be viewed in such selfish terms, however. If it is in the interest of the black community to elect black representatives to city councils, state legislatures, and the U.S. Congress, then residential dispersal might be viewed with some qualms. Currently, for example, most Negro public officials are members of city councils (they number 382), school boards (304), and state legislatures (172) [Congressional Quarterly, 1969]. It has been proven much easier for big-city Negro voters to elect congressmen than mayors.

In addition, the black community is placing growing emphasis on the campaign to achieve control over the institutions that affect the lives of black people—public schools, police departments, welfare agencies, for example. Though in part this trend is a result of despair over the prospects for integration in American society, it is no doubt also independent of such disappointments. To the extent that community control will make institutions more responsive to the needs and desires of the black population, the disappearance of the community through integration may be considered a mixed blessing.

The net effect of segregation on the political power of a minority group is thus difficult to assess. Ignoring the more subtle benefits of meaningful communication between black and white citizens, it would seem that as black voters become increasingly organized they have more and more to lose in terms of crude political "clout" from the potential erosion of their urban geographical bases.

Segregation and Social Problems

It is well known (Pinkney, 1969) that the reported incidence of such measures of social deviance as crime and delinquency, mental illness, and drug addiction are higher among the black (and Spanish-surnamed) population than among the majority. Despite biases in the way the figures are reported, and in addition to the role of economic and social disadvantage in generating such deviance, it is also likely that segregation itself plays some role. The notion behind this assertion is that there exists a neighborhood "contagion" effect. How powerful this effect may be has not been subject to systematic empirical appraisal. Another line of argument, rarely made explicitly because of what may be considered as racist overtones, holds that ghetto culture encourages attitudes that are at variance with majority norms and, therefore, inhibit conventional achievement of children raised in a segregated milieu. Ghetto Negroes are alleged to have deviant attitudes on the delay of gratification, the importance of decorum, the sanctity of work, marriage, and private property.

For what it may be worth, a recent quantitative study of juvenile delinquency (Fleisher, 1966: 61-62, 108) found that the availability of targets for crime, indexed by means of the local dispersion in income distribution, had a significant and positive effect on the propensity to commit offenses among crime-prone youth.[13] Integration could serve then to exacerbate delinquency by reducing the economic distance between potential delinquents and their targets.

Segregation and the Configuration of the Metropolis

Given the fact that segregation exists and is enforced by means of the bigotry premium incorporated in the self-segregation model or in other ways, what impact is it likely to have on the general configuration of metropolitan areas? Earlier we discussed the possible effect on business locations which stems from the penning of the poor, minority residents in the central city. The reduction in the tax base and increased demand for services (e.g., welfare, compensatory education, protection) can serve to drive traditional residents as well as firms outward beyond the central-city limits. The result, of course, is a speed-up in the process of suburbanization with all of the costs in service provision inefficiency and fragmentation in political and social life that this process is alleged to entail.

RESIDENTIAL SEGREGATION
AND PUBLIC POLICY

The lack of knowledge about the effects of segregation on other important social phenomena, which the foregoing review has pointed out, compounds the problem of framing public policy. Statements about the necessity for fostering free choice in housing location pretty much exhaust the list of objectives about which any general consensus exists. Not knowing the consequences of segregation prevents the specification of public goals as to racial settlement pattern since there is no clear perception of what an optimal pattern might be. As Lowry (1968: 23) notes in a recent call for research on this question: "Despite the prominence of the issue, there has been virtually no discussion or analysis of the spectrum of possibilities between the polar cases of the monolithic ghetto and salt-and-pepper integration."

Clarification of objectives for housing integration constitutes the necessary beginning in a process of assessment and evaluation of alternative program instruments. The devising of programmatic approaches depends in turn on the development of predictively powerful explanatory models of segregation. I hope that the self-segregation model will be considered a useful first step in that direction.

But, even in the absence of unambiguous objectives and fully developed models, some current policy research is needed. There exists, after all, a disparate set of programs that were designed to further integration or are thought to have an important effect on it. These can be classified by type of approach and should be compared on the basis of success achieved.

(1) Laws guaranteeing open occupancy and equitable treatment in the housing market probably constitute the most common approach. What have such laws accomplished? For instance, are variations among the states in enforcement procedures, remedies, and penalties associated with differential performance in breaking down segregation? Related federal policies with respect to housing at government installations should be examined for effectiveness at the same time.

(2) Though not a part of governmental policy, in many cities, volunteer groups have taken up the task of improving information as a way to promote integration. That is, the typical Fair Housing Council seeks to find majority house sellers (or landlords) with low bigotry levels and bring them together with similarly disposed minority house seekers. Admittedly there would be substantial problems of data assembly in assessing such voluntary efforts, but it might be a task worth attempting.

(3) Among indirect approaches to the reduction of segregation are programs that raise the socioeconomic status of minority group members. More knowledge is needed on the impact of confluence in economic conditions between the races, where it has occurred, and confluence in the pattern of residential-site choices. Particularly important would be investigations of the effect on locational outcomes of rent supplements and other housing subsidies to the disadvantaged.

(4) Finally, we could design a program for attacking residential segregation in a very direct fashion. A scheme similar to the one offered here was first proposed by Lowry (1968: 24-25). It could, if properly organized, provide an extremely interesting experiment in social policy. The idea, essentially, is to compensate people for behaving in a socially desirable way, when it is counter to their inclinations to do so. One version of the scheme calls for the establishment of a metropolitan agency that would subsidize, at some predetermined percentage of the value of the transaction, any sales or rental transaction that would cause a convergence between the fraction of, say, blacks in a neighborhood and the fraction of

blacks in the metropolitan area as a whole. Thus, for example, when a black family moved from an apartment in the ghetto to one in the suburbs, the agency would provide a subsidy of x percent of the rent. And when a white family sold its home in a white-segregated neighborhood and purchased one in a neighborhood with a higher-than-average black residential population, there would be financial assistance in making the second transaction. To the extent it is thought that prejudice on the part of sellers and landlords is effective in preventing transactions that foster integration, steps could be taken to insure that both the mover and the seller (or landlord) would share in the subsidy. An extra advantage of this plan is that the probability of neighborhood-tipping is reduced since the availability of subsidies ends when neighborhood racial balance is achieved.

In the logic of the self-segregation model, the public treasury is in effect here paying the bigotry premium. Some may find such an idea objectionable. If that is the case, we can call the subsidy an integration bonus. Then it will be seen that it is not so different in principle from the payment of subsidies to firms that hire minority workers to cover, ostensibly, "extra costs of training" when, in reality, the subsidy may be compensation in exchange for violation of an existing prejudice. In reality, of course, a long list of government programs contains subsidies (often in kind) to people to encourage them to act in socially desirable ways, e.g., tuition-free education, free public health facilities, below-cost sanitation services, and the like.

Currently, next to nothing is known about the relative efficacy of various policies for achieving housing integration. An effort in preliminary research which systematically compared the options I outline here would be well worthwhile.

In the longer pull, however, a much more fundamental analysis needs to be done. Until we know how and why segregation exists on its current scale, and what its social impacts are, we can accomplish little in healing the breach between the two societies, black and white, which increasingly characterizes America.

NOTES

1. In an extremely thorough appendix on the measurement of segregation, Taeuber and Taeuber (1965: 195-245) argue for the convenience and meaningfulness of the index

$$D_i = \frac{1}{2} \left[\sum_{i=1}^{n} \left(\frac{N_i}{N} - \frac{W_i}{W} \right) \right],$$

where N_i (W_i) is the number of Negro (white) households on block i and N(W) is the number in the entire city. They show how D_i is related to other commonly used segregation indexes and why it is generally superior. Interesting properties of D_i are its independence from the citywide fraction Negro, i.e., (N/N+W) and its values which— between 0 and 100—can be interpreted as the fraction of Negroes who have to change their block of residence if the city were to achieve an even overall distribution of Negroes. Henceforward, in this paper references to the degree of segregation in a city or metropolitan area will refer to the phenomenon which a formula such as that for D_i would index.

2. Reservations follow on the interpretation of the Sudman et al. findings. Even if accepted, however, they do not depart very far from those of the Taeubers' except with respect to how "small" and "large" are defined. The latter find that an average of about eighty-six percent would have to change their block of residence in the larger United States cities in 1960 to make the cities unsegregated. (Shifting from block to neighborhood and from city to metropolitan area and weighting individual city indexes by city populations would lower this number.) They seem to feel that eighty-six percent is a large number, while Sudman et al. suggest that eighty-one percent is small.

3. Such an approach does *not* depend on the assumption that racial differences in household characteristics are random. The problem is to separate, for analytical convenience, the effects of prejudice in housing markets from the prejudice operative in other spheres, e.g., employment or education.

4. The steps, as best I have been able to make out, consist in first subtracting from the explained variance the ratio of the sum of the squared deviations from a line forced through the origin (and with predetermined slope) to the sum of the squared deviation from the mean of the observations on the dependent variable. The remainder of the explained variance is compartmentalized by a procedure that is used on problems of variance analysis with two principles of classification but is inappropriate to simple correlation problems (Taeuber and Taeuber, 1965: 87-90).

5. The gap between Taeubers' analysis of racial housing patterns in Cleveland (1965) and in a subsequent study (1968) is difficult to bridge. In Taeuber and Taeuber (1965: 188), he shows a strong negative relationship between the occupational and educational status of the blacks in a tract and the fraction of blacks in that tract, which means that the higher the economic status of a black, the more likely he is to live in an integrated neighborhood, while he denies (1968) that raising black incomes in Cleveland will more than negligibly reduce segregation in that city.

6. The white tenants' preference for living in the building, other things being equal, could be expressed as a function of the number or the proportion of black tenants. Where race prejudice exists, it would be a negative function. An attempt is made in this

study to more fully explicate the way in which racial feelings may be expressed in housing choices. It will be seen that one reasonable interpretation of such an expression will actually work to reinforce the landlord's economic disincentive to rent to blacks.

7. There is no convenient word to refer to persons against whom prejudice is directed. Apologizing in advance, I will employ the terms, bigots and bigees, to describe prejudiced people and the objects of their prejudice, respectively. In the general case, a bigot may be white or black, rich or poor, and similarly for bigees.

8. Thomas C. Schelling (1969), in an extremely imaginative though also highly abstract paper on the theory of segregation, constructs a set of analytical devices to which the model explicated here owes some filial debt. In general his concern is to show that, given a preference—negative or positive—regarding some characteristics of one's neighbors such as skin color, the final pattern of clustering by color will depend on such variables as strength and variation in preferences, population ratio of the colors, size of neighborhood, initial pattern of color clustering, and like factors. He deliberately excludes from his analysis segregation that results from socioeconomic separation and from organized collective action, and seems rather uninterested in generating empirically testable hypotheses. Particularly interesting is Schelling's exploration of the determinants of stability in cluster patterns—i.e., the problem of neighborhood-tipping.

9. I am currently in the process of working out a formal specification of the model described in these pages. It will appear in a forthcoming publication.

10. It should also be noted that the value the household attributes to the site may be purely a function of its distance from concentrations of low-income people for the kinds of reasons listed on pages 10-12. Since, deplorable fact that it is, there is a high correlation between being Negro and having a low income, the two situations may be indistinguishable in practice.

11. Kain (1969: 98) contends that black buyers are charged a "discrimination premium" for homes outside the ghetto, but offers no evidence to support his contention. He admits, in fact, that there is no unanimity among researchers on this point and that, "No one has been able to carry out the standardization in sufficient degree to demonstrate conclusively that measured price differences are not simply the systematic differences in the housing [services] consumed by whites and blacks."

12. Municipal boundaries within metropolitan areas often appear to serve effectively as racial and ethnic dividing lines. Why this should be so is not clear. Perhaps it is simply a side effect of municipal desires to screen out the poor for fiscal reasons, e.g., through zoning ordinances, structure and maintenance codes, and the like. There may also be informal threat mechanisms that can be applied within the limits of small municipalities, though it is difficult to specify how these might operate. Finally, city limits may help provide information to bigots on where they can find clusters of like-minded people, so that "Cicerdale" will attract white racists and "Marinoville" anti-Semites, thereby decreasing the probability that Negroes will be found in the former or Jews in the latter.

13. It should be noted that findings from the same study (Fleisher, 1966) suggest that raising the income of the population from which crime-prone youth come—e.g., fatherless families in the inner city—has an even larger *negative* effect on delinquency.

REFERENCES

ABRAMS, C. (1955) Forbidden Neighbors: A Study of Prejudice in Housing. New York: Harper & Row.

ALLPORT, G. W. (1958) The Nature of Prejudice. Garden City: Doubleday.

BECKER, G. S. (1967) The Economics of Discrimination. Chicago: University of Chicago Press.

BOWLES, S. and H. LEVIN (1968) "The determinants of scholastic achievement—an appraisal of some recent evidence." Journal of Human Resources, 13 (Winter): 3-24.

CARROLL, J. D. (1949) "Some aspects of the home-work relationship of industrial workers." Land Economics, 25 (November): 414-422.

COLEMAN, J. S. et al. (1966) Equality of Educational Opportunity. Washington, D.C.: Government Printing Office.

Congressional Quarterly (1969) "Detroit win reflects growing black political strength." Volume 27 (September 12): 1681-1684.

DRAKE, S. and H. R. CLAYTON (1945) Black Metropolis: A Study of Negro Life in a Northern Ghetto. New York: Harcourt, Brace & World.

DuBOIS, W. E. B. (1899) The Philadelphia Negro. Philadelphia: Publications of the University of Pennsylvania.

DUNCAN, B. and O. D. DUNCAN (1957) The Negro Population in Chicago. Chicago: University of Chicago Press.

DUNCAN, B. and P. M. HAUSER (1960) Housing a Metropolis—Chicago. New York: Free Press.

DUNCAN, O. D. and S. LIEBERSON (1959) "Ethnic segregation and assimilation." American Journal of Sociology, 64 (January): 364-374.

FARLEY, R. and K. E. TAEUBER (1968) "Population trends and residential segregation since 1960." Science, 159 (March): 953-956.

FLEISHER, B. M. (1966) The Economics of Delinquency. Chicago: Quadrangle Books.

GLAZER, N. and D. McENTIRE (1960) Studies in Housing and Minority Groups. Berkeley: University of California Press.

HARRIS, B. (1961) "Some problems in the theory of intra-urban location." Operations Research, 9 (September-October): 695-721.

HERBERT, J. D. and B. H. STEVENS (1960) "A model for the distribution of residential activities in urban areas." Journal of Regional Science, 2 (Fall): 21-36.

HOOVER, E. M. (1948) The Location of Economic Activity. New York: McGraw-Hill.

HYMAN, H. H. and P. B. SHEATSLEY (1964) "Attitudes toward desegregation." Scientific American, 211 (July): 16-23.

ISARD, W. (1956) Location and the Space Economy. New York: MIT Press and John Wiley.

KAIN, J. F. (1969) "The effect of housing segregation." Savings and Residential Financing Conference Proceedings. Chicago: U.S. Savings and Loan League.

——— (1968) "Housing segregation, Negro employment, and metropolitan decentralization." Quarterly Journal of Economics, 82 (May): 175-197.

——— (1964a) "The effect of the ghetto on the distribution and level of non-white employment." Paper presented at the Annual Meetings of the American Statistical Association, Chicago, December 1964. (Published as P-3059 by the RAND Corporation, Santa Monica, California, January 1965.)

——— (1964b) "A contribution to the urban transportation debate: an econometric model of urban residential and travel behavior." Review of Economics and Statistics, 46 (February): 55-64.

——— (1962) "The journey-to-work as a determinant of residential location." Papers and Proceedings of the Regional Science Association, 9.

LAURENTI, L. (1960) Property Values and Race: Studies in Seven Cities. Berkeley: University of California Press.

LIEBERSON, S. (1963) Ethnic Patterns in American Cities. New York: Free Press.

LOGUE, E. (1968) "Prospects for the metropolis." Pp. 17-23 in A. H. Pascal (ed.) Contributions to the Analysis of Urban Problems. (Published as P-3868 by the RAND Corporation, Santa Monica, California.)

LOWRY, I. S. (1968) "Housing." Pp. 4-26 in A. H. Pascal (ed.) Cities in Trouble: An Agenda for Urban Research. (RM-5603-RC). Santa Monica: RAND Corporation (August).

McCALL, J. (1968) Economics of Information and Job Search. (RM-5745-OEO). Santa Monica: RAND Corporation (November).

McENTIRE, D. (1960) Residence and Race: Final and Comprehensive Report to the Commission on Race and Housing. Berkeley: University of California Press.

MEYER, J. R., J. F. KAIN and M. WOHL (1963) The Urban Transportation Problem. Cambridge: Harvard University Press.

MILLER, H. (1969) "Urban crisis: prospects not all black." Los Angeles Times (July 13): 11, 4-8.

MOONEY, J. D. (1969) "Housing segregation, Negro employment, and metropolitan decentralization." Quarterly Journal of Economics, 83 (May): 299-311.

MOSES, L. (1958) "Location and the theory of production." Quarterly Journal of Economics, 73 (May): 259-272.

MUTH, R. F. (1968) "Urban residential land and housing markets." In H. S. Perloff and L. Wingo, Jr. (eds.) Issues in Urban Economics. Washington, D.C.: Resources for the Future, Inc.

——— (1962) "Housing prices and race in Chicago." Unpublished manuscript.

——— (1961) "Economic change and rural-urban land conversions." Econometrica, 28 (January): 1-23.

MYRDAL, G. (1944) An American Dilemma. New York: Harper & Row.

Newsweek (1969) "Report from black America." Volume 83 (June 30): 17-39.

PARK, R. (1962) Human Communities: The City and Human Ecology. New York: Free Press.

PASCAL, A. H. (1967) The Economics of Housing Segregation. (RM-5510-RC). Santa Monica: RAND Corporation (November).

PINKNEY, A. (1969) Black Americans. Englewood Cliffs, N.J.: Prentice-Hall.

SCHELLING, T. C. (1969) Models of Segregation. (RM-6014-RC). Santa Monica: RAND Corporation (May).

SCHNORE, L. F. (1954) "The separation of home and work." Social Forces, 32 (May): 336-343.

SHEATSLEY, P. (1968) "White attitudes toward the Negro." Daedalus, 95: 217-238.

SUDMAN, S., N. M. BRADBURN and G. GOCKEL (1969) "The extent and characteristics of racially integrated housing in the United States." Journal of Business, 42 (January): 50-87.

TAEUBER, K. E. (1968) "The effect of income distribution on racial residential segregation." Urban Affairs Quarterly, 4 (September): 7-14.

——— and A. F. TAEUBER (1965) Negroes in Cities: Residential Segregation and Neighborhood Change. Chicago: Aldine.

U.S. Bureau of the Census (1962) U.S. Census of Population and Housing: 1960, Final Report PHC (1). Washington, D.C.: Government Printing Office.

U.S. Bureau of Labor Statistics (1968) Handbook of Labor Statistics—1968. Washington, D.C.: Government Printing Office.

U.S. Bureau of Labor Statistics and Bureau of the Census (1967) Social and Economic Conditions of Negroes in the United States. B.L.S. Report 32, CPS Report 24. Washington, D.C.: Government Printing Office (October).

U.S. Office of Policy Planning (1965) The Negro Family: The Case for National Action. Washington, D.C.: U.S. Department of Labor (March).

WEAVER, R. (1948) The Negro Ghetto. New York: Harcourt, Brace & World.

15

The Urban Economy and Public Problems

RICHARD MUTH

☐ CITIES TODAY have many problems—two of the most pressing are those commonly called urban blight and suburban sprawl. By urban blight is meant the poor or deteriorating condition of all kinds of real estate, but especially residential structures in the central city. Suburban sprawl is the disparaging expression frequently used for urban population decentralization—the tendency for population to grow more rapidly in the outer parts of urban areas than in their inner, older zones. These problems, like the so-called ghetto problem or the residential segregation of Negroes, are intimately related to the workings of urban housing and residential land markets. Others, urban transportation and central city fiscal problems in particular, are closely though less obviously related to forces that importantly affect these markets.

While urban housing and residential land markets are of crucial importance for today's urban problems, their workings are not well understood. Most diagnoses of urban problems attribute them to market failures. In particular, it is usually said that urban problems result primarily from the lack of planning and detailed public regulation over residential real estate. Such diagnoses frequently reflect little more than the all too common, vulgar belief that unregulated individual behavior of any sort in economic matters produces chaotic and undesirable results. More deserving of serious attention are arguments that rest upon barriers of some sort, often unspecified, to the efficient operation of housing and residential land markets or to so-called external effects—the failure of individual participants in these markets to take into account the effects of their own actions upon others.

It is surprisingly difficult to find in either popular or professional writing on urban problems a reasonably complete and logically consistent statement of how the failure of housing and residential land markets has produced today's urban problems. Such a statement, however, might run somewhat like the following summary. For one or more of many possible reasons, the demand for housing in the older, central parts of urban areas has fallen. This might have happened because of their age and age-associated obsolescence, the lack of proper planning and regulation when these areas were initially developed, and the failure of local governments to supply the proper kinds and amounts of local services. The automobile is also often held to be the villain of the piece, both because of the noise and congestion it produces and because it has made more distant residential locations relatively cheaper than they once were. Another set of initiating forces often mentioned are external effects produced by the intrusion of manufacturing plants or even small retail shops into once good neighborhoods, the lack of loans for the purchase or improvement of residences in older neighborhoods, and a variety of features of real property and income taxes which are alleged to inhibit the redevelopment of older properties and neighborhoods.

Whatever the reason cited, the effects of the decline in housing demand in the older parts of cities are the same. Because the decline in demand reduces the earnings received by owners of existing structures in the affected areas, these owners reduce the expenditures they make on maintenance and repair. This reduction leads to a further deterioration of the affected neighborhoods and to a further decline in housing demand. Along with the decline in demand, the previous middle- and upper-income residents of the affected neighborhoods move out of them and into suburban areas. Their places are taken by lower-income households, frequently members of racial minorities, who, because of their more limited means, have less of an aversion to living in poorer quality neighborhoods than the original residents.

Arguments such as that which I have just sketched are used to justify a wide variety of government interventions, both federal and local, in housing and residential land markets. Such measures are aimed at undoing the mistakes of the past and preventing their future occurrence. Public housing and rent subsidy schemes are aimed at providing the decent housing to lower-income groups that the private market has failed to provide. Urban renewal attempts to lure middle- and upper-income persons and business firms back into the central city or, at least, to prevent their future flight. More strict enforcement of local building and occupancy codes is urged to slow deterioration of the inner city, while the building of new towns, properly planned and federally subsidized, is urged to prevent future deterioration.

In my judgment, the analysis set out above is almost totally incorrect. Housing and residential land markets might be made to operate more efficiently and external effects might be of greater importance in them than in many other markets. My study of the spatial aspects of the urban housing market and of the determinants of housing quality, however, has convinced me that so-called urban blight, suburban sprawl, and related problems arise for quite different reasons (Muth, 1968: 285-333; 1967). The problem of poor housing quality is largely the result of the low incomes and hence the low housing expenditures of its residents. Urban decentralization has occurred mostly because of improvements in automobile transportation, and to a lesser extent because of the growth of urban populations. Neither urban housing quality nor decentralization has been influenced very much by physical or by social conditions in the central city, except to the extent that the latter have influenced the incomes of lower-income population. In addition, as I interpret the data I have examined, there is little evidence that Negroes pay higher prices for housing of the same quality than whites or that Negroes are confined by segregation to areas of poor housing quality. As a result, I am convinced that many of the measures currently taken, or urged, to ameliorate urban problems do little to combat them and, in many cases, intensify these problems.

In the next three sections of this paper I will discuss my own analysis of problems associated with urban housing and residential land markets more fully. The first section is principally concerned with the problem of urban decentralization, the second with the determinants of housing quality, and the third with the residential segregation of Negroes. In the concluding section I will analyze briefly what I feel to be the shortcomings in current public policy on these problems.

THE PROBLEM OF
URBAN DECENTRALIZATION

Economists have never paid much attention to intraurban population distribution or to the pattern of population densities in different parts of cities. Population density is fairly easy to analyze in economic terms, though, for it is merely the output of housing per unit of land divided by the per capita consumption of housing. The latter two quantities are examples of variables with which economists regularly deal. I will begin this section by considering how, in allocating their incomes between expenditure on housing, transportation, and other commodities, the location of households in cities is determined and the spatial pattern of housing prices is established. I will then show how variation in housing prices induces variation in land rentals. Next I will consider the variation of the output of housing per unit of land or the intensity of residential land use. Finally I will indicate the implications of this variation for urban decentralization.

First, however, it is necessary to make explicit a few definitions of terms. An understanding of my vocabulary is especially important, for I have found it analytically convenient to attach meanings to certain words generally consistent with their technical economic definitions but different from their common usage. *By housing,* I refer to that which is produced by residential structures and by the land upon which they are located, unless modified the term housing refers to the flow of services or satisfactions produced rather than to the assets themselves. More important, by the *quantity of housing,* I mean something more than the services of a certain number of dwellings. The latter differ in a variety of ways, some producing more satisfactions per unit of time than others. It is probably sufficient in most of what follows to suppose that dwellings differ only in size measured in square feet of floor space. The quantity of housing consumed per capita is then floor space per dwelling divided by persons per dwelling; the quantity of housing produced per unit of land is simply total floor space in a neighborhood divided by its area in square miles. The major modification to the definition just noted is that needed when discussing housing of different quality. For this purpose it is sufficient to suppose there are two distinct housing qualities, good and poor.

In speaking of the *price of housing* I mean the amount of money paid per month, say, for a given number of square feet of floor space. It is important to keep this usage of the term price firmly in mind, for in common usage price frequently means the total amount spent for the services of a dwelling for a certain period (or even for the structure itself and its site, the latter being equal, of course, to the capitalized value of all future services rendered by the structure and by its site). In my vocabulary, the total amount spent for the services of a dwelling for a month is expenditure and is equal to price per square foot multiplied by the number of square feet (of services) purchased.

In a similar vein, in referring to the amount spent on land I will use the phrase, expenditures for land. By rental of land I mean the amount spent per month per square foot of land. I use the term rental for the price per unit of the services of land rather than the term rent because, though the latter started out to mean payments for the use of land, its present technical meaning in economics is quite different. Finally, when speaking of land or of housing, I will suppose in my analysis, unless noted otherwise, that they are homogeneous, except for location, and perfectly divisible. By so doing I neglect features of real estate such as durability, hetero-geneity and immobility which seem to many person to be the essence of urban housing problems. I am convinced, however, that such problems arise for quite different reasons and that urban blight and suburban sprawl can be understood quite readily with-out appealing to these seemingly distinguishing characteristics of real estate.

The one differentiating feature of housing and land, whose presence I assume throughout, is location. Since other quality differences are generally ignored, location is important only insofar as it affects the costs of households or business firms located on it. I also neglect variations in the prices of consumer goods other than housing and producer's inputs other than land. Thus, location is significant only because of differences in transportation costs for consumers and land rentals for producers that are associated with it.

Because households with members working in the central busi-ness district (CBD) of the city incur greater transportation costs the greater their distance from the CBD, the price per unit of housing paid must decline with distance. If it did not, a house-hold located, say, five miles from the CBD could, by moving closer, reduce its transportation costs, consume the same amount of housing and all other commodities as it did before, and have something left over. It would thus be willing to offer more per unit of housing at the closer than at the more distant location. Indeed, it is fairly easy to demonstrate that if a household cannot improve itself by moving, then at its location, the relative rate of decline in housing prices per mile must be equal to the marginal cost of transportation to it (that is, the increase in the cost of transportation which is incurred by moving a mile farther from the CBD) divided by its expenditures on housing, both being expressed on a per month basis (Muth, 1969: 21-23).

It follows, then, that the relative rate of price decline must become smaller if transport costs decrease. If all households were in their best locations prior to the decline in transport costs, the reduction in the expenditure necessary to purchase the quantity of housing purchased would be just equal to the additional transport costs incurred for a short move. Hence, with a decline in marginal transport costs, any household could, by a move further from the CBD, purchase the same quantity of housing and other commodities it would have previously and, by incurring the now smaller additional transport costs, have income left over. Households would thus seek to locate farther from the CBD than they otherwise would. By so doing they would bid housing prices up at greater distances from the CBD and bid them down at shorter distances.

The decline in housing prices with distance from the CBD in turn causes residential land rentals to decline. For if rentals did not, firms producing housing (including owners, whom I treat as if they were landlords selling housing to themselves as tenants) would incur the same costs of producing housing but receive lower prices for it at greater distances from the CBD. Since the net incomes of firms would thus decline with distance, it would be in the interest of more distantly located firms to offer higher rentals for land closer to the CBD. If, then, firms were to have no incentive to shift their locations, land rentals would have to decline with distance from the CBD.

Of course, the rate of decline of land rentals would be greater, the greater the rate of decline of housing prices. Expenditures on land, however, are only a fraction—about one-tenth to one-twentieth in recent years—of all expenditures on housing. To equalize the net incomes of firms producing housing in different locations, residential land rentals would therefore have to decline at a rate ten to twenty times as fast.

The decline in residential land rentals, coupled with construction costs which are the same everywhere, produces differences in the kinds of structures used to produce housing. Toward the center of a city, high-rise apartment buildings predominate. These, of course, yield housing using relatively little land. Toward the outskirts of an urban area, housing is relatively cheaper to produce in the form of single-family, detached structures, frequently one story on slab foundations. In the outskirts, then, the intensity of housing output per square mile of land is lower than in areas of the city where apartment buildings predominate.

The rate of decline in the output of housing per square mile of land depends upon the decline in land rentals and, hence, housing prices. The responsiveness of housing output to a given price change varies directly with the ease of substitution of land for structures in producing housing, and inversely with the fraction of expenditures on housing paid out for land. The latter is the case because, residential land being in relatively fixed supply, the larger the fraction of housing expenditure made for land, the more difficult is the substitution of additional structures for land with an increase in housing output. Now, toward the CBD where expenditures per square foot of residential land are relatively high, the fraction of housing expenditures spent on land is also relatively high. The response of housing output to price changes is thus greater in the outer parts of an urban area than it is toward the center.

I am now in a position to explain why population growth and improvements in transportation lead to population decentralization in urban areas. First, consider the effects of population growth. A growth in a city's population, of course, implies a growth in housing demand and thus in the demand for land and for structures to produce housing. While the inputs into the construction of structures—mainly materials and labor—probably flow quite freely into a city from other uses, increases in residential land result only from bidding the fixed amount of land at any location away from other uses. Indeed, increases in land used to produce housing probably come mainly from local agriculture, implying a growth in a city's land area, since increased population also results in an increase in the demand for land for roads, grocery stores, and bakeries.

The increased residential land rentals, in turn, result in higher housing prices to consumers in all parts of the city. It appears, though, that consumers cut down the quantity of housing they purchase as its price increases by just about the amount required to keep their housing expenditure constant. For this reason the relative rate of decline in housing prices is not affected, at least so long as population growth does not have other effects (especially on transport costs). Because the value of housing produced responds more rapidly to the increase in housing prices in the outer parts of cities, and expenditures per person remain largely unchanged, population must increase more rapidly in the outer parts of cities.

With transportation improvements, which in recent years have resulted primarily from expressway construction financed largely by federal funds, the cost of traveling an additional mile by car has declined. Perhaps the most important element of this decline is in the time people spend in travel. Those of us who have reached middle age can readily appreciate that time is not in unlimited supply; hence, it has value, a value that increases as we grow older. The decline in transport cost, in turn, causes the rate of decline in housing prices with distance to become smaller. The latter, in turn, means that housing output and population grow in the outer parts of cities relative to their inner parts.

In my study of urban population distribution I have compared statistically the relative rate of decline in population densities in 1950, the fraction of an urbanized area's population residing in the central city, and the former's land area both for 1950 and 1960 with a variety of different measures. Of the determinants of urban population distribution and land area, population and car registrations per capita, which I interpret as a proxy for transport costs, have been of overwhelming importance in explaining changes taking place in the postwar period. The increase in car registrations per capita of 0.26 to 0.35 from 1950 to 1960 by itself would account for a decline in the fraction of an urbanized area's population residing in the central city from 7 to 14 percent. The latter compares well with an actual decline of 9 percent for the areas I studied. Similarly, the increase in land area of the urbanized areas studied averaged about 82 percent from 1950 to 1960. Of this, the increase in car registrations would account for an increase in land area of about 45 percent, while the urbanized area's population growth would account for a land area growth of about 25 percent (Muth, 1969: 180-183).

While I did find certain other variables associated with urban population and land area, notably income and the spatial distribution of manufacturing employment, I found virtually no evidence that population distribution in cities is appreciably affected by the physical condition of the central city. Among the indicators of physical condition examined were the age of the central city and its dwelling units, the fraction of the latter which were substandard—dilapidated and/or without private bath—the average population density of the central city, and the proportion of its employment which is in manufacturing. Of these, the fraction of substandard central city dwellings did tend to be inversely associated with the relative rate of decline in population density, as arguments such as those outlined in my first section would imply. The fraction substandard, however, showed no clear association with the central city's share in urbanized area population or the latter's land area. Furthermore, the fraction of substandard dwellings declined markedly during the fifties—from 0.20 to 0.11 in the ones studied, so changes in dwelling condition would imply recentralization.

If anything, the fraction of the central city's population which is Negro showed a somewhat clearer inverse association with the relative rate of population density decline within the central city than did dwelling condition; like the latter, though, it showed no consistent association with the central city's population fraction or with land area. With these two exceptions, though, none of the indicators of physical or social conditions in the central city showed any association with population decentralization. Thus, *not only can one account for most of the suburbanization of population and spread of urban areas that occurred during the fifties by transportation improvements and population growth, but there is virtually no evidence that these phenomena are in any way a "flight from blight."*

DETERMINANTS OF HOUSING QUALITY

Most discussions of the so-called slum problem and the theory outlined in the first section attribute the *spatial* spread of slums in recent years to increase in their supply (thoughout this paper I use the term slum and poor-quality housing to mean the same thing and identify them empirically with substandard housing, as defined previously). Some of the forces stressed in these discussions may have some empirical support. Such arguments, though, conflict rather strongly with three rather simple but important empirical considerations. In beginning this section I wish first to set forth these inconsistencies. After discussing an alternative theory which is more consistent with these empirical considerations, the section will conclude with a discription of the major findings of my statistical investigations into the determinants of housing quality.

Forces affecting supply, such as age and obsolescence, the spread of industrial and commercial firms into residential neighborhoods, or a shortage of funds for investment in residential real estate, which are commonly stressed, tend to increase the number of poor-quality dwellings. These forces by themselves, however, have no effect upon demand for such housing, or the willingness of people to inhabit such dwellings. Thus, in order for additional substandard housing to be *inhabited,* the price of this poor-quality housing would have to decline relative to that of good-quality housing, as would the relative earnings of firms producing poor-quality housing in other parts of the city. For reasons that are essentially the same as those discussed in connection with the variation of residential land rentals with distance in the preceding section, the rentals of land (plus existing structures) in poor-quality housing areas would thus decline. The output of poor-quality housing per square mile of land would thus tend to fall relative to that of good-quality housing.

Therefore, if conventional arguments were correct, one would expect that the production of poor-quality housing would be a rather unprofitable business and that land in so-called slum areas would be less intensively utilized than in comparably located areas of better housing. Empirical observations are at variance with the implications of conventional wisdom, both with respect to the unprofitability and to the lower-density conclusions. First, if anything, it is widely believed that slums are especially profitable to their owners. It is quite difficult to obtain reliable data on the gross earnings of producers of slum housing and, even if one could, there are several sensible reasons why high gross earnings would be consistent with normal net earnings. Second, I have found that, if anything, population densities and measures of housing output per unit of land tend to be higher, not lower (other factors such as location being held constant), the higher the fraction of dwellings that are substandard (Muth, 1969: 240).

Closely related to the above-noted shortcoming in conventional theory is the fact that urban renewal projects lose money and require government subsidy. Indeed, the receipts from the resale of cleared land themselves are typically much smaller than site acquisition costs (Anderson, 1964: 19-23). Yet according to traditional arguments, if urban renewal projects were properly planned and executed—particularly if they corrected the deficiency that led to the initial decline in demand for good-quality housing—one would expect them to earn a surplus. While urban renewal projects might find justification on other grounds, experience with them to date would seem to be inconsistent with most arguments for the existence of slums.

Most important, the forces affecting supply discussed earlier, almost without exception, would lead one to expect continued deterioration of central city housing in the postwar period. Central cities and the majority of their dwellings continued to age, local governments became progressively more hard pressed in supplying services to their residents, the building of urban expressways greatly facilitated automobile transportation, and so forth. Indeed, it is widely believed that urban housing has continued to deteriorate. It is important, though, to carefully distinguish between the spatial spread of a city's relatively poorest housing, on the one hand, and the average (absolute) level of quality of the entire housing stock. It is undoubtedly true that, as urban populations have grown, the number of units comprising and land area occupied by the poorest, say, fifth of its housing stock have both increased. At the same time, however, census data make it abundantly clear that the fraction of substandard urban housing that is below a given *absolute* level of quality declined dramatically during the fifties (Muth, 1969: 124-125, 280).

The three sets of considerations which seem to me to be inconsistent with traditional arguments are readily explainable by shifts in the effective demand for poor-quality housing. I stress the adjective effective since few persons would prefer to live in poor housing simply because they are slothful or ignorant. Many, however, have such limited incomes that to spend more for better housing would necessarily cut sharply into already meagre expenditures for food, clothing, and medical care. The income of a family is certainly the principal determinant of its expenditures on housing. Households with small incomes make small expenditures on housing, and small expenditures buy not only fewer feet of floor space or rooms per person but poorer quality as well. The latter is the case because it is relatively cheaper for the private housing market to convert existing dwellings, especially older ones in multi-unit structures, than to build new ones when meeting the low-income demand for housing. Such conversions are accomplished partly by cutting larger apartments up into smaller ones, and partly by allowing them to deteriorate in quality through reducing expenditures for maintenance and repair.

As cities grew during the 1920s and 1930s, their growth brought about an increase in the demand for all kinds of housing including lower-income housing. For reasons already discussed in the second section, the growth in housing output and population took place disproportionately in the outer parts of urbanized areas. These, of course, tended largely to be outside the central city. At the same time, higher-income households tend to live primarily in newer dwellings. The latter may be partly due to preferences for newness as such, but mainly, I suspect, it is because it is relatively cheaper for firms producing higher-income housing to do so in newer dwellings. The residual effect, of course, was that the central city came to house disproportionately large numbers of an urban area's lower-income households.

Within the central city, then, the demand for poor-quality relative to good-quality housing increased. Indeed, despite the increase in average income during the interwar period, the per capita consumption of housing showed little increase, on the average. This was primarily because housing prices increased relative to other prices. Thus, the fraction of the housing stock produced by substandard units probably (though we have no data) showed little decrease in the whole of the urban area and may actually have increased within the central city. In the postwar period, by way of contrast, the evidence suggests that the per capita consumption of housing increased markedly, both because incomes grew more rapidly than the average rate from 1920 to 1940, and because housing prices increased less rapidly in relation to other prices (Muth, 1969: 128-129).

The considerations just discussed explain both the spread of poor-quality housing within the central city and the improvement in average quality during the postwar period. To account for the apparently high price of slum housing in relation to its quality, it is necessary to consider the conditions of its supply as well. The output of poor-quality housing is increased as its price rises relative to that of good-quality housing, with a growth in demand, partly by more intensive utilization of poor-quality structures and the land upon which they are situated, and partly by the conversion of better quality housing, as discussed earlier. If all structures were equally easy to convert to the production of poor-quality housing, their conversion would, given sufficient time for adjustment, return the relative price to the level that prevailed prior to the growth in demand.

It is probably the case, however, that due to the great variety of structures used to produce housing, structures differ in their ease of conversion or in their relative earnings in producing poor- vs. good-quality housing. If this latter is the case, with a growth in poor-quality demand, recourse would be to dwellings that are successively more expensive to convert. Even after full adjustment to the change in demand, the relative price of poor-quality housing, and thus the earnings of properties used to produce it before the change in demand, would be higher than previously. Previously existing slums, then, become more profitable to their owners as the land area occupied by slums grows. At the same time, urban renewal projects that, in effect, take land from a higher- to a lower-priced market incur losses.

I have related the fraction of dwelling units that are substandard to a variety of measures statistically, both for census tracts within a given central city for 1950 and 1960 and among different central cities for 1950. The one variable most strongly and consistently related to this fraction is the income of its inhabitants. In addition, the fraction is highly responsive to changes in income. On the average, with a ten percent increase in the average income of the residents of a census tract or of a central city, the fraction of its substandard dwelling units would fall by about one-third. Furthermore, using my estimates of the response of this fraction to changes in income and construction costs made from cross-section data and observed changes in real median family income and construction costs of three and one percent per year, respectively, from 1950 to 1960, one calculates an expected decline in the fraction substandard of fifty percent. This, of course, compares quite closely with the fall from 0.20 to 0.11 cited toward the close of the second section.

Now one might be tempted to argue that these findings could also be explained by the effects of dwelling-unit condition on the kinds of households that locate in them. This is admittedly a problem with comparisons within a single city. As to the comparisons among different cities, it is a bit of a strain on the argument to suggest that the better housing quality in, say, Chicago over Atlanta has attracted higher-income households to the former. It is also difficult to account for the consistency in changes over time for the nation as a whole on this basis. However, using data for the South Side of Chicago, I have tried to separate out the effects of income and dwelling-unit condition on the location of households using appropriate statistical methods. It turns out that this alternative association is quite small quantitatively. Hence, theories that attribute the spread of slums to increases in supply cannot account for the association between the substandard fraction and income.

Even though slums may exist principally because of low-income housing demand, other forces may contribute to their presence and the private (supply) market might overreact to this demand. In fact, I have found little empirical support for either proposition. In regard to the first contention, a variety of measures similar to those indicators of physical and social conditions mentioned in the second section were included in my comparisons. The only ones that showed up at all strongly in the proper direction were the age of dwellings and a measure of population turnover, both in 1950 but not in 1960. The difference in the results for the two years may have been the result of rent control in 1950. When apartment rents are held down by rent control, landlords have less incentive to spend to maintain their properties. The reduction in their expenditures for maintenance is likely to have greater effects on older housing and upon housing with a higher turnover of tenants, the latter because the landlords' costs are higher. With regard to the second contention, as indicated earlier in this section, if the private market had produced too much poor-quality housing relative to the low-income demand for it, land would be used less intensively in slum areas than in comparably located areas of good-quality housing. In fact, it appears that in South Chicago in 1960, at least, the reverse was true.

RESIDENTIAL SEGREGATION
OF NEGROES

The residential segregation of Negroes and other minority groups is a third urban problem intimately related to the workings of urban housing and residential land markets. Despite widespread attention, civil rights activity, state, local, and (most recently) federal legislation seeking to curb it, such segregation has shown itself to be remarkably resistant to change. As with other urban problems, the apparently intractable nature of segregation may be partly due to our failure to understand it. In this section I first consider reasons for segregation. I then turn to a consideration of whether the residential separation of Negroes necessarily implies higher housing prices for them and whether, in fact, they pay them.

The most common explanation for the segregation of Negroes is that landlords and real estate agents have a unique aversion, not shared by the rest of the community, to dealing with them. There are several shortcomings to this explanation, but it has one overwhelming one. If, because of their own preferences, landlords refused to rent, or real estate agents refused to sell to Negroes, they would earn smaller incomes than they would otherwise. Landlords and real estate agents would thus profit from selling their businesses to others who do not share their aversions. Segregation is also sometimes attributed to a grand conspiracy on the part of landlords, real estate agents, mortgage lenders, and perhaps others to profit from higher housing prices to Negroes. (Conspiracy by the "interests" has long been a popular explanation for all kinds of economic and social phenomena.) Such a collusive agreement though, which would necessarily involve hundreds of firms in cities of even moderate size, would be quite difficult to organize. If organized and successful in raising housing prices to Negroes for a time, any member or outsider could profit by selling at a lower price than the collusive one. Such an agreement, then, would almost certainly be short-lived.

A much more satisfying explanation for the residential segregation of Negroes is Becker's (1957: 59) consumer preference hypothesis, namely that whites have a greater aversion to living among Negroes than do other Negroes. If so, whites would live in the vicinity of Negroes only if they paid a lower price for housing than elsewhere. Thus, if initially some Negroes were living in a predominantly white area, it would be in the interest of the owners of the occupancy rights to their properties to sell them to whites who, by hypothesis, would pay more for occupancy. Whites and Negroes would thus gravitate over time to disjointed residential areas.

As Bailey (1966: 215-220) has pointed out, though, if housing prices were the same or lower in the interior of the white, relative to the Negro area (i.e., sufficiently far from the boundary so that the presence of the other group does not affect the prevailing housing price) along the boundary separating them, prices in white occupancy would be lower than those in Negro occupancy. Consequently, it would be in the interests of owners of occupancy rights to white residences to sell them to Negroes, the Negro area would tend to grow relative to the white area, and housing prices paid by Negroes would fall relative to those paid by whites. Consequently, under stationary conditions of relative population and income, only if housing prices were higher in the interior of the white area than in the Negro area could relative prices remain stable over time. However, if the Negro population, and thus housing demand, were growing just enough relative to that of whites, housing prices in the interior of the Negro area could remain above those in the white area. Clearly, though, as Becker (1957: 59) pointed out over a decade ago, their segregation need not imply that Negroes pay more for housing of given quality than whites do.

The Becker hypothesis readily explains the persistence of landlords and real estate agents who refuse to deal with Negroes. By renting to Negroes, a landlord would either have to reduce the rents to whites or lose his white tenants. During the transition to Negro occupancy he would suffer a reduction in income while some units were vacant. Similarly, a real estate agent stands to suffer the loss of future business from white residents of an area in which he arranges a sale to a Negro. The Becker-Bailey analysis also explains how so-called blockbusters make money by buying occupancy rights from whites and selling to Negroes. Assuming the greater aversion of whites to Negroes than Negroes to other Negroes, so long as the Negro area is expanding, the price paid for housing by whites along the boundary is below that paid by Negroes. If, as seems likely, whites cannot contact Negro buyers and established real estate agents refuse to handle interracial transactions, blockbusters are able to buy at the lower white price and sell to Negroes at a higher price. Note, too, that Negroes are not harmed monetarily by such transactions since they always have the alternative of purchasing from other than the blockbuster in the interior of the Negro area. Because of their contiguity to the Negro residential area, however, white sellers cannot sell at the price prevailing in the interior of the white area.

Good evidence of the price paid for housing by Negroes is difficult to obtain and evaluate. Certainly census data rather clearly show that at given income levels the expenditures for housing by nonwhites are greater than those for whites. Expenditure, however, is price multiplied by quantity purchased. It would appear that the quantity of housing purchased is responsive enough to price changes so that, with a rise in price, expenditure either remains unchanged or, perhaps, declines. Thus, higher housing expenditures by nonwhites are not necessarily consistent with higher housing prices. My comparisons of population densities and related magnitudes on the South Side of Chicago for 1950 and 1960 (Muth, 1969: 238-239, 280) provide little evidence of any appreciable price differential. Not only are population densities very similar, as between otherwise comparable nonwhite and white census tracts, the fraction of dwelling units in one-unit structures, which I would interpret as a measure of the physical intensity of housing output per unit of land, was also only very slightly different. In addition, if Negroes paid higher prices for housing than whites, one would expect the former's per capita consumption of it to be smaller. Holding income and other factors constant, though, the fraction of households living in dwellings with more than one person per room, and the fraction of substandard dwellings are no higher for nonwhite areas than for white areas. This last finding is also inconsistent with the common assertion that segregation restricts Negroes to areas of the worst housing quality. Finally, in comparing changes in census-reported rents in census tracts switching from white to nonwhite occupancy from 1950 to 1960 with changes in tracts remaining in white occupancy, I found negligible differences (Muth, 1969: 11). Negroes are, of course, more poorly housed than whites, but I submit that it is because of their lower incomes, not because of a racial differential in housing prices.

SHORTCOMINGS IN CURRENT PUBLIC POLICY

Toward the close of the introductory section, I asserted that, probably because the reasons for urban blight and suburban sprawl are not well understood, government programs directed against them may very well intensify these problems. Having presented my own analysis, I would now like to explain my reasons for this assertion. I shall discuss, in turn, programs affecting the quality of central city housing, programs affecting urban decentralization, and those affecting the residential segregation of Negroes. In each instance I first will discuss the shortcomings of current programs. Following this I will describe briefly the kinds of programs I think would be more useful in attacking current problems.

Measures designed to improve central-city housing may seek to demolish poor-quality structures, increase the cost to private firms of producing poor-quality housing, or reduce the price of better housing to lower-income groups. Demolitions, as under the federal urban renewal program, may initially reduce the fraction of dwellings that are substandard. They thus reduce the supply of poor-quality housing but do nothing to affect the demand for it. Consequently, they increase the price of poor- relative to good-quality housing. Some households that might otherwise have lived in poor-quality housing are induced by the price change to inhabit better housing, while others continue to inhabit poor-quality housing but consume less of it. The latter is accomplished through larger numbers of persons per dwelling, further deterioration of remaining poor housing, or both.

The long-run effects of demolitions, however, may be quite different. Due to the rise in the price of poor- as compared with good-quality housing, some owners of the latter who would otherwise have maintained their properties now have the incentive to convert them to poor quality. As properties are converted, the fraction of substandard dwellings increases again and the relative price of poor-quality housing falls. If residential structures did not differ appreciably in their costs of conversion from good to poor quality, demolitions would have no long-run effect on housing quality. Rather they would merely relocate the slums. It seems more likely to me, though good evidence is lacking, that because structures differ in their costs of conversion, demolitions would still, in the long run, result in some net reductions in the fraction that is of poor quality. However, demolitions would also raise the price of poor-quality housing. Thus they reduce the housing opportunities for the poor and increase, not reduce, poverty.

While many have come to realize that demolitions are harmful to the poor, few appreciate that measures which increase the cost of producing poor-quality housing have similar effects. Such measures include stricter enforcement of building and occupancy codes, the rent strike, and public receivership of slum dwellings. The imposition of any such measures reduces the earnings of poor-relative to good-quality housing. While perhaps beneficial to lower-income persons in the short run, in the long run fewer dwellings would be converted to poor quality as the lower-income population grows, and some units might even be reconverted to good-quality ones. Thus, in the long run, cost-increasing measures also reduce the housing opportunities for the poor. Unlike demolitions, though, this will be the case regardless of the ease of conversion, so long as it is possible at all.

Unlike demolitions and code enforcement, measures such as the provision of public housing and rent subsidies do increase housing opportunities. They probably do so inefficiently for two reasons. First, a favorite government means to reduce the price of something is to provide loans for its production at below-market interest rates. Doing so may reduce the price of housing, but it encourages producers of housing to use more capital expenditure than they otherwise would, and less expenditure for maintenance and for operation. Indeed, many of the difficulties now being experienced by public-housing projects may be traceable to this source. Structures in which such housing is located are frequently very well built and durable but often poorly maintained. Interest rate subsidies would be desirable on economic grounds only if barriers to the investment in urban housing resulted in higher interest rates on such loans than on other uses of capital, allowance being made for differences in administrative cost and risk of making loans. Otherwise, interest rate subsidies are more costly to society in achieving a given increase in housing consumption than housing price or rent subsidies.

Second, rent subsidies, while the most efficient way to increase housing consumption, may be inefficient in improving the well-being of the lower-income population. This group, while poorly housed, is also poorly fed and clothed. If the equivalent of the rent subsidy were given as an income subsidy, the recipients would undoubtedly increase their expenditures on housing. But they would also spend more on food, clothing, and medical care as well. Recipients of the subsidy, though, would always have the option of spending the whole of it on housing if they wished. If they did not, one presumes they consider themselves better off under the expenditure pattern they actually select for themselves. Housing subsidies might be preferable to income subsidies only if, by inducing lower-income groups to consume more housing and less of other things, other persons would be made better off as well. This last issue is far too complicated to explore here, but there is little good evidence that increased housing consumption produces such external benefits.

As I argued earlier, the low-income housing demand is the basic cause of poor housing. It would thus seem sensible to attack the problem at its source through income subsidies. One of the principal reasons I prefer income subsidies as opposed to urban renewal is that the former avoid the harmful side effects of other such programs. In addition, the fraction of housing which is of poor quality is highly responsive to income, having fallen by about half during the decade of the fifties alone because of the income growth that occurred then. Indeed, I have estimated that an income subsidy would have cost no more per substandard dwelling eliminated than public housing did during the 1950s (Muth, 1969: 334-335).

The major contributor to urban decentralization during the postwar period has undoubtedly been the building of urban expressways, largely financed with federal funds. Expressways are rarely built primarily as a device for reducing decentralization though it is sometimes argued that by improving the access of outlying parts of urban areas to downtown employment and shopping, they prevent decentralization of business firms. Actually, by producing dispersion of residential locations, they may well contribute to the decentralization of business as well.

Other programs that have contributed, or would contribute, to decentralization are home-ownership subsidies to middle- and upper-income groups and new towns. I have estimated that the federal income tax advantage to home-ownership and federal mortgage programs by as much as thirty-five percent. In addition, I would judge that the land area of cities may be as much as seventeen percent larger than they otherwise would be because of such programs and the central city's fraction of the urbanized area population three to seven percent smaller (Muth, 1969: 101-104, 319-322). Since they tend primarily to benefit middle- and upper-income groups, such subsidies ought to be eliminated. Federal subsidies for the building of new towns are frequently urged to inhibit future decentralization by avoiding the undesirable results of past urban development. Yet as I stressed previously, there is little or no evidence that our current problems result from past mistakes. And, by reducing housing prices in the outer parts of urban areas, subsidized new towns would certainly contribute to decentralization currently.

Actually, with one exception, it is quite difficult to make much of a case for trying to inhibit or reverse urban decentralization. Largely because of improvements in transportation, the central cities of urban areas have lost part of the advantage they once possessed as places of residence and work. Not to take advantage of such improvements is about as sensible as would be forbidding farmers to use tractors in trying to solve farm problems.

The one exception is the fiscal problem of central cities. Because central city, lower-income populations have grown, the demand for many kinds of public services has risen while yields of taxes levied and collected in the central cities have declined. Raising taxes would only tend to encourage further decentralization. Seeking to bring population and business firms back into the central city, though, is much like bringing the mountain to Mohammed. A much simpler and less costly solution to the problem, in principle, is for a higher level of government to levy taxes throughout the whole of the urban area, or an even broader region, and make increased payments to central cities. Federal revenue sharing would be an excellent device for so doing.

Finally, I would like to comment briefly on the problem of the residential segregation of Negroes. As I stressed in the fourth section, such segregation need not reduce the housing consumption of Negroes and, indeed, the evidence suggests to me that it has not done so. Rather, to the extent that Negroes are more poorly housed than whites, it is largely due to the fact that their incomes are smaller. Segregation may well be undesirable for other reasons. However, I would not anticipate that so-called open-occupancy legislation, even if enforced much more vigorously than such legislation has been in the past, would appreciably reduce segregation. While such legislation could, of course, lead recalcitrant landlords and real estate agents to rent and sell to Negroes, it would not prevent whites from moving away from the immediate vicinity of Negroes. Similarly, schemes to colonize Negroes in suburban areas would probably result in enclaves of Negroes, many examples of which already exist. To prevent the development of new, all-Negro neighborhoods, quota systems that would limit the number of Negro families have sometimes been suggested. Such systems could perhaps prevent the development of new ghetto areas, but they would certainly work to limit the housing opportunities of Negroes.

The basic difficulty to be overcome in eliminating the residential segregation of Negroes is the greater aversion of whites to living among Negroes than that of other Negroes. Admittedly, I have no suggestions for overcoming these aversions. It is certainly better, however, to face up to the situation than to incur further alienation and frustration by promising to correct the problem by means that are bound to fail.

REFERENCES

ANDERSON, M. (1964) The Federal Bulldozer. Cambridge: MIT Press.

BAILEY, M. J. (1966) "Note on the economics of residential zoning and urban renewal." Land Economics 42.

BECKER, G. S. (1957) The Economics of Discrimination. Chicago: University of Chicago Press.

MUTH, R. F. (1969) Cities and Housing. Chicago: University of Chicago Press.

——— (1968) "Urban residential land and housing markets." In H. S. Perloff and L. Wingo, Jr. (eds.) Issues in Urban Economics. Baltimore: Johns Hopkins Press for Resources for the Future, Inc.

HOW THE "SYSTEM" AFFECTS THE POOR

16

Tax Structures and
Their Impact on the Poor

DICK NETZER

☐ THE TAX STRUCTURES that are relevant to this paper include
those of both state and local governments. In the aggregate, the
states collect somewhat more in taxes from urban residents than
do the local governments, and state government taxes are rising
more rapidly. In a very real sense, we function with fifty-one
distinctive tax structures, each applicable to the whole collection
of governmental units operating within the boundaries of the fifty
states and the District of Columbia. The tax policy options open
to individual urban local governments are circumscribed by the
statewide structure. This is not only because state legislative
approval is typically required for major local innovations, but also
because the composition and level of state taxation effectively
determine, by preemption, what local governments can do. The tax
burdens perceived by local residents clearly include state income,
sales, and excise taxes.

Every study of the incidence of state and local taxes by income
class, whether its geographic frame is the nation, a state, or a city,
has found state-local tax structures to be regressive in their incidence (University of Wisconsin Tax Study, 1959; Brownlee, 1960;
Musgrave and Daicoff, 1958; Graves, 1964; Rostvold, 1966; Netzer,
1966, ch. 3; Donheiser, 1966; Gillespie, 1966: 122-186). The
simple fact of regressivity need not be grounds for social concern,

given that the federal tax structure is rather progressive in incidence and that the benefits from public expenditure are decidedly "pro-poor," that is, expenditure benefits amount to large percentages of the money incomes of low-income households and much smaller percentages of the incomes of the high-income households.

For many types of taxes, regressivity results largely because the tax amounts to small *percentages* of the income of the rich. Where this is the case—for example, with highway-user taxes—the regressivity is essentially technical in nature. The real problem occurs when the regressive nature of the tax or tax structure shows up mainly at low-income levels, that is, in heavy tax burdens on the poor. Unfortunately, that is the general characteristic of state-local tax structures in the United States; sharp regressivity at the very low incomes combined with a roughly proportional incidence over a broad middle-income range. Typically, the finding is that state-local taxes as a percentage of income for households with incomes below $5,000 are more than twice as high as for incomes above $5,000; for incomes below $3,000, the percentage is usually three times (or more) as high as for those above $5,000.

However, it is quite clear that, on the average, state-local public expenditure benefits to the poor far exceed the tax burdens. In one study, for the nation as a whole in 1960, benefits less taxes for income classes below $3,000 exceeded twenty-five percent of family money income (Gillespie, 1966: 162). But this does not obviate concern for heavy tax burdens on the poor, for governments can use alternate forms of taxation to finance the same income-redistributive expenditure program. Moreover, the imposition of taxes has political and economic costs. If the same objective can be achieved by raising less in taxes from the poor and spending less to benefit them, these costs are in effect totally unproductive ("deadweight losses") and to be avoided. It is at least conceivable that, in some instances, taxing the poor less may be the most efficient way of aiding them.

Another reason for concern is connected with the fragmented structure of subnational government in the United States. Most of the incidence studies do not examine all taxes and expenditures applying within a single political jurisdiction, but instead look at the fiscal situation aggregated over a multijurisdictional geographic area. This tends to understate the regressivity of the tax structure and overstate the "pro-poor" incidence of public expenditure. As I have written elsewhere:

Aggregation tends to reduce the regressivity of the principal local revenue, the property tax, by combining high-housing-consuming suburbanites with low-housing-consuming city dwellers, the former with high incomes and the latter with lower incomes. Effective property tax rates are usually higher in the larger central cities, but these differences are not reflected in the aggregate studies, which apportion property taxes in some relation to housing consumption. In any case, the evidence suggests that the property tax is far more regressive within individual cities than it is on a statewide or nationwide basis. It seems likely that, in reality, the fiscal systems of the large metropolitan areas produce almost no redistribution between the central city poor and the suburban rich, aside from that not insignificant element resulting from the operation of state and federal grant programs [Netzer, 1968a: 439].

Before further examination of the sources of regressivity, a word of caution is in order, in connection with the very bottom group in any set of income distribution statistics. Such statistics invariably indicate the existence of a significant group of households whose income is virtually zero. For example, it is estimated that as of January 1968, nearly 40,000 households in New York City had incomes of less than $1,000 (Gordon, 1969: 68). Since public assistance or other transfer payments usually exceed such levels, many of the households in the very bottom group are not the "ordinary" poor. Instead, they are individuals and families with very low temporary incomes, surviving on private gifts or past savings—college students, ordinarily well-off families with volatile incomes (for example, professional artists and writers, and commodity traders), persons between jobs, and the like. Since they continue to consume taxed goods and services in amounts far in excess of these temporarily low incomes, their apparent tax burdens may amount to huge percentages of current incomes. But it is clearly inappropriate to adjust public policy to the temporary circumstances of households with good long-term income histories and prospects.

Since so many of the households in the lowest income category are not truly poor, the data themselves are misleading. Therefore, in the following paragraphs, the emphasis is on family income levels of $2,500 and $3,500, levels that spell poverty for families of four or more persons in large American cities today. True, there are temporarily poor households at these levels, but they amount to minor portions of the total numbers at such income levels.

STATE-LOCAL TAX SYSTEMS

The extent to which state-local tax structures burden the urban poor in a given community depends upon the particular combination of tax instruments applicable in that community and the features of each of the major tax instruments specific to the city and state in question. For the country as a whole, state-local tax collections in 1966-1967 are classified in five groupings, each with distinctive incidence characteristics, in Table 1; these groups are the framework within which the rest of this paper is organized. A major distinction is made between personal and other taxes, principally taxes on some facet of business activity.

TABLE 1

MAJOR SOURCES OF STATE-LOCAL TAX REVENUE, 1966-1967
(in billions of dollars)

Source	State Governments	Local Governments	Combined
Total	31.9	29.3	61.2
Personal Taxes	21.9	15.1	37.0
Property Taxes on Housing[a]	.1	12.7	12.8
General Sales Taxes[b]	7.1	1.0	8.1
Individual Income Taxes	4.9	.9	5.8
Other Personal Taxes[c]	9.8	.5	10.3
Other Taxes (mainly on business receipts, profits, property and purchases)	10.0	14.2	24.2

SOURCES: U.S. Bureau of the Census. Governmental Finances in 1966-1967 (1968); State Government Finances in 1967 (1968).

[a] Estimated at 50 percent of local property tax collections and a small part of state property tax collections.

[b] Estimated at 80 percent of total collections; the remainder is on intermediate business purchases.

[c] All licenses and excises on alcoholic beverages, tobacco, parimutuels, amusements, hunting and fishing; death and gift taxes; 70 percent of highway-user taxes; 50 percent of insurance and utility taxes.

There is by no means universal agreement on the theory of tax incidence. Indeed, years of theoretical dispute and empirical testing have left us with tenaciously held but diametrically opposed hypotheses regarding the shifting of nearly every important form of taxation. Nonetheless, for the major personal taxes, there is a clear-cut majority viewpoint. In brief, the conventional wisdom is as follows:

(1) Property taxes on housing (one-third of total state-local personal taxes). The portion of the tax which falls on the value of the land underlying the building cannot be shifted from the owner who (in the United States) is legally liable for the tax. This element of the tax has the effect of reducing the value of the bare land. It does so because it reduces the net income after tax, yielded by the land; this in turn occurs because the supply of sites is more or less fixed and thus lower returns cannot eventually drive up prices by reducing the supply, as can occur with regard to commodities or reproducible assets like buildings. Moreover, individual landowners will not withhold their land from use in response to the tax; if they are rational, they will do quite the opposite—they will attempt to use the site as intensively as possible, so as to have more cash from which to pay the tax.

The building portion of the tax will be borne by the occupants, whether homeowners or renters. For homeowners this is obvious, since there is no one to whom to shift the tax. For tenants this is less obvious, for there can be one of two immediate reactions to the imposition of new or higher taxes on rental housing. If market conditions permit (i.e., vacancy rates are low), owners will raise rents promptly, thus shifting the tax forward to tenants. If vacancy rates are high, this will not occur. Instead, the new tax will reduce the profitability of investing in housing. With lower profitability, prospective owners are less likely to invest in housing vis-à-vis other types of assets not burdened with the property tax, other things being equal. Existing owners will disinvest over time, through the simple expedient of reducing outlays for maintenance of rental properties. The reduction in the rate of new construction and the increased loss of housing units through insufficient maintenance will tend, over time, to force rents up in existing buildings as tenants bid against each other for the smaller number of available apartments. Thus, in the long run, tenants will bear the burden of the tax in any case. For instance, a recent econometric study of property taxation and housing investment in New York City suggests that, during the 1960s, for each one percentage point increase of the effective property tax rate, new investment in rental housing in Manhattan was reduced by well over $100 millions annually. Beginning in the mid-1960s, rent levels in New York City, especially in Manhattan, soared and vacancy rates plummeted; the data indicate that the increase in property taxes on apartments free of rent control during the 1960s was more than fully shifted forward to tenants in the form of higher rents.

Even within the conventional reasoning, there are some qualifications to this conclusion. For example, some low-quality central-city housing occupied by the poor might have very high vacancy rates if rents were raised, even if the housing supply contracted; the quality of this part of the housing stock might be so low that families would double up in better housing units rather than pay higher rents. If there is no other alternative use of the sites (which is often the case for such property), a building owner who is also the landowner would tend to absorb the tax. Thus, part of the tax would be shifted backward to reduce land values (for further discussion of this, see Netzer, 1966: 37-39). But on balance, the qualifications do not change the general conclusion.

(2) Taxes on other consumer expenditures (a little less than half of total state-local personal taxes). The usual assumption is that these taxes are mainly shifted from the producer or distributor bearing the legal liability to consumers. The reasoning is similar to that applying to taxes on buildings, but the time required for adjustment is much briefer, since the life span of consumer goods and services other than housing is relatively short. If the producer or distributor must bear the tax, the scale of output will decline since it will be less profitable to produce the taxed commodity than to produce others. The reduction in supply will lead to higher prices and thus consumers eventually bear the burden.

The extent to which this happens in individual cases depends on a number of factors. One is the sensitivity of consumers to price increases for the taxed commodity and their willingness to substitute nontaxed goods and services for the now higher-priced taxed ones (that is, the price elasticity of demand for taxed and nontaxed goods). The greater the sensitivity of consumers to price increases with respect to taxed goods, the less the tax will be shifted forward (and the less the revenue produced by the tax). Implicit recognition of this by governments over many decades has led them to tax heavily those goods and services where consumption levels are thought to be relatively insensitive to price changes, like liquor, tobacco, gasoline, utility services, insurance premiums, and gambling expenditure.

Obviously, the wider the net of the tax—the more goods and services subject to the tax—the less possibility there is for consumers to substitute nontaxed for taxed goods and services. But no American consumption taxes are truly comprehensive. Even the general sales tax does not reach a large fraction of consumer expenditure. It does not cover housing expenditure or expenditures for many services, ranging from medical care to foreign travel. Nonetheless, the evidence we do have suggests that prompt forward shifting to consumers in the form of higher prices is the rule rather than the exception. For example, when federal excise taxes were reduced or eliminated on a wide range of goods (mainly consumer durables) in the 1950s and again in 1965, the tax reductions were quickly reflected in lower prices to consumers (Brownlee and Perry, 1967: 235-249).

(3) Taxes on individual income and wealth, that is, income, death, and gift taxes (about one-sixth of total state-local personal taxes). There is little doubt that these taxes fall on the persons or wealth taxed, reducing the after-tax income of income recipients and the value of estates transferred by death or gift. It is sometimes alleged that heavy taxation of personal incomes may reduce the supply of labor or the willingness to engage in risky ventures, and heavy death taxes may discourage saving, but the empirical evidence on this indicates that such effects are very small indeed.

(4) Taxes on business receipts, profits, property, and purchases. The incidence of business taxes is by no means clear; here the conventional wisdom provides no guidance. Both economic theory and empirical inquiries lead to conflicting answers. For example, whether corporate profits taxes are substantially shifted forward in higher prices or largely serve to reduce the income and wealth of stockholders is a hotly debated subject. It is generally believed that the employer share of payroll taxes tends to result in lower wage levels, but there is no agreement on the strength of this tendency. If business taxes are largely shifted forward to consumers or backward to employees, they will be more or less regressive; if they are largely borne by stockholders, the opposite will be true.

The picture is cloudy enough for nationwide business taxes. It is more obscure still for state and local taxes on business, which differ greatly in nature and level. It is a fair presumption that significant elements of state-local business taxes are not shifted forward to consumers, but borne by the factors of production—land, labor, and capital—to the extent that these factors cannot escape to lower-taxed jurisdictions or to activities subject to lower taxes in the same jurisdiction. This would suggest some elements of progressivity in state-local business taxes, especially to the not inconsiderable extent that they fall on landowners, and some elements of regressivity.

Because all of this is so obscure, it would be misleading indeed to attempt any quantification. Therefore, in the remainder of this paper, we will be concerned entirely with the major forms of personal taxation, accounting for roughly two-thirds of state government tax revenue and slightly over half of local government tax revenue.

Before going further, it should be noted that there is a major dissent to the conventional wisdom. This is a line of reasoning which holds that all commodity taxes, including those on buildings, are really shifted backward to the factors of production. To the extent that a tax has partial coverage, this argument goes, it is borne by the occupants, the tax is a decidedly regressive one land that cannot be readily used to produce nontaxed goods and services. If the tax is a general one, the factors of production will bear the tax depending on the factor proportions in different activities differentially burdened by that form of taxation. Thus, the burden of the property tax will be mainly on capital and land and other types of taxes will burden labor more heavily. This reasoning depends on a set of assumptions which may or may not hold. This is not the place to investigate them; suffice it to say that this remains a minority viewpoint among theorists and has limited empirical support.

THE PROPERTY TAX ON HOUSING

To the extent that the burden of the property tax on housing is borne by the occupants, the tax is a decidedly regressive one and, indeed, the principal factor in the regressivity of total state-local tax systems, that is, in the relatively high tax burdens borne by the urban poor. This is simply a reflection of the fact that housing expenditure is so high a fraction of the current incomes of the poor, despite the low quality of much of the housing they occupy.

This in turn is partly explained by the "threshold" aspects of housing expenditure and housing quality. Even the very poorest housing that building and housing codes permit to be inhabited at all in large American cities has significant economic costs; we do not allow the squatters' settlements that characterize housing for the poor in many large cities in Latin America. The minimum-standard housing that is legal may very well require monthly rentals of no less than $20 a room. If a family of six occupies four rooms (not luxurious by U.S. standards), the annual rental payments of $960 will be nearly one-third of a $3,000 annual income. If they occupy six rooms, the annual rent will absorb over one-third of the income earned by a wage earner fully employed for fifty-two weeks at a $2.00 an hour wage.

Better-off American families on the average do not spend anything like one-third of their incomes for housing. In middle-income ranges, the housing expenditure/income ratio is usually below twenty-five percent for homeowners and below twenty percent for renters. The result is the pattern of housing tax incidence shown in Table 2.

Like most recent studies of the incidence of housing taxes, the estimates in Table 2 apply to years centering on 1960, the year in which a Census of Housing was taken. If anything, such estimates probably understate the prevalent situation today. During the 1960s, housing costs rose more rapidly than incomes, by a wide margin. The margin by which central city property tax rate increases exceeded income increases was greater still. There is good

TABLE 2

ESTIMATES OF HOUSING PROPERTY TAXES AS A PERCENTAGE OF INCOME, BY INCOME CLASS

Income Class	United States, 1959[a]		New York City, All Private Housing[b] (percent)
	Single-Family Homeowners (percent)	Renters (percent)	
Less than $2,000	6.4	8.5	8.6
$ 2,000 - 3,000	4.5	3.9	5.6
3,000 - 4,000	3.7	3.0	4.1
4,000 - 5,000	3.3	2.5	3.4
5,000 - 7,000	3.1	2.1	2.8
7,000 - 10,000	2.8	1.8	2.4
10,000 - 15,000	2.6	1.6	2.2
Over $15,000	2.3	1.4	2.7

[a] From Netzer (1966, Tables 3-8).

[b] Adapted from Donheiser (1966: 177).

reason to believe that these disparities have been especially large for lower-income households in the larger central cities; this is one of the factors explaining the frequent central role of housing grievances in the disorders afflicting the larger cities.

The geographic coverage in Table 2 leaves something to be desired. As noted earlier, geographic aggregation, as in the nation-wide estimates, tends to appreciably understate the regressivity of the property tax. New York City, because of the existence of rent control and other factors, has a rather unusual pattern of housing consumption. However, estimates of the incidence of housing taxes done for other areas on somewhat different bases also display the pattern of severe regressivity and disproportionately heavy burdens on the poor.

It is sometimes alleged that, on a lifetime basis, housing expenditure tends to be more nearly proportional to income and thus taxes on housing are less regressive when considered in this light. This is no doubt true, to a considerable extent. Housing consumption decisions are made on the basis of long-term income prospects and when the prospects change, for better or for worse, changes in housing consumption patterns do not occur at all promptly. However, this is not necessarily relevant to short-term public policy considerations. Governments continually face the need to increase tax rates or impose new taxes. A decision to raise the needed revenue by increasing property tax rates will disproportionately burden those families that are poor here and now.

Moreover, there are substantial numbers of low-income urban families that will remain poor for years to come, in the absence of massively increased income transfers from the government. For many of these, housing costs including the property tax component of rental payments may not be of concern since they are public assistance recipients whose assistance payments are adjusted when rents rise. But, for the working poor, housing costs are a serious concern; indeed, increases in housing costs can be a deterrent to work efforts, by making living on public assistance relatively more attractive. In this connection, it should be noted that the official definitions of poverty are based on the costs of food, not housing. In the urban United States, there are millions of households not considered poor in the official statistics, but who are truly poor in the sense of being unable to afford adequate housing (Downs, 1968: 147).

OTHER TAXES ON CONSUMPTION

Nearly all urban Americans live in states and/or cities imposing a general sales tax. In addition, all American states and many local governments impose selective sales or excise taxes on particular types of goods and services, particularly liquor, tobacco, gasoline, motor vehicles, utility services, and insurance premiums. Assuming that the burden of such taxes is borne by consumers, the regressivity of a consumption tax depends on how the consumption of the taxed item changes as family income levels rise. In general, there is some degree of regressivity in nearly all consumption taxes, since there are few types of things for which expenditure rises as rapidly as income throughout the range of incomes. Consumption as a whole rises less rapidly than income, since the proportion of income saved rather than spent for current consumption tends to rise with income.

The real public policy issue in this regard is the degree of regressivity. As noted at the outset of this paper, for some taxes the regressivity shows up only at high-income levels. For the very rich, expenditure as a percentage of income is likely to be relatively low even for true luxury goods. This is hardly a matter for great concern. In other cases, where the regressivity begins at low points on the income scale, the dollar amounts of tax may be so small that the regressivity of the particular tax, considered by itself, is not of great moment.

The general sales tax tends to be a regressive one. However, the regressivity is very substantially mitigated if food is exempt from the tax, as is the case in one-third of the states with general sales taxes. This is illustrated in Table 3, where two states with five percent sales tax rates in 1968 are compared—Illinois, where food is subject to tax, and Rhode Island, where it is not. Roughly half of the regressivity displayed by the Illinois tax is eliminated in Rhode Island by the food exemption.

It is easy to understand why a general sales tax with food exempt should be only mildly regressive. Such a tax excludes the major items in the budget of the poor—food, housing, and public transportation fares. Its base consists largely of house furnishings, apparel, and automobiles, the consumption of which rises almost as fast as incomes, if the higher-income brackets are ignored. In fact, a careful study for New York City suggests that the sales tax there has no significant regressivity (Donheiser, 1966: 169).[1]

Since by now most states have both income and sales taxes, there is available a better means of mitigating sales tax regressivity than the exemption of food. The food exemption complicates the administration of a sales tax. More importantly, its ability to reduce regressivity depends heavily on family composition; for small households and especially for older people, food consumption is a relatively small part of the family budget. In lieu of this, six states, beginning with Indiana in 1963, have adopted provisions for credits or cash rebates, administered with the state income tax, to offset the regressivity of the sales tax. In Indiana, for example,

TABLE 3

EFFECT OF FOOD EXEMPTION: ESTIMATED SALES TAX PAYMENTS AS A PERCENTAGE OF INCOME, 1968, ILLINOIS AND RHODE ISLAND, FOR FOUR-PERSON FAMILIES[a]

Adjusted Gross-Income Level	Illinois (percent)	Rhode Island (percent)
$ 2,500	4.2	2.3
3,500	3.6	2.1
5,000	3.0	1.8
7,500	2.5	1.6
10,000	2.2	1.5
15,000	1.9	1.3

[a] Interpolated from Internal Revenue Service 1968 Optional State Sales Tax Tables. In each of the states, a total sales tax rate of five percent applied, and the IRS calculations in both cases included the burden of a five percent tax on utility services. Food is exempt in Rhode Island but not in Illinois.

there is a credit of $8 per person; since the sales tax rate is two percent, in effect the first $400 per capita of purchases subject to the sales tax in any year is tax-exempt (Advisory Commission, 1968). Were a $15 per capita credit applied in Illinois (with its higher sales tax rate), the regressivity of its sales tax would be eliminated, for all practical purposes.

For typical families at different income levels, the burden of the usual package of state-local selective consumption taxes is not particularly regressive, that is, on the average, spending for liquor, tobacco, gasoline, utility services, insurance premiums, and amusements rises about as rapidly as income, well into the middle-income brackets. However, this statistical finding conceals some major deviations from the average. In particular, it is in part a measure of deprivation, of the failure of the poor to consume many of the things that are part of the American standard of life. Nearly all American families with incomes of $5,000 or more have telephones, automobiles, and life insurance and indirectly pay the consumption taxes imposed on these services, but sizable fractions of the urban poor—a third or more—do not include these items in family consumption. For those among the low-income population who do spend their modest incomes for these items, the taxes imposed are indeed regressive.

Moreover, it can be argued that public policy should encourage the urban low-income population to spend their money on some consumption items that are now heavily taxed. For example, it has been demonstrated that increased ownership of more and newer cars by residents of central-city ghettos may be a relatively economical way of improving their access to expanding suburban job opportunities, from society's standpoint. But we tax automobiles and automobile insurance (which may be the critical factor here) fairly heavily. Similarly, it can be shown that the lack of adequate insurance coverage—for example, fire coverage for household goods—is a significant factor in periodically reducing the working poor to dependency status.

No doubt this is not sufficient reason for abandoning the selective consumption taxes commonly used. It surely may be sufficient reason for central cities to hesitate to tax more heavily these conventional and obvious targets of state-local taxation.

PERSONAL INCOME TAXES

Among those concerned about the regressivity of tax structures, personal income taxation strikes an immediate and favorable response. However, the contribution of personal income taxes to offsetting the regressivity of much of the rest of the state-local tax structure is heavily dependent on the particular form of personal income taxation employed. Indeed, most of the local government income taxes now used are regressive, although only at higher income levels.

Essentially, there are three major forms of income taxation utilized in the United States, with each of the forms used with a variety of tax rates. One is the prototype progressive income tax form, covering all major types of personal income and applying graduated rates to net taxable income, that is, income in excess of personal exemptions (and, in most cases, personal deductions). This, of course, is the form of the federal income tax, nearly all state income taxes, the New York City income tax and the income tax used by Baltimore and counties in Maryland.

The second form also applies to all types of income and permits personal exemptions, but applies a flat tax rate to net taxable income; this is the kind of tax used at the state level in Indiana, Michigan, and Massachusetts and by cities in Michigan.[2]

The third form is a flat-rate tax on gross income, without any exemptions, but covering only earned income (that is, excluding interests, dividends, rents, and capital gains). This is the prevalent local government form of income tax, used in Ohio, Pennsylvania, and Kentucky, where local income taxation is very widespread, and by Kansas City and St. Louis in Missouri.

It would be possible for a state or city, by granting high personal exemptions and utilizing a steeply graduated rate structure, to have an income tax that is much more steeply graduated than is the federal tax. Few do so, in part no doubt because of apprehension that very high rates on the rich will cause them to move to lower-tax-rate jurisdictions. However, a number of them do achieve a significant degree of progressivity in the lower half of the income distribution, but this is largely achieved not by a graduated rate structure but by personal exemptions and tax credits (like the sales tax credit used in six states).

This is illustrated by Table 4, in which three hypothetical income tax structures, all designed to yield roughly the same revenue in an urban state or local jurisdiction, are compared. The first is the typical local government tax, a flat-rate tax on gross earned income. Because property income usually increases as a fraction of personal income as income levels rise, this tax becomes somewhat regressive at higher-income levels. Clearly, most of the progressivity shown by the other two tax forms, up to gross incomes of $10,000, can be attributed to the exemptions.

Indeed, a flat-rate tax with both exemptions and credits (as in Indiana) will ordinarily be less burdensome to low-income households, especially large ones, than the more conventional state income tax, even if the latter has relatively steep graduation. Obviously, graduation can be combined with a system of tax credits designed to provide cash rebates to large low-income fam-

TABLE 4

TAX LIABILITY UNDER THREE HYPOTHETICAL

INCOME TAX STRUCTURES, FAMILY OF FOUR[a]

		Tax Liability		
		I	*II*	*III*
			Flat-Rate Tax	*Graduated-Rate*
		Flat-Rate Tax	*on All Income*	*Tax on All Income*
Gross	*Earned*	*on Gross*	*After Exemptions*	*After Exemptions*
Income	*Income*[b]	*Earned Income*[c]	*and Deductions*[d]	*and Deductions*[e]
$ 4,000	$ 4,000	$ 40	$ 18	13
5,000	5,000	50	32	27
7,500	7,500	75	65	67
10,000	9,000	90	99	112
15,000	13,000	130	167	202

[a] The tax structures are designed so that each would produce roughly the same revenue from residents of a typical large city.

[b] That is, excluding income from dividends, interest, rents, and capital gains.

[c] One percent rate on gross earned income.

[d] Exemptions of $600 a person and deductions equal to ten percent of gross income; 1.5 percent rate on net taxable income.

[e] Exemptions and deductions as in (d); rate of 1.0 percent on first $1,000 of net taxable income, 1.5 percent of next $2,000, and 2.0 percent on net taxable income above $3,000.

ilies; this is done in such states as Hawaii, Iowa, Nebraska, and Colorado. In Hawaii, under provisions in effect in 1968, a family of four with an income of $2,500 ordinarily would receive a cash rebate of $110, equal to 4.4 percent of gross income, while a four-person family with a $10,000 income typically would pay $330 in income tax, or 3.3 percent of gross income.

There are, in short, three policy implications from this discussion of personal income taxation. First, the common flat-rate local income tax without exemptions does not significantly help to redress the burdensome nature of other parts of the state-local tax structure for the poor. Second, exemptions and credits are the most important features of income taxation in this connection. Third, since any income tax with personal exemptions provides significant relief for the poor, the more important such income taxation is in a state-local tax structure relative to other forms of taxation, the less regressive will that structure be.

THE "PACKAGES" IN SEVEN LARGE CITIES

The impact of a tax structure on families at various income levels in a given city can be assessed only by examining family tax liabilities under all the major forms of taxation used by state and local taxing units with jurisdiction in that city. A crude attempt at this is shown in Table 5, for seven large American cities chosen to illustrate the different tax "packages" in effect in the larger cities more generally. In all seven cities, the state-local tax burden on the poor is high in percentage terms—in no case less than seven percent for families of four with a $2,500 income—but there are substantial differences in both the level of tax on the poor and the degree of regressivity.

TABLE 5

MAJOR STATE-LOCAL PERSONAL TAXES AS
PERCENTAGES OF INCOME, FAMILES
OF FOUR IN SELECTED LARGE CITIES[a]

| City | Family Income Level | | | |
	$2,500 (percent)	$3,500 (percent)	$10,000 (percent)	Ratio: Column 3 to Column 1
Chicago	10.5	8.8	5.8	.55
Cleveland	8.0	6.8	4.9	.61
Philadelphia	10.6	9.1	7.0	.66
Los Angeles	7.2	6.3	5.6	.78
Detroit	8.1	7.0	6.4	.79
New York	9.2	7.9	8.1	.88
Minneapolis	7.2	6.9	8.5	1.18

[a] These are crude estimates based on a series of statistical manipulations, short-cuts and assumptions of varying validity. They are not to be construed as indicators of intercity differences in tax level, but instead are measures of differences in tax structure. The taxes analyzed include only those identified as personal, rather than business, taxes, on the assumption that the incidence of personal taxes can be described with some confidence. The personal taxes included fall in four groups: property taxes on housing; general retail sales taxes; personal income taxes; and other consumer-type taxes, mainly liquor, tobacco, gasoline, amusements, and the like. The property tax data are as of 1966; the other data apply to 1968. See the Appendix for a description of the derivation.

In Minneapolis, the level of the tax burden of the poor is low for this group of cities, even though state-local taxes per capita and as a percentage of personal income are very high overall in Minnesota. This is accomplished by imposing relatively high taxes on middle-income families. At the other extreme, taxes on the poor are high, for this group, in Chicago despite a level of state-local taxation that is low overall. The explanation is the regressivity of the tax structure applicable in Chicago, as severe as one can find in large cities. (These comparisons antedate the adoption by Illinois of a personal income tax in 1969.)

The underlying explanation for the degree of regressivity is, in turn, found in Table 6, where the characteristics of the tax systems operating in the seven large cities are indicated. Reading Tables 5 and 6 together, the important role of properly designed income taxation becomes clear. Important reliance on income taxation that treats the poor well can do a lot to offset high property taxes on housing (Minneapolis and New York) or a high sales tax that covers food (Detroit). Even if sales and property taxation is relatively low, combining these taxes with a modest income tax or a flat-rate gross income tax (Los Angeles and Cleveland, respectively) will produce a regressive tax system. And, of course, there is the

TABLE 6

CHARACTERISTICS OF STATE-LOCAL TAX SYSTEMS IN SELECTED LARGE CITIES[a]

Tax and Character[b]	Cities
Property Tax on Housing	
Relatively High	Philadelphia, New York, Chicago, Minneapolis
Relatively Low	Cleveland, Los Angeles, Detroit
Sales Tax	
Relatively High, Food Taxable	Chicago, Detroit
Relatively High, Food Exempt	New York
Relatively Low	Los Angeles, Philadelphia, Cleveland, Minneapolis
Income Tax	
None or Low Effective Rates	Chicago, Los Angeles
City Tax Only, Flat Rate on Gross Income	Philadelphia, Cleveland
State and/or City Tax with Progressive Structure	Minneapolis, New York, Detroit

[a] Based on property tax data for 1966, from U.S. Census of Governments, 1967, and sales and income tax provisions in effect in 1968, from Advisory Commission on Intergovernmental Relations tabulations and Internal Revenue Service Optional State Sales Tax Tables.

[b] For each major tax, characteristics with most regressive effect are listed first.

pre-1969 Chicago system in which nothing mitigates the lot of the poor—no income tax, high sales tax with coverage of food, and high property taxes on housing.

CONCLUSION

This analysis suggests that if we are concerned about the substantial burdens of state-local taxes borne by the urban poor (especially larger families among the poor), two major policy emphases are required. The first is to reduce, or at least avoid further increasing, property taxes on housing in central cities. There are also good reasons for this, connected with the impact on the central-city housing stock (see Netzer, 1968b). The second is to accelerate the historic process of relying more heavily on sales and income taxation, but, in so doing, to concern ourselves with the specific design of these taxes, for a sales tax that covers food and an income tax without exemptions will do little to assist the poor. However, comprehensive sales taxation in combination with a refined system of credits and exemptions in the income tax (the Hawaii model is a relatively good one), can be the cornerstone of a state-local tax structure that does impose mimimum burdens on the poor.

One implication of this is heavier reliance on taxes imposed at the state government level; local government sales and income taxes usually do not have the refinements needed to avoid regressivity, and it often would be difficult to operate a local-level system with such refinements. Moreover, there are good reasons on other counts to emphasize increased, relative importance of state-level taxation to finance urban services, including the effects on intrametropolitan tax differentials or the location of residents and economic activity, the need to tap the income and wealth located outside central cities, and the need to organize as well as finance many services over a geographic area broader than that of the central city itself. To this author at least, concern over the income distribution consequences of state-local tax systems is the capstone of the case for a substantial realignment of responsibilities among the levels of government—much more for state governments and much less for local governments, *especially* the central cities (Netzer, 1969).

APPENDIX

Derivation of Tax Incidence Data in Table 5

(1) *Property taxes on housing.* These estimates represent a combination of two other sets of estimates. One is an estimate of the distribution of housing property taxes by income class for a single city; it is assumed that the interclass relationships apply to the other large cities in this table. The estimate used applies to New York City in 1960-1961; it was used because it is by far the most careful analysis of tax incidence in a single large central city yet done (Donheiser, 1966: 153-208).

The 1967 *Census of Governments,* Volume II, *Taxable Property Values* (1968) provides data on 1966 median effective tax rates for single-family houses in many large cities and on assessment ratios for these houses and for "ordinary real estate" in these cities. These data permit us to crudely estimate effective tax rates for both single-family houses and for other than luxury apartment buildings. These effective rates can then be expressed as an index of the effective rate implicit in the Donheiser study. The Donheiser incidence values, adjusted to the illustrative family income levels used here, were then multiplied by the index for each city. It was assumed that low-income families are apartment dwellers and $10,000 familes are homeowners, except in New York and Chicago (where the $10,000 families are treated as apartment dwellers) and Minneapolis (where the low-income families are treated as owners or renters of single-family houses).

(2) *State and local sales taxes.* Estimates of dollar tax payments by the illustrative families were interpolated from the Internal Revenue Service's Optional State Sales Tax Tables for 1968, included in the Instructions for Form 1040 for 1968.

(3) *State and city income taxes.* Effective rates for state income taxes in 1968 were taken from Advisory Commission on Intergovernmental Relations, *State and Local Finances: Significant Features, 1966 to 1969* (1968), Table 33. Effective rates for city income taxes in Detroit, New York, Cleveland, and Philadelphia were computed by the author on the assumption that all income of the illustrative families are subject to the tax.

(4) *Other personal taxes.* These taxes include the following: all receipts from sales and license taxes on alcoholic beverages, tobacco, amusements, and parimutuels; seventy percent of highway-user taxes and licenses; fifty percent of taxes on insurance companies and utilities; and death and gift taxes. For state governments, the data were derived from the Census Bureau's *State Government Finances in 1968* and were expressed as a percentage of statewide personal income for each of the states containing the selected cities. Rough adjustments were made for the relatively minor city taxes of this type, significant only in New York and Chicago.

An incidence series was constructed for the urban United States as a whole, from the data in Fabian Linden, editor, *Expenditure Patterns of the American Family* (National Industrial Conference Board, 1965). That is, expenditure by income class was tabulated for alcoholic beverages, tobacco, gasoline, insurance, utilities, and dues and fees for sports (a proxy for amusements, parimutuels, and so forth). These data were weighted by the approximate tax rates typical of major areas in the United States. The incidence series and tax level series described in the preceding paragraph were combined as in the case of property tax, by multiplying the nationwide incidence values by an index number of these taxes as a percentage of personal income, for each selected city.

NOTES

1. New Yorkers do have peculiar consumption patterns; in particular, car ownership in the lower half of the income distribution is relatively infrequent.

2. The Massachusetts tax applies a higher rate to property income than to earned income and therefore has a crude graduation of rates. All state and local income taxes apply only to the earned income of nonresidents, not to their property income.

REFERENCES

BROWNLEE, O. H. (1960) Estimated Distribution of Minnesota Taxes and Public Taxes and Public Expenditure Benefits. Minneapolis: University of Minnesota.

——— and G. L. PERRY (1967) "The effects of the 1965 federal excise tax reduction prices." National Tax Journal (September): 235-249.

DONHEISER, A. D. (1966) "The incidence of the New York City tax system." Financing Government in New York City. New York: Graduate School of Public Administration, New York University.

DOWNS, A. (1968) "Moving toward realistic housing goals." In K. Gordon (ed.) Agenda for the Nation. Washington, D.C.: Brookings Institution.

GILLESPIE, W. I. (1966) "Effect of public expenditures on the distribution of income." In R. A. Musgrave (ed.) Essays in Fiscal Federalism. Washington, D.C.: Brookings Institution.

GORDON, D. M. (1969) "Income and welfare in New York City." Public Interest (Summer).

GRAVES, L. F. (1964) "State and local tax burdens in California." In California Assembly Interim Committee on Revenue and Taxation, 1964 Report.

MUSGRAVE, R. A. and D. W. DAICOFF (1958) "Who pays the Michigan taxes?" Michigan Tax Study Staff Papers. Lansing: Michigan Legislative Tax Study Committee.

NETZER, D. (1969) State-Local Finance and Intergovernmental Fiscal Relations. Washington, D.C.: Brookings Institution.

——— (1968a) "Federal, state and local finance in a metropolitan context." In H. S. Perloff and L. Wingo, Jr. (eds.) Issues in Urban Economics. Baltimore: Johns Hopkins University Press.

——— (1968b) Impact of the Property Tax: Its Implication for Urban Problems. Washington, D.C.: National Commission on Urban Problems, Joint Economic Committee Print (90th Congress, Second Session).

——— (1966) Economics of the Property Tax. Washington, D.C.: Brookings Institution.

ROSTVOLD, G. N. (1966) "Distribution of property, retail sales, and personal income tax burdens in California: an empirical analysis of inequity in taxation." National Tax Journal 19.

U.S. Advisory Commission on Intergovernmental Relations (1968) State and Local Finances: Significant Features, 1966 to 1969.

University of Maryland (1965) Maryland Tax Study. College Park: University of Maryland.

University of Wisconsin Tax Study Committee (1959) Wisconsin's State and Local Tax Burden. Madison: University of Wisconsin.

17

Metropolitan Employment and Population Distribution and the Conditions of the Urban Poor

ROGER NOLL

☐ DURING THE PAST twenty-five years several rather obvious—and quite significant—changes have occurred in the spatial structure of large urban complexes in the United States. Together these changes constitute the general phenomenon of the erosion of the dominance of the central-core city in the economic and political affairs of the metropolis. Specifically, population growth has come to be virtually completely confined to suburban areas; jobs and business activity have increased far more rapidly outside central cities; growing jurisdictional fragmentation (as the population becomes spread over a wider area) has undermined the ability of the central city—or of any other local government—to deal effectively with larger metropolitan problems; and finally the balance of political power in many states and in the nation, having switched from rural areas and small towns to big cities one to three generations ago, is shifting from the cities to the suburbs.

AUTHOR'S NOTE: *This monograph draws heavily on earlier research done jointly by the author and Benjamin I. Cohen, presented in their manuscript (Cohen and Noll, 1969). Although responsibility for errors remains with the author, many of the ideas in this paper were initially suggested by Professor Cohen.*

Recently an impressive body of literature has developed in which changes in the spatial structure of metropolitan areas have been cited as a major obstacle to improving the social and economic conditions of the urban poor (see, for example, Meyer, Kain and Wohl, 1965; Kain, 1967; Ginzberg, 1968). This literature characterizes the central city as caught in a permanent downward trend in employment opportunities, business investments, and middle- and upper-income population, coupled with a permanent upward trend in low-income, disadvantaged residents. Together these trends produce an eroding financial capacity to deal with a rapidly increasing need for government social services. Furthermore, job growth in suburban areas is said to be inaccessible for reasons of physical distance to the central-core poor, thus creating a paradox of excessive underemployment in the central city, and labor shortages in the suburbs (Kain, 1968; Mooney, 1969).

The preceding characterization of the plight of the American urban poor has led to several policy proposals. A suggested interim proposal is the subsidization of transportation of low-income central-city residents to suburban jobs. (In 1968, the federal government sponsored some experimental projects for transporting central-core poor to suburban jobs in Los Angeles, St. Louis, and Boston.) Longer-run solutions involve changing the trends in the spatial distribution of population and employment opportunities. Some propose subsidies of private enterprise to encourage more investment in central cities, while others propose strategies to transplant significant portions of urban ghettos to the suburbs. Finally, to slow the growth of low-income city population, rural development assistance—particularly in the South—is also suggested.

These policy proposals all can be defended on other grounds beside their ability to alleviate problems associated with spatial characteristics of urban growth: one need not support deeper federal subsidies of urban mass transit solely because suburban jobs can be made more accessible to central-core ghetto residents, nor higher welfare benefits in Mississippi only as a means of slowing the migration of southern Negroes to large northern cities. Yet resources for dealing with urban problems—or for any purpose—are limited, so that the importance of spatial development patterns in determining the conditions of the urban poor affects the allocation of resources among policy operations. Assigning major significance to spatial structure reduces the relative policy emphasis placed upon education, training, income maintenance, political action, and other strategies for reducing urban poverty.

This chapter seeks to present information and analysis pertaining to the effect of population and employment distribution on the conditions of the poor. The first section sketches the recent trends in population growth in metropolitan areas, including changes that have occurred in the characteristics of central-city residents. The second section deals with patterns of employment distribution, and how they relate to the employment problems of the poor. The final section deals with policy implications of the changing urban structure.

THE CENTRAL CITY POPULATION

The spatial distribution of population within a metropolitan area can affect the conditions of low-income households in several ways. Individuals in predominantly low-income neighborhoods are likely to face higher retail prices (see U.S. Federal Trade Commission, 1968; 1969) as a consequence of the fact that establishments in their vicinity have only low-income customers. Poverty-area schools are likely to be less effective, partly because more qualified teachers generally avoid teaching children from low-income families, partly because facilities in low-income areas are likely to be older and poorer, and partly because of the impact upon motivation arising from a homogeneous group of students from low-income households (Coleman, 1965; Bowles and Levin, 1968). Furthermore, while the literature on the so-called "culture of poverty" that appeared in the 1960s (for example, see Glazer and Moynihan, 1965) probably overstated its case, the attitudes and behavior developed by individuals to adapt to the conditions in poverty areas are probably not conducive to success in escaping poverty (see Valentine, 1968). Finally, if higher-income, more successful residents of poverty areas move out, an important source of information about available job opportunities is lost to the remaining less skilled, more irregularly employed residents. A particularly important vehicle for finding work, especially among lower-income and minority-group individuals, is employment information gained through informal contacts with friends and relatives (see Shepard and Belitsky, 1966).

The distribution of the poor among political jurisdictions can also affect the conditions of the poor through its impact upon the provision of services by government. The costs of maintaining any given standard in income-redistribution programs (welfare, public health, food stamps, public housing) are directly proportional to the size of the low-income population, and achieving a given effectiveness of education costs more for children from low-income families. In addition, in high-population-density areas, such as central-city ghettos, pressures will be exerted upon local government to provide more police protection, more recreational facilities, and more waste control services per capita. Consequently, political jurisdictions with a higher-than-average incidence of poverty will face a higher per capita cost of maintaining any given standard for most local government services, yet in the absence of a substantially above-average commercial property tax base they will also find themselves with fewer resources than communities with a lower incidence of poverty. If jurisdictions of high- and low-poverty incidence are contiguous, the problem is more acute since the low-poverty unit can offer the prospect of lower taxes to finance caring for the poor as an inducement to upper-income families and businesses to locate in that jurisdiction. To the extent that locational decisions are altered by this consideration, the tax base of the high-poverty area is further eroded, and the services provided to the poor are commensurately less. If the low-poverty community can prevent resettlement of the poor within its boundaries, the patterns of growth will produce an increasing dispersion of incomes between the two areas, and a similarly increasing dispersion in the level and quality of government services.

The following paragraphs examine the extent to which metropolitan population distribution has produced this type of situation.

CHANGES IN CENTRAL CITY POPULATION

The 1960 Census confirmed the emergence of a new trend in the population of the largest American cities. For the first time, a majority of the nation's largest cities lost population between decadal Censuses. Only during the 1930s had a significant number of large cities lost population, but even then—when the growth in total urban population was the slowest in the history of the United States and when, for the first time since 1810-1820, the urban population grew only about as fast as the rural—only four of the thirteen cities exceeding 500,000 population experienced a decline.

Large central cities apparently fared even worse during the 1960s. Sample survey data for 1960 and 1968 (Table 1) indicate that while central cities in the very largest metropolitan areas have continued their very slow aggregate growth, central cities in smaller metropolitan areas experienced precipitous declines in population growth rates. Population growth in the suburbs was much less rapid in the 1960s, but the suburbs still grew very rapidly in comparison to other areas.

By the mid-1960s, the central city population may have begun to decline. Five-month averages of Current Population Survey (CPS) data centered on April show an estimated decline of 380,000 in central-city population between 1966 and 1968. The significance of these results is difficult to assess: they may reflect a

TABLE 1

POPULATION CHANGES IN METROPOLITAN AREAS

Area	Population[a]		Population[b]		Percent Change	
	1950	1960	1960	1968	1950-1960	1960-1968
		(figures in millions)				
All SMSA's[c]	89.3	112.9	112.4	128.0	26.4	13.9
Central Cities	52.4	58.0	57.8	58.2	10.7	0.7
Suburbs[d]	36.9	54.9	54.6	69.9	48.5	28.0
Largest SMSA's[e]	49.6	61.6	61.2	70.0	24.2	14.4
Central Cities	29.8	30.7	30.2	30.4	3.0	0.7
Suburbs	19.8	30.9	31.0	39.6	56.1	27.7
Other SMSA's	39.7	51.1	51.1	58.0	29.2	13.5
Central Cities	22.6	27.3	27.6	27.8	20.8	0.7
Suburbs	17.1	24.0	23.5	30.2	40.4	28.5

[a] Decadal Census tabulations. Compiled from Bureau of the Census, Current Population Reports Series P-23 Number 23, Statistical Abstract 1965, and County and City Data Book 1962.

[b] Sample data. 1960 figures from one-in-a-thousand sample of 1960 Census; 1968 data from March, 1968 Current Population Survey. Data published in Bureau of the Census Current Population Reports Series P-23 Number 27.

[c] As defined in 1960; Standard Metropolitan Statistical Areas (SMSA) contain at least one city (or twin cities) of over 50,000 population, the county in which the central city is located, and any other counties dominated by suburbs of the central city.

[d] All SMSA territory not in the central city. Of course, since counties are the building blocks for SMSA's, many SMSA's also include substantial rural areas and some smaller cities that are too distant from the central city to be appropriately classified as suburbs.

[e] Exceeding one million population in 1960.

temporary response to the civil disturbances in many cities during the mid-1960s, a secular spreading of the trend of population decline from the largest cities to smaller ones, or perhaps a sampling error in the CPS (possibly connected to the known, extremely large Census undercount of young men from low-income urban areas). In any event, the data do show that the relative position of the central city in the population of the metropolitan area is declining rapidly, and that central-city residents are now a minority of all metropolitan residents.

SOURCES OF THE DECLINE IN CENTRAL CITY GROWTH

Suburban areas have accounted for most population growth in large metropolitan areas for many years—in many cases since before the beginning of this century. The population of Manhattan Island peaked in 1910, but even before then its share in the total population of present-day New York City had begun to decline. Between 1900 and 1930, while the population of Manhattan peaked and then fell back to its level at the turn of the century, the population of Brooklyn doubled and that of Queens and the Bronx sextupled. The six largest cities surrounding Boston grew in population by sixty-seven percent during the same period, while the city of Boston grew thirty-nine percent.

Many large American cities have grown less rapidly than their natural rate of increase for decades. In 1900, eleven cities[1] exceeded 300,000 population; since 1930 all but one has grown less rapidly than its natural rate of increase, and six had fallen below the natural rate by 1920. This means that for approximately fifty years a majority of the largest cities have lost more migrants than they have gained.

The most important source of population growth for most larger cities has been annexation of surrounding territories; because of the importance of annexation, census population statistics seriously understate the relative stagnation of population in the central core. Table 2 shows that the population of the central cities of thirty large metropolitan areas would have declined had the 1950 political boundaries remained unchanged until 1960; however, annexations added nearly 1.5 million residents to these cities. By the 1950s annexation activity had all but stopped in the fifteen largest metropolitan areas. Some relatively slow population growth—less than one-half of one percent per year—occurred within the 1950 political boundaries of central cities in the next size group of metropolitan areas.

TABLE 2

CENTRAL CITY POPULATION CHANGES: 1950-1960

Central City Group	Population Changes		Net Change in Population
	Within 1950 Boundaries	Due to Annexation	
	(000)	(000)	(000)
Thirty Large SMSA's[a]	- 75	+1,460	+1,385
Fifteen Largest	-476	+ 15	- 461
Remaining Fifteen	+401	+1,445	+1,846

SOURCE: Bureau of the Census (1961) U.S. Census of Population: 1960, Vol. 1, Characteristics of the Population. Part A: Number of Inhabitants. Washington, D.C.: Government Printing Office.

[a] Includes thirty of the thirty-five most populous SMSA's; deletions from the latter group were made on the basis of central-city population.

A possible explanation of the further slowing of central-city population growth in the 1960s among other than the very largest cities could be a diminished ability to annex suburban areas, as opposed to a fundamental change in population distribution trends within constant geographical areas. If this is the case, then the following characterization of the broad changes in metropolitan-area population distribution would be appropriate:

(1) The central geographic areas of most urban areas have had roughly stable populations for several decades (exceptions are largely confined to a few new, extremely rapidly growing urban centers in the Southwest and Far West).

(2) Annexations of adjacent vacant land in advance of population growth (such as the San Fernando Valley by Los Angeles or Queens, Bronx, and Richmond by New York City) or of suburban cities (such as Brooklyn by New York City) have been the principal source of population growth for most central cities, and this source has diminished significantly over the past twenty years, at least among the largest cities.

(3) Although only within the last decade has a majority of the metropolitan population lived outside central cities, suburban areas have been growing much more rapidly than central cities throughout the twentieth century; while the shift of the majority of the population to suburban areas may have been a critical turning point politically and socially for the cities, it is the culmination of a long period of widely differing growth rates between central cities and suburbs.

POPULATION CHARACTERISTICS

The preceding paragraphs deal only with broad trends in total population distribution. Another important issue regarding the structure of metropolitan areas is the social and economic status of residents in particular areas. Approximately a decade ago social scientists and others began to warn, as did Grodzins (1959): ". . . many central cities . . . are fast becoming lower class, largely Negro slums." Although the distinction is rarely made, two quite separate hypotheses are woven into Grodzins' statement: first, that central cities are becoming increasingly populated by Negroes; and second, that they are becoming largely slums. More recently Ginzberg et al. (1968) has stated the conventional assumptions behind forecasts such as Grodzins': "[Racial] designations provide a shorthand for other critically important variables that will largely determine . . . the present economic condition and future prospects of the metropolis. . . . [A] substantial proportion of . . . Negroes and Puerto Ricans are poorly educated, have only modest skills, . . . are condemned to poverty and deprivation, and . . . represent a drag on the vitality and growth of the city's economy."

The first half of the conventional characterization of central-city population is certainly accurate: the fraction of central-city residents who are Negroes is increasing. In 1968, Negroes accounted for one-fourth of the population of central cities in metropolitan areas of over one million population, compared with eighteen percent in 1960. In smaller metropolitan areas, the fraction of Negroes in the central-city population increased from fourteen to sixteen percent during the same period. Recent survey data indicate some changes in both white and Negro population-growth rates: the annual decline in central-city white residents is estimated to have tripled in 1966-1968 over the 1960-1966 rate, while the annual increase in Negro central-city residents was apparently only about one-third as large in the later period (and probably fell below the rate of natural increase). Apparently both Negroes and whites are, on balance, migrating away from central cities, although among the former the net outward movement is not sufficient to counterbalance all of the natural population growth. Because changes among both whites and Negroes in natural population growth and resettlement patterns have occurred in the same proportion, the rate at which the racial composition of central cities is changing has remained approximately the same.

In the absence of economic growth and other improvements in social conditions, an increase in the relative number of Negroes in central cities would produce a decline in average income and educational attainment. Yet median income (corrected for price changes) rose sixteen percent in central cities between 1959 and 1967; other indexes of social conditions also showed significant gains over the same period (Table 3). These improvements occurred despite rapid changes in the racial composition of cities because of rapid increases in income and educational attainment among central-city Negroes: median income of Negro families in central cities increased nearly twice as rapidly (twenty-eight percent) as for whites (sixteen percent).

Rising real income in the central city is evidence to make one question the aptness of Ginzberg's notion that the social characteristics of minorities allow one to use racial composition statistics as surrogates of social and economic conditions—and particularly of change, or the capacity for change, in those conditions. Of course, other phenomena could offset the gains in the indicators cited in Table 3. Certainly the nature of the housing market (discussed in chapters 14 and 15 of this volume) could be creating deterioration in central-city neighborhoods despite rising real incomes, thus validating the hypothesis of growing slums. But if this is the case, the difficulty lies in the social institutions that cause limited, inferior housing to be provided to minorities, not in an inability of members of minority groups to improve their economic conditions. For the fact remains that if slums are growing in central cities, the median income of slum residents must be increasing rather rapidly —more rapidly, in fact, than the increase in median income for the whole city.

TABLE 3

SOCIAL AND ECONOMIC CONDITIONS IN CENTRAL CITIES

Characteristic	All Residents		White		Negro	
	1960	1968	1960	1968	1960	1968
Median Years Schooling (age 25-29)	12.4	12.5	12.5	12.6	11.4	12.2
Percent High School Graduates (age 25-29)	62	73	66	77	43	61
Percent Families with Income[a] Under $4000 (1967 prices)	22	19	18	16	45	33
Percent Families with Income[a] $4000 to $5999 (1967 prices)	20	15	20	14	25	21
Percent Families with Income[a] Over $10,000 (1967 prices)	23	34	27	37	7	18
Unemployment Rate	5.8	4.3	4.9	3.5	10.7	7.8
Males Over Age 20	5.6	3.1	4.8	2.5	9.9	6.0
Females Over Age 20	5.2	4.0	4.3	3.5	10.1	5.9
Percent Female Workers in:						
Domestic Service	8	6	3	3	34	20
Professional Occupations	17	18	19	20	8	11
Percent Male Workers in:						
Laborer Jobs	7	6	5	5	22	17
Professional Occupations	24	27	26	30	6	9
Incidence of Poverty[b]						
Number Poor (millions)	11.3	8.3	3.1	2.3	4.1	3.5
Percent Poor	20	14	15	10	43	30

SOURCE: Bureau of the Census (1969) Current Population Reports: Special Studies. Series P-23, Number 27 (February 7).

[a] Income data for 1959 and 1967, in constant (1967) prices.

[b] Poverty as defined by Social Security Administration Index, which takes account of residence, family size, and age.

WHERE ARE THE POOR?

Two strong trends in the distribution of the poor are indicated by income data for the 1960s. First, reflecting national figures, the number and incidence of the poor in central cities fell, particularly in metropolitan areas of under one million population. At the same time, the fraction of all poor living in the larger central cities increased. Both trends can be seen in Table 4.

The incidence of poverty in suburban areas was one-half of that in the central cities by 1967, but this difference was mainly due to the relatively small Negro population in suburbs. The incidence of poverty among suburban Negroes is nearly as high as in central cities, although nearly four times as many Negroes (total and in poverty) live in the central city. Among whites, roughly the same number are poor in both central cities and suburbs, but, because of the larger total white suburban population, the incidence of poverty among whites is considerably lower in the latter areas.

An interesting contrast can also be drawn between large and small metropolitan areas: in the latter the incidence of poverty has apparently fallen below that in the former, but only because the latter have a relatively smaller Negro population—the incidence of poverty among Negroes is higher (and of whites approximately the same) in the smaller metropolitan areas. The slightly lower overall poverty incidence in the smaller areas stands in contrast to income levels. Larger central cities have considerably higher median income, indicating a much more unequal income distribution.

Within central cities, changes in the composition of neighborhoods in which the poor reside—particularly minority-group poor—appear to be occurring. Special censuses in 1965 of low-income areas in Los Angeles and Cleveland (U.S. Bureau of the Census, 1966; 1967) revealed that in the worst slum ghettos, income (corrected for price changes) had actually declined since 1960, along with population, employees in higher-status occupations, housing quality, and a number of other indexes of social and economic conditions. The incidence of poverty among Negroes living in poverty areas in the three largest cities increased between 1959 and 1967. Meanwhile, in all metropolitan areas the number of families living in poverty areas declined nearly as much between 1967 and 1968 as it had in the preceding seven years. Among Negroes, net migration from central-city poverty areas totaled 138,000 families between 1967 and 1968, compared with 16,000 from 1960 to 1967.

TABLE 4

A. THE DISTRIBUTION OF POVERTY
(percentage of U.S. total)

Area	Total		White		Negro	
	1959	1967	1959	1967	1959	1967
Central Cities	28	32	25	26	40	44
In Largest SMSA's[a]	13	17	11	13	19	26
In Other SMSA's	16	15	14	13	21	18
Suburban Areas	18	19	20	22	12	11
In Largest SMSA's	7	9	8	11	4	4
In Other SMSA's	11	10	12	11	8	7
Outside Metropolitan Areas	54	49	55	51	48	45
Total U.S.	101[c]	100	100	99[c]	100	100

B. THE INCIDENCE OF POVERTY
(percentage of area population)

Central Cities	20	14	15	10	43	30
In Largest SMSA's	17	15	13	10	36	28
In Other SMSA's	23	14	17	10	54	33
Suburban Areas[b]	13	7	11	6	52	28
Outside Metropolitan Areas	32	19	27	15	77	55
All of U.S.	22	13	18	10	55	31

SOURCE: Bureau of the Census (1969) Current Population Reports Special Studies. Series P-23, Number 27 (February 7).

[a] Exceeding one million population, 1960.

[b] Breakdown by SMSA population not available.

[c] Errors are due to rounding.

    The central-city poor, particularly in the largest cities, are apparently becoming a smaller, more isolated minority. As rising incomes and inroads on discrimination make some middle-income residential neighborhoods in central cities and suburbs more accessible to minority groups, the urban ghetto becomes more homogeneously poor. For the reasons cited in the introductory paragraphs of this section, this phenomenon is likely to cause at least some social and economic conditions in the ghetto to worsen for those remaining there. The consequences could be especially disastrous for poverty-area children as the quality of neighborhood life declines, the educational system deteriorates, and contacts with stable, successful families diminish.

EMPLOYMENT DISTRIBUTION

Changes in the distribution of jobs within large metropolitan areas are often held to be important in determining the economic conditions of the urban poor. According to the Kerner Commission, high ghetto unemployment rates are due in part to "a net loss of jobs in [many] central cities" which has resulted from "the outflow of manufacturing and retailing facilities" (National Advisory Commission on Civil Disorders, 1968). In a more detailed study, Kain (1967) concludes that "the most central parts of metropolitan areas are losing employment to outlying areas, and . . . this process is, if anything, accelerating." Finally, Mooney (1969) has estimated that a one-percentage-point decline in the percent of metropolitan area jobs located in the central city will cause a fall in the employment rate of Negro males of .19 of 1.0 percent, and of Negro females, .15 of 1.0 percent.

CHANGES IN JOB LOCATIONS

Unfortunately, almost no data are available which would present an unambiguous picture of the distribution of jobs in relation to the residences of the poor over the recent past. The lone exception is estimates of employment in retail trade in central business districts and other major retail centers, published in the *Census of Business.* Between 1958 and 1963, the total payroll of retail establishments in the central business districts of the central cities of thirty populous metropolitan areas declined about two percent, indicating an employment decline of roughly ten percent. Of course, the decline was not experienced by all central business districts: New York, Boston, San Francisco and Washington CBD's, all in metropolitan areas ranking among the ten largest, showed significant growth in retail payrolls. Yet in all thirty metropolitan areas, the CBD's share of the metropolitan-area retail payroll declined.

Employment data for several sectors can be obtained either for entire cities or for counties. Table 5 shows employment data for thirty large metropolitan areas for retail trade, wholesale trade, manufacturing, and "selected services"—a category that includes most of the lower-wage, business and personal services marketed

TABLE 5
EMPLOYMENT DISTRIBUTION IN THIRTY LARGE METROPOLITAN AREAS

| | 1958 | | 1963 | | Change Percentage | | | |
	Central City	Suburbs	Central City	Suburbs	Central City	Suburbs	Central City[a]	Suburbs[b]
	(000) (1)	(000) (2)	(000) (3)	(000) (4)	(000) (5)	(000) (6)	(7)	(8)
Populous SMSA's								
Manufacturing	4,258	2,935	3,994	3,328	-263	393	- 6.2	13.4
Wholesale	1,289	302	1,241	445	- 48	143	- 3.8	47.5
Retail	2,133	1,216	2,241	1,772	108	556	5.1	45.7
Services	1,252	460	1,415	665	173	205	13.9	44.7
Total	8,924	4,913	8,891	6,210	- 33	1,297	- 9.0	26.4
New York SMSA								
Manufacturing	998	186	927	220	- 71	34	- 7.1	18.6
Wholesale	311	27	293	43	- 18	16	- 5.8	55.8
Retail	437	127	469	180	32	53	7.3	42.1
Services	330	48	378	70	48	22	14.5	45.3
Total	2,076	388	2,067	513	- 9	125	- 0.4	32.2
12 Larger SMSA's								
Manufacturing	2,212	2,158	2,008	2,422	-203	263	- 9.2	12.2
Wholesale	603	215	557	315	- 45	100	- 7.5	46.5
Retail	1,049	825	1,074	1,177	25	353	2.4	42.8
Services	595	311	661	449	66	138	11.1	44.4
Total	4,459	3,509	4,300	4,363	-157	854	- 3.2	24.3
17 Large SMSA's								
Manufacturing	1,048	591	1,059	687	11	96	1.0	16.2
Wholesale	376	59	391	87	15	27	4.0	46.2
Retail	648	265	698	415	51	150	7.8	56.5
Services	317	101	376	146	59	45	18.5	45.0
Total	2,389	1,016	2,524	1,335	136	318	5.7	31.3

SOURCES: Bureau of the Census, 1958 and 1963 Census of Business (Area Statistics); 1963 Census of Manufactures (Area Statistics).

[a] Column 7 is derived by dividing column 5 by column 1.

[b] Column 8 is derived by dividing column 6 by column 2.

through business establishments. About half of all employment is in these four sectors. Excluded are domestic service, professions, public utilities, finance, construction, and government. Between 1958 and 1963, employment in the four sectors declined slightly in the central cities of the largest metropolitan areas, while employment in suburban areas rose roughly one-fourth. Central cities lost jobs in manufacturing and wholesale trade, showed some slow growth in retailing, and experienced reasonably rapid growth in services.

Declining employment was by no means characteristic of all central cities; in fact, in precisely half of the thirty metropolitan areas employment increased. Furthermore, the employment declines were largely concentrated among the largest metropolitan areas—ten of the thirteen most populous metropolitan areas contained central cities with declining employment in the four sectors.

Government employment has shown significant growth in central cities. In the same thirty metropolitan areas, employment by central-city governments increased by 52,000 between 1957 and 1962 (roughly eight percent), which more than offsets the decline in employment in the four sectors shown in Table 5. Employment by all local governments and by the federal government is available by counties only, but in eight metropolitan areas the central city is an entire county. Table 6 shows that employment declines in the four private sectors were offset by increases in government employment in these eight central cities.

No adequate data exist for estimating employment changes in the remaining sectors; however, one would expect central cities to have done better in the excluded industries than in the four private sectors shown in Table 5. Total employment data are available for New York City, and they confirm this expectation; between 1958 and 1963 employment gains in construction, government, and finance offset declines elsewhere, raising total employment within the city by over 50,000 jobs.

TABLE 6

EMPLOYMENT IN EIGHT CITIES (in thousands)

Sector	1958		1963		Change		Percentage Change	
	Central City	Suburbs	Central City	Suburbs	Central City	Suburbs	Central City[d]	Suburbs[e]
Wholesale Trade	514	150	485	135	- 29	16	- 5.6	10.7
Retail Trade	941	410	979	596	38	186	4.0	45.4
Selected Services	570	147	647	223	80	75	14.0	51.0
Manufacturing	1,744	712	1,608	890	-136	128	- 7.8	18.0
Subtotal	3,769	1,419	3,719	1,844	- 48	404	- 1.3	28.5
Local Government[a]	405	191	445	267	40	77	9.9	40.3
Federal Government[b]	438	143	478	175	40	32	9.1	22.4
Total[c]	4,612	1,753	4,642	2,286	32	513	0.7	29.3

SOURCE: Compiled from Bureau of the Census, Census of Business, Census of Manufactures, and Census of Governments for relevant dates. (For details, see Cohen and Noll, 1968.)

[a] Data for 1957 and 1962.

[b] Data for 1958 are average of 1957 and 1959 figures.

[c] May not add due to rounding.

[d] See note a, Table 5.

[e] See note b, Table 5.

Part of the change in the central-city employment reflects annexation of new territory, but unlike Census population figures, the Census Bureau does not estimate the contribution of annexation to central-city employment totals. The annexation effect is important for two reasons. First, annexed areas, being on the boundaries of central cities, are probably only slightly more accessible to central-city poverty areas than other, unannexed, suburban areas. Thus, a measure of changes in the proximity of jobs to poverty areas should exclude jobs annexed to the central city. Second, the ability to annex suburban areas can be very important to central cities in adding to their tax base, and consequently their ability to provide services to central-city poor. Furthermore, annexation and growth are likely to increase employment by the central-city government, and these jobs are more likely to be located in the central core of the city—and to be open to members of minority groups—than local government jobs in the suburbs.

Some estimates have been made of employment trends within central cities, correcting for effects of annexations (see Meyer, Kain and Wohl, 1965; Kain, 1967), but the assumptions underlying these estimates are arbitrary (see Cohen and Noll, 1968). In any event, most central cities in large metropolitan areas cover a wide geographical territory, containing many areas not physically close to central-core poverty districts. Even when corrected for annexation, central-city employment data over the past two decades are not a very accurate measure of job opportunities located near the poor.

With these caveats in mind, the existing data seem to indicate that if all industries are included, most central cities are probably experiencing some relatively slow growth in total employment, even with adjustments made for annexation. However, in some cities—probably several of the older and very largest—employment may be declining. Furthermore, within the cities with expanding job opportunities, much of the growth could be occurring in suburban areas inside the limits of the central city but not physically near central-core poverty neighborhoods. Certainly in the retail trade sector the slow growth in total central-city employment is occuring despite declining employment in the central business district, although this loss conceivably could be offset by growing employment in service industries and government.

DETERMINANTS OF EMPLOYMENT DISTRIBUTION

Business locations are normally regarded as the result of rational choices based upon the profit potentials implied by various market signals and technological possibilities seen by firms. The list of factors that could influence location decisions is long, and will not be summarized fully here (for more complete treatments, see Hoover and Vernon, 1959; Chinitz, 1964; Schnore and Fagin, 1967). The following discussion is focused upon factors that may have been especially important in determining the pattern of metropolitan employment distribution that has emerged over the past two decades.

Population and Jobs

Population and employment distribution are closely connected. Dispersion of the metropolitan population has pulled consumer-oriented service industries (retail trade, personal services, local government) and, indirectly, some other industries providing business services to the suburbs. Conversely, suburbanization of employment generates some further dispersion of population as both employees and, in the case of consumer-service industries, customers find the convenience of suburban residences increased.

Not all consumer-service industries are equally attracted to the suburbanizing population. Firms with markets so small that they must attract customers from the entire metropolitan area continue to find central locations more attractive; examples of this category are specialized retail stores, concert halls, and live theaters. In addition, because of the larger central area market, retail stores carrying a large inventory of a wide variety of items in a particular line of merchandise will also find central locations attractive. Consequently, the share of the central city in retail sales and employment should always remain above its share in total population. Nevertheless, an acceleration in population dispersion can be expected to lead to an acceleration in the suburbanization of consumer-service industries.

The Transportation System

Changes in the metropolitan transportation system have undoubtedly affected the pattern of metropolitan development. Automobiles, trucks, and freeways have made intrametropolitan transportation cheaper (although partly because of subsidies, and partly because some of the other costs such as air pollution are never paid) and more flexible, thereby increasing the relative attractiveness of suburban locations.

The importance of transportation in the suburbanization of employment is easy to overemphasize; certainly much suburbanization would have occurred had the internal combustion engine not been invented. In the absence of trucks and freeways, workers would tend to live closer to their place of work and in more densely settled neighborhoods, railroads would have developed suburban passenger and freight service far more extensively (including a much more complex system of sidings to suburban industrial areas), and smaller suburban firms would exhibit a greater tendency to locate in compact clusters. New York and Chicago had both become very large and rather extensively suburbanized before trucks and autos dominated intrametropolitan transportation (although the suburban areas were within the political limits of the central city), and provide examples of the more dense pattern of development that could have occurred elsewhere.

The Age of the Central Core

The period in which the central-core area of a large city was developed will affect the attractiveness of the core to businesses. Because of the long life of buildings, the central areas of most large cities contain a large number of structures built when the city initially grew to prominence, reaching a quarter to a half a million population. These old structures embody the tastes, business practices, and technology of the period in which they were built; the older the central city, the more obsolete the buildings in its central core.

As a rough approximation, firms will operate in the same location until the additional profits to be expected from moving into a newer, more modern facility justify incurring the costs, or until the firm can no longer cover its operating costs. The more obsolete the structure, the more likely it is that the firm will face one of these conditions.

Most American cities fall into one of three age groupings. A few cities were populous and extensively developed at the turn of the century, and therefore contain many commercial and industrial buildings constructed in the nineteenth century. Another group of cities grew to prominence between World War I and the Great

Depression, and a final group has emerged since the beginning of World War II. The incidence of employment losses in the four sectors included in Table 5 is highly correlated with the age group to which the city belongs. Of the eleven cities with populations exceeding 300,000 in 1900, ten showed declines in employment in the four sectors between 1958 and 1963. Ten of the thirty cities studied reached 300,000 population between 1900 and 1930, of which four lost employment in the four sectors between 1958 and 1963.

The Availability of Labor

Business location decisions are also affected by labor-market conditions. The argument that ghetto unemployment is caused by shifts in employment distribution assumes that the metropolitan area is comprised of several somewhat independent labor markets. It further assumes that because more employment growth is occurring in the suburbs, labor markets must, perforce, be tighter there. But such differences in labor-market conditions will affect firm-location decisions. A disproportionately large share of the metropolitan labor force works in the city (the ratio of employment to population is fifty percent higher than in the suburbs). It is possible that a much larger number of suburbanites working in the central city would prefer suburban jobs than can be filled by suburban employment growth, and that firms are locating in suburbs in response to tighter central-city labor markets.

Some fragmentary evidence suggests that central-city labor markets have considerable excess demand for workers. A study of ghetto employment in Boston found high job-vacancy rates in the central core, especially in low-skill occupations (see Doeringer, 1969).

Tighter central-city labor markets should be detectable through higher wages and lower unemployment (corrected for skill differentials in labor-force composition). A rough indication of wage differentials by skill groups can be obtained by observing wages in certain specific industries in which firms do not exhibit wide differences in the skill composition of their work force. In retail trade, five types of establishments appear to be somewhat more uniform in labor-force requirements than the others (although even these are far from homogeneous): automobile dealers, dime stores, shoe stores, restaurants and women's ready-to-wear stores. For these five types of stores in the thirty metropolitan areas covered by Table 5, data are available for 129 of the 150 possible observations on wage differentials. Of these, 101 shows higher wages in the central business district, twenty-one show higher wages elsewhere, and seven show equal wages. Eleven of the twenty-eight cases of higher or equal wages outside the central business districts are accounted for by automobile dealers, which also is the industry among the five in which the central business district has by far the lowest share of metropolitan area sales and employment. Removing automobile dealers from the sample, of the 105 observations of differentials in the remaining four types of stores, the central business district wages were higher in eighty-eight cases.

Table 7 shows wage differentials for several manufacturing industries between Manhattan and the rest of the New York metropolitan area, and between Philadelphia and the rest of the Philadelphia metropolitan area. These industries were selected partly because of a greater expected homogeneity among firms in labor-force requirements. They also show generally higher wages in the central city. These data are in contrast to the differential in average wages for all manufacturing: in the fifteen largest metropolitan areas, fourteen central cities (including New York and

TABLE 7

**WAGE DIFFERENTIALS IN MANUFACTURING
NEW YORK AND PHILADELPHIA
(1963)**

| | Average Hourly Wage of Production Workers | | | |
| | New York | | Philadelphia | |
Industry	Manhattan	Entire SMSA	City	Entire SMSA
All Manufacturing	$ 2.37	$ 2.42	$ 2.53	$ 2.69
Baking	2.46	2.71	2.56	2.50
Knitting Mills	2.18	2.08	1.88	1.89
Dresses, Blouses	2.27	2.08	1.85	1.83
Yarn and Thread	1.60	1.57	1.83	1.76
Footwear	2.09	2.00	1.94	1.93
Metalworking Machine	2.92	2.80	2.90	n.a.
Men's Suits, Coats	2.29	2.33	2.22	2.20
Bookbinding	2.41	2.36	2.21	2.17
Purchased Glass	2.11	1.96	2.30	2.11
Cutlery, Tools, etc.	2.14	2.13	2.41	2.34
Jewelry, Silverware	2.46	2.39	n.a.	n.a.

SOURCE: Bureau of the Census, 1963 Census of Manufactures (Area Statistics).

Philadelphia) have lower average wages in manufacturing than their suburbs. Among the next fifteen metropolitan areas, ten central cities have lower wages. The opposite result for specific industries compared to all manufacturing indicates that *a much greater concentration of lower-skilled manufacturing jobs in the central city accounts for the lower average wage of the central-city employee.*

Unemployment rates are another index of labor-market conditions. Except for adult white males, no statistically significant differences exist in age, sex, and race-specific unemployment rates between residents of central cities and suburbs (see Flaim, 1968). These rates are not corrected for different skill compositions of the labor forces. One would expect lower unemployment among suburban workers because of their above-average occupational skills.

All of these findings indicate that jobs, particularly for the less skilled, are easier to find in the central city. They also suggest that suburbanization of employment may be a response to labor-market conditions rather than a cause of unemployment.

Business Cycle Sensitivity

For reasons related to labor-market conditions, to the age of the central core, and to the preferences of businessmen, central-city employment is likely to be more sensitive to changes in the business cycle. When economic contraction occurs, older, less productive capital is withdrawn from production first. Since the central-city capital is, on the average, older, central-core employment falls more, relatively. In addition, recessions of the postwar era have been characterized by widely differing impacts by occupational class. Professionals, managers, and highly skilled labor have experienced less variance in unemployment than have operatives, laborers, and other unskilled or lower-skilled workers. Central cities contain proportionally more lower-skilled workers, while professionals and managers are the most highly-suburbanized occupational group (Wheeler, 1968). Consequently, during a recession, the income of the central city is affected more than are suburban incomes. Retail sales and other consumer services should therefore be more affected in the central city. Firms relatively less touched by the recession and seeking to expand employment will find the recruitment of lesser-skilled labor and production workers far less difficult than the recruitment of managers and professionals. Locations in the suburbs—with a larger pool of the scarcer occupational group—will then be relatively more attractive.

The commercial and industrial composition of a city also affects its sensitivity to the business cycle. Northeastern and Midwestern cities have historically been the home of investment-goods manufacturing. A slowing of economic growth causes cutbacks on plans of firms to expand capacity and, therefore, a decline in investment-goods sales. Since these manufacturing cities are also generally the oldest, they are doubly damaged by a recession.

The Physical Structure of a City

The incredible durability of buildings and the difficulties of assembling land hinder employment expansion in central cities. The value in current use of a plot of land and the building upon it can fall considerably below the value of the land in best use without inducing conversion simply because of the expense of clearing the old structure. Land assembly presents difficulties in large part because inefficient, low use-value structures are normally not all lined up neatly in a row, but are interspersed with structures generating income adequate to justify their continued existence. Consequently, a new investment covering several tracts of land must be able to recover the costs of purchasing some profitable assets—and then destroying them—as well as the costs of purchasing and clearing obsolete structures. Furthermore, the assembling of several tracts of land must overcome attempts by owners of individual plots to capitalize on the necessity for including their own property in a larger development by setting a monopoly price.

The assembly and clearing problems are likely to be more difficult, the older the central city. Through time, the minimum efficient size of establishments apparently has increased, so that one is likely to be required to assemble more places of business and individual plots of land for a given investment, the older the neighborhood to be redeveloped. Pre-World War I cities, constructed before the era of automobiles and trucks, normally exhibit much more intensive use of land. More compact, larger buildings and less open space mean more expense in land assembly and clearing.

The problems of land assembly and clearing do not represent a permanent block to redevelopment of the central city. What must occur is a fall in land and structure prices to a level much below that which, in the absence of these problems, would induce replacement; replacement then comes later than would otherwise be the case. This is true even of government decisions regarding urban renewal of a neighborhood. A government is willing to bear the costs of assembly and clearing for some sufficiently high return (considering both social effects and changes in tax revenues). If assembly and clearing were cheaper, neighborhoods would be cleared when land and structure values in current use were higher. The more deteriorated the neighborhood, the cheaper the site prices will be, and the higher the social benefits from renewal as viewed by the city.

An Accelerating Decline?

The preceding remarks indicate that the years from the end of the Korean War to the mid-1960s may have been unusually difficult ones for older central cities. This period saw the convergence of two depressing influences on central-city employment. First, many structures became obsolete in central-core areas within those central cities that had become prominent commercial and industrial centers in the nineteenth century. However, their value had not declined sufficiently to induce business or governments to undertake the costs of demolishing and clearing them. Second, the economy softened during this period, passing through a ten-year period of slow growth and high unemployment. The older the central city, and the more dependent the central city on manufacturing, the more detrimental was the impact of a sluggish national economy.

If this analysis is correct, the return to higher employment in the mid-1960s ought to have improved the economic performance of older central cities substantially, as the more recent data seem to indicate. But obsolescence of the central core may continue to be a growing problem in the middle-aged cities that experienced most of their development during the 1920s. Built after the beginning of the era of the automobile, these cities are generally less densely developed, contain lower structures, and use land less intensively. Consequently, the land-assembly and structure-demolition problems are likely to be less difficult. Maintenance of high employment will also lessen the difficulties of redevelopment.

A longer-range problem may be posed by the increasingly intensive use of central-city land. High assembly and clearance costs lead to much larger structures replacing older ones, which raises the question of what will happen when the immense structures built since 1930 begin to become obsolete? In fifty years the Empire State Building may be as much of an albatross to New York City as buildings constructed during the 1880s are today (although office buildings may obsolesce more slowly than manufacturing plants). The costs of dismantling and removing such a structure are likely to be monumental, and the chances of spreading the cost over units in a much larger structure seem remote.

IMPLICATIONS FOR THE URBAN POOR

Employment opportunities are relevant to the elimination of poverty only among the poor who are employable. Poverty among the disabled, the aged, and the very unproductive will not be significantly affected, at least in the short run, by expanding employment.

The reductions in poverty in recent years have been heavily concentrated among families with an employable male head. Between 1959 and 1967, poverty among families with a male head declined forty-one percent in central cities—and declined equally rapidly for Negroes and whites. By 1967 only sixteen percent of Negro families headed by a male were in poverty in central cities; among whites only six percent were poor. Clearly the economic expansion of the 1960s has benefited working-age males.

During the same period, poverty in central cities among Negro families headed by a woman *increased* twenty-five percent (among all Negro families poverty declined fifteen percent). Among whites, poverty among families headed by a woman declined fourteen percent—still small compared with the decline in male-headed families. Some of this shift in the composition in poverty is probably due to the consequences of an expanding welfare system, closed to men, which provides increasing incentive for family dissolution in order to qualify children for assistance. But much of the shift is due to increased employment at higher pay among lower-skilled, less productive, or discriminated-against adult males.

The pattern of metropolitan employment growth does not explain the recent trends in poverty incidence and unemployment among females. Mooney makes the point that the types of jobs growing reasonably rapidly in the central city are those most likely to be filled by women. As a result, the physical access to jobs must be getting relatively better for central-city women than men, yet the relative distribution of employment and poverty is moving in the opposite direction.

Little progress has been made during the 1960s in reducing unemployment among teen-agers, whether white or Black, in either central cities or suburbs. Probably a large fraction of jobs open to teen-agers, particularly those seeking part-time employment, are in retail-trade and service firms. Both types of employment are growing rapidly in the suburbs, and the latter is growing rapidly in the central city. Again, job access would not appear to be the problem. Perhaps the fundamental reason is the large increase in the relative number of teen-agers over the past twenty years as the postwar baby boom generation has reached maturity. Increases in the minimum wage also probably affect teen-agers significantly more than any other segment of the labor force. By raising the cost to the employer of labor previously below a minimum wage, the employer will tend to purchase fewer units of such labor.

PUBLIC POLICY CONCLUSIONS

The most important conclusion to be derived from the arguments and evidence presented in this paper is that maintaining full employment is an especially important policy for reducing urban poverty and preserving the economic viability of the central city. This conclusion remains whether or not one assigns high importance to the spatial distribution of employment opportunities in lowering ghetto unemployment, for the connection between intrametropolitan job distribution and national full employment is particularly strong.

Another important conclusion is that the "pure growth" strategy of curing poverty by encouraging private investment and expanded employment within the central city is inadequate. Easy job access and high job-vacancy rates are not relevant to the problems of several large—and either growing or very slowly declining—categories of the poor. Concentration on employment expansion at the expense of other policies actually exacerbates the problems of these groups, since the more successful residents of low-income areas are more prone than ever to move to better neighborhoods, leaving the cen-tral-core low-income district with an increasing concentration of hard-core poverty cases.

Because of the wide divergence in the magnitude of the central-core deterioration problem between cities, general strategies to improve employment conditions in large cities are inevitably ineffi-cient, directing too few resources into those cities with the greatest problems. While this is a particularly strong argument against pro-grams providing relatively automatic incentives for all who invest in lower-income or older areas of central cities (such as tax incentives), it is likely to hold to some degree for almost any federal program. Cities with really serious central-core declines are relatively few in number—perhaps fifteen metropolitan areas in a dozen states. While the number of people living in these areas is very large, they still constitute a rather small minority of the population and of the elected representatives in Congress. The federal government, with its tendency to try to spread the benefits of federal programs over the maximum number of congressional districts, seems unlikely to apply the proper concentration of funds. The Model Cities program is a particularly good example of this tendency: initially conceived as a program to do a rather complete face lifting of a few cities—perhaps a half dozen—as a demonstration of how cities could be redeveloped, the programmed number of cities to be included grew to over two hundred by the time the first money for construction was appropri-ated. Yet the requested annual commitment of funds to the program remained roughly constant.

The nature of the federal decision-making process, as illustrated by Model Cities, suggests that much greater reliance will have to be placed upon the states for funding redevelopment programs in the oldest cities. In this connection, it is interesting that the policy con-clusions in the final chapter of the Kerner Commission Report relate only to local and national governments, neglecting the states. Perhaps the federal government can supply funds ample for treating the prob-lems of the most seriously declining cities indirectly, through assum-ing a greater federal financial responsibility for other, more univer-sally distributed problems, such as education, public health, welfare, and others connected specifically to poverty. Many cities that have significant problems caring for low-income families do not suffer from a *general* loss of employment. A fully federalized welfare system, for example, would free roughly $4 billions of state and local government expenditures for other uses.

An area deserving further investigation for possible policy overtones (particularly with regard to state constitutions) is the politics of annexation. Why are some cities far more able than others to annex surrounding suburban areas? Why does the ability of some cities to annex change through time: for example, why was New York City able to annex Queens and Bronx, but not Great Neck and Yonkers? Annexation would appear to be a very powerful potential response to the problems of a declining central core as a means of retaining a tax base and expanding central-city government employment. Even today, a number of large cities include a very large proportion of the suburban communities in the metropolitan area (e.g., Houston, Memphis, San Antonio, San Diego). How has this affected the ability and the willingness of these cities to deal with central-area problems?

Further investigation and experimentation in modular structures, as well as continued attacks on restrictive building codes and union rules, are also suggested, with the emphasis upon developing structures that, while durable, can also be more easily removed. Modular construction could significantly raise the scrap or salvage value of structures to be cleared, in addition to reducing dismantling costs. One could imagine the eventual development of a market for used modular units. This would greatly facilitate the conversion of urban land from one use to another, and would provide some incentive to owners to maintain structures in order to preserve the resale value of the modules.

The analysis of this chapter provides little support for the development of urban mass-transit systems as vehicles to reduce poverty. The experimental programs for busing ghetto residents to suburban jobs in several large metropolitan areas were not particularly successful, largely because lack of participation by both employers and workers prevented the programs from holding the promise of reaching significant scale. Of course, the case for urban mass-transit systems does not depend upon its effect on ghetto unemployment—its value will be determined by the tastes and density of the population of an urban area and the necessity for reducing automobile use for reasons of pollution and central-area congestion. The argument here is simply that *urban mass transit cannot be expected to produce serious inroads against the poverty problem.*

Finally, the depopulating, declining ghetto amidst rapidly increasing prosperity elsewhere in the central city is probably the most dangerous current social trend. Great emphasis should be placed upon adequate income support and educational opportunities in these areas. In addition, renewal programs should be more oriented toward the *original* Model Cities concept, integrating a wide variety of social services into redevelopment, as opposed to the urban renewal-public housing emphasis on physical aspects of housing and business activity. Furthermore, building on the fact that informal contacts with employed persons are an especially important vehicle for finding jobs among the unemployed, the importance of providing employment services and job information in the ghetto is increasing to the extent that more of the regularly employed are moving elsewhere.

NOTE

1. New York, Chicago, Philadelphia, St. Louis, Boston, Baltimore, Cleveland, Buffalo, San Francisco, Cincinnati, and Pittsburgh. All still are among the twenty-one most populous cities, although Buffalo and Cincinnati are at the bottom of that list.

REFERENCES

BOWLES, S. and H. LEVIN (1968) "The determinants of scholastic achievement." Journal of Human Resources (Winter).

CHINITZ, B. (1964) City and Suburb. Englewood Cliffs, N.J.: Prentice-Hall.

COHEN, B. I. and R. G. NOLL (1968) "Employment trends in central cities." Social Science Discussion Paper 69-1. Pasadena: California Institute of Technology.

COLEMAN, J. (1965) Equality of Educational Opportunity. U.S. Department of Health, Education and Welfare. Washington, D.C.: Government Printing Office.

DOERINGER, P. (1969) "Manpower programs for ghetto labor markets." Proceedings of the 21st Annual Winter Meeting of the Industrial Relations Research Association: 257-267.

FLAIM, P. O., (1968) "Jobless trends in twenty large metropolitan areas." Monthly Labor Review (May).

GINZBERG, E. et al. (1968) Manpower Strategy for the Metropolis. New York: Columbia University Press.

GLAZER, N. and D. MOYNIHAN (1965) Beyond the Melting Pot. Cambridge: MIT Press.

GRODZINS, M. (1959) The Metropolitan Area as a Racial Problem. Pittsburgh: University of Pittsburgh Press.

HOOVER, E. M. and R. VERNON (1959) Anatomy of a Metropolis. Cambridge: Harvard University Press.

KAIN, J. F. (1968) "Housing segregation, Negro employment and metropolitan decentralization." Quarterly Journal of Economics, LXXXII (May): 175-198.

––– (1967) "The distribution and movement of jobs and industry." In J. Q. Wilson (ed.) The Metropolitan Enigma. Washington, D.C.: Chamber of Commerce of the United States.

MEYER, J. R., J. F. KAIN and M. WOHL (1965) The Urban Transportation Problem. Cambridge: Harvard University Press.

MOONEY, J. D. (1969) "Housing segregation, Negro employment and metropolitan decentralization." Quarterly Journal of Economics, LXXXIII (May): 299-311.

National Advisory Commission on Civil Disorders (1968) Report. New York: Bantam Books.

SCHNORE, L. and H. FAGIN (1967) Urban Research and Policy-Planning. Beverly Hills: Sage Publications.

SHEPPARD, H. L. and A. H. BELITSKY (1966) The Job Hunt. Baltimore: Johns Hopkins Press.

U.S. Bureau of the Census (1967) "Characteristics of selected neighborhoods in Cleveland, Ohio: April 1965." Current Population Reports, Series P-23, No. 21. Washington, D.C.: Government Printing Office (January 23).

––– (1966) "Characteristics of the south and east Los Angeles areas: November 1965." Current Population Reports, Series P-23, No. 18. Washington, D.C.: Government Printing Office.

U.S. Federal Trade Commission (1969) Economic Report on Installment Credit and Retail Sales Practices of District of Columbia Retailers. Washington, D.C.: Government Printing Office.

––– (1968) Economic Report on the Structure and Competitive Behavior of Food Retailing. Washington, D.C.: Government Printing Office.

VALENTINE, C. A. (1968) Culture and Poverty: Critique and Counter Proposals. Chicago: University of Chicago Press.

WHEELER, J. O. (1968) "Residential location by occupational status." Urban Studies 5 (February): 24-32.

Part IV

URBAN SERVICES
AND THEIR DELIVERY

Part IV

URBAN SERVICES AND THEIR DELIVERY

Introduction

In Part II, we discussed the process by which public expenditure decisions were made. These decisions, in fact, allow public agencies to provide particular goods and services in amounts which do not cost more than can be financed in the agencies' budgets. Seldom specified is how these goods and services are distributed to the population, in terms of the quantity, quality, or method of delivery. Also seldom specified is how goods and services get distributed to population groups. What are garbage pickup rates, whose street gets repaired first, and so on? Is public opinion at all sensitive to variations in service levels over time or over the city? This is the topic of the Aberbach and Walker paper. Simply put, their conclusion is that within the structure of the existing delivery system and within the (considerable) range of service levels provided in Detroit, public opinion is pretty insensitive to public agency actions.

Eisinger focuses on some attempts to place the delivery of public goods and services under closer control of the consumers of these goods and services. An examination of how six antipoverty organizations devised their own strategies for combating poverty in their own neighborhoods leads to some insights into how many of the proposed decentralized decision-making and self-determination schemes put forth recently might work in action. A detailed examination of the implications of government sponsorship (legal and financial) of agencies which are, on occasion, also government antagonists is presented.

Finally, Reiss critically examines the existing mechanisms for the delivery of local government services. He points out, for example, the fact that the only municipal services available on a twenty-four hour basis and delivered to the customer's door are police and fire protection. He also draws attention to the public bureaucracy's strong preference for what he calls "preprocessed individuals"—those individuals who have the appropriate forms already filled out, who fit neatly into the categories and standard operating procedures of the agency, and who in general can be dealt with routinely. Individuals who do not know how the system works and how to package themselves in such a way as to receive favorable treatment, Reiss argues, are precisely those who need help the most. Several imaginative proposals are offered.

—*J. P. C.*

THE PROVISION
AND DELIVERY
OF URBAN SERVICES

18

The Attitudes of Blacks and Whites Toward City Services: Implications for Public Policy

JOEL D. ABERBACH
JACK L. WALKER

◻ DURING THE 1960s, urban problems moved to the top of the American agenda of public debate. The "urban crisis" was the subject of countless newspaper articles, books, television documentaries, and public speeches. Central cities were afflicted with racial hostility and frequent outbreaks of violence and, while tax bases were declining, crime rates were rising with alarming rapidity. In northern cities, with growing black populations and histories of violent racial conflict, there are few areas of public policy which are not touched by the race issue in some important way. The response of many people to these conditions has been to leave the city for new homes in the surrounding suburbs. At the same time,

AUTHORS' NOTE: *This study was supported by the National Institute of Mental Health, the Horace H. Rackham Faculty Research Fund, and the Institute of Public Policy Studies, The University of Michigan. James D. Chesney and Douglas B. Neal assisted in the data analysis, and Joan Aberbach read the earlier draft and made valuable suggestions for revision.*

because of changing economic conditions, industries in larger numbers have begun locating their new plants in open country outside the cities, depriving established urban areas of both residents and tax revenues. This spiral of urban decay has begun to cause increasing alarm among many observers of American life.

This widespread concern over the future of cities has led to the creation of many plans for urban renewal or rehabilitation. Schemes abound for improving the public schools, reforming the welfare system, creating new parks and recreation programs, and professionalizing the police. Some believe that attention should be concentrated on regaining public confidence in government by eliminating rigid or insensitive bureaucracies. In response to these desires, plans for police review boards, community advisory committees and administrative decentralization are being actively considered and debated in almost every large city in the country. Others believe that many of the urban problems we face can be solved through massive infusions of public money to create new housing, more adequate educational facilities, hospitals, and social agencies. Some groups press for racial integration of all aspects of urban life, while others believe that blacks are primarily interested in controlling their own communities and advocate a form of racial separation. The political and administrative leaders of city government are being pressed on all sides with new proposals and demands for social change.

The problems of choice faced by city officials could easily be resolved if they could simply do everything at once. If there were enough resources, and the programs did not conflict with each other, no hard decisions might every have to be made. Of course, no city official has ever, or will ever, face such circumstances. Even if city governments were in much better financial condition than they are, important choices between competing solutions would have to be made. In the contemporary setting, where all resources are in short supply and racial tensions are more intense than ever before in our history, it is extremely important that urban policy-making be as informed and effective as possible.

One of the principal goals of policy-making in urban areas must be to increase public satisfaction with city services, and, in most contemporary American cities, the reduction of racial tensions must also be a major objective. Of course, the quality of city government and the services it provides are not the only elements which affect the attractiveness of urban life. Contemporary racial tensions also are rooted in aspects of the American social environment which are, for the most part, far beyond the reach of city hall. Nevertheless, city governments are not powerless to influence the attitudes and behavior of their citizens. Before this potential influence can be exercised, however, city officials ought to be aware of the current reactions of the public to the existing services of city government. They also ought to know what characteristics of government and which services seem to be producing the most dissatisfaction and to have some understanding of the causes of the dissatisfaction. Before planning new policies or adopting elaborate programs of bureaucratic reform, they should understand the desires, expectations, and basic orienting values of both the black and the white communities.

This paper contains some of the knowledge of public reactions to city services which is prerequisite to effective decision-making. The paper begins with a general review of the public's evaluations of city services in a major American city; it includes an analysis of the correlates of dissatisfaction among both black and white citizens; it concludes with a review of the policy implications suggested by our findings. If policy makers are to avoid possibly disastrous unanticipated reactions to their decisions, they must gain a firm grasp of the current public mood. By both describing and analyzing the shifting focus of urban discontent, we hope to provide a solid foundation for the formation of humane and successful public policies.

THE DATA

The data upon which our analysis is based were collected in two surveys of Detroit, Michigan. The first was completed in the fall of 1967 and the second in the winter of 1968.

In the 1967 survey, 855 respondents were interviewed (394 whites and 461 blacks); in all cases, whites were interviewed by whites, blacks by blacks. Approximately sixty-two percent of the respondents were chosen in a random sample of the households in the entire city; the rest represent a supplementary random sample of the zone within the city in which Detroit's 1967 riot took place. Each of these original respondents was asked to supply his name so that he could be reinterviewed in the future. Approximately eighty-two percent did so, and of this group, 295 respondents were reinterviewed in 1968 (154 whites and 141 blacks). Before resampling, the original respondents who agreed to be reinterviewed were stratified according to race, section of the city, age, and sex.

EVALUATIONS OF CITY SERVICES

In order to measure the level of satisfaction with Detroit's city services, we asked our respondents: "What about the quality of parks and playgrounds for children in this neighborhood—are you generally satisfied, somewhat satisfied, somewhat dissatisfied, or very dissatisfied?" The same kinds of questions were asked about public schools, sports and recreation centers for teen-agers, police protection, and garbage collection. The results of these questions are reported in Table 1.

TABLE 1

DEGREE OF SATISFACTION WITH CITY SERVICES, BY RACE [a] (in percentage)

	Whites			Blacks		
	Satisfied	Dissatisfied	Don't Know	Satisfied	Dissatisfied	Don't Know
Public schools	55	19	26	55	25	20
Parks	59	33	8	40	46	14
Teen centers	43	28	29	24	48	28
Police	81	18	1	51	43	6
Garbage	91	5	4	74	23	3

[a] Answers to the questions on satisfaction with services were dichotomized for ease of presentation. "Generally satisfied" or "somewhat satisfied" were classified as satisfied and "somewhat dissatisfied" or "very dissatisfied" as dissatisfied.

Blacks express more dissatisfaction than whites in regard to all five city services included in Table 1.[1] More whites are satisfied than dissatisfied in every category of the table and, generally, are much more contented with existing city services than blacks. The whites express their highest levels of satisfaction with garbage collection and the police, their lowest levels of satisfaction with recreational facilities for teen-agers, but in all categories, more white respondents are satisfied than dissatisfied with the services they are now receiving. Among our black respondents, however, more dissatisfaction than satisfaction is expressed regarding both parks and recreational facilities for teen-agers. In fact, twice as many black respondents express dissatisfaction than satisfaction with teen centers. About half of the black community is satisfied with the police, but this is almost thirty percent lower than the white community. As with the whites, garbage collection draws the fewest complaints and the most satisfaction among blacks, but again, the level of satisfaction is about seventeen percent lower among blacks than whites. Only the public schools receive similar ratings from the two racial groups, and even in this category, blacks express more dissatisfaction about the schools than whites. Overall, there is a clear difference between whites and blacks in Detroit in their evaluations of the services of city government.

Even though blacks are more dissatisfied with city services than whites, surprisingly, most do not believe their services are much different than those provided in other parts of the city. As we can see in Table 2, when we asked our respondents to compare the services they receive in their neighborhood with those provided in other parts of the city, there was little difference in the answers of the two racial groups. Fewer blacks felt they received better treatment and three times as many felt they were getting worse treatment, but in both racial groups, more than seventy percent of the respondents believed their neighborhoods were being treated about the same as all other parts of the city.

The picture changes somewhat when we asked whether our respondents expected city officials to make a positive reply to complaints about poor service. As Table 3 demonstrates, blacks were decidedly less confident about this prospect than whites. A sizable number of respondents from both racial groups do not believe they could get a positive response to a complaint, but the feeling is much more widely held in the black community.

Table 4 demonstrates, however, that even though blacks are less optimistic about getting action from city officials when they complain, they report having made complaints about as frequently as do the whites. About one-third of our respondents in each racial group, a strikingly large number, report having made some kind of complaint to the city about poor service.

TABLE 2

RELATIVE RANK OF NEIGHBORHOOD CITY SERVICES, BY RACE (in percentage)

Question: "Thinking about city services like schools, parks and garbage collection, do you think your neighborhood gets better, about the same, or worse services than most other parts of the city?"

	Whites	*Blacks*
Better	19	11
Same	72	71
Worse	5	18
Don't know	4	0
	100	100

TABLE 3

BELIEF THAT CITY OFFICIALS WILL RESPOND TO COMPLAINTS, BY RACE (in percentage)

Question: "If you have a serious complaint about poor service by the city, do you think you can get city officials to do something about it if you call them?"

	Whites	*Blacks*
Yes	68	51
No	29	45
Don't know	3	4
	100	100

TABLE 4

PERCENTAGE OF POPULATION HAVING REGISTERED COMPLAINTS, BY RACE (in percentage)

Question: "Have you ever called the city with a complaint about poor service?"

	Whites	*Blacks*
Yes	35	33
No	64	66
No answer	1	1
	100	100

In summary, our data indicate that blacks are much more dissatisfied with existing city services than whites, but neither group believes it is receiving unusual or special treatment. Blacks are decidedly more pessimistic than whites, however, about the prospects of getting city officials to respond to their complaints. Even though there are differences in satisfaction and expectations between the two racial communities, and even though many blacks believe the city government to be unresponsive, about the same proportion of each group report having made complaints about poor services. It seems that the relative satisfaction of the whites or the dissatisfaction of the blacks has not restrained either group from frequently communicating their feelings to city officials.

ATTITUDES TOWARD THE POLICE

Responses to our general question on the adequacy of city services revealed a great difference of opinion between the two racial groups regarding the police. Since the conduct of local police has so often been the spark for violent protests or riots in black neighborhoods during the last five years, we asked a further series of questions about the citizens' evaluations of the police force. Only eighteen percent of the white community expressed any dissatisfaction about the quality of the police work in their neighborhood when asked for their general evaluation of city services. When we followed up this general question with a series of questions on specific abuses charged against the police, we see an even greater difference of opinion between the two racial groups on these potentially explosive issues. The results of these questions, displayed in Table 5, show that whites are almost unanimous in denying that the police in their neighborhood lack proper respect for the rights of citizens, stop or frisk people without cause, or use unnecessary force. It is important to note that a majority of our black respondents *also* deny that the police are guilty of such tactics, but over one-third of the blacks agree with the charges in each case. The black community obviously is not unified in its outrage at the police, but a very large minority expresses dissatisfaction. Only a small minority of whites express any negative feelings about the police on these questions. Our data illustrate the

TABLE 5

EVALUATIONS OF POLICE BEHAVIOR,
BY RACE (in percentage)

Question: "Some people say policemen lack respect or use insulting language. Others disagree. Do you think this happens in this neighborhood?"

	Whites	Blacks
Yes	10	34
No	89	62
Don't know	1	4
	100	100

Question: "Some people say policemen search and frisk people without good reason. Others disagree. Do you think this happens to people in this neighborhood?"

	Whites	Blacks
Yes	1	38
No	90	56
Don't know	9	6
	100	100

Question: "Some people say policemen use unnecessary force in making arrests. Others disagree. Do you think this happens to people in this neighborhood?"

	Whites	Blacks
Yes	7	34
No	91	60
Don't know	2	6
	100	100

TABLE 6

EVALUATION OF THE PERFORMANCE OF THE POLICE,
BY RACE (in percentage)

Question: "Some people say the police in Detroit do not do as good a job as they could in protecting people from burglars, purse snatchers, and other law breakers. Others disagree. What do you think?"

	Whites	Blacks
Do as good as they could	64	32
Do not do as good as they could	33	57
Don't know, no answer	3	11
	100	100

deep divisions existing between the two racial communities over these important issues.

The attitudes of both whites and blacks toward the conduct of the police force, however, is more ambiguous than their responses to the items in Table 5 would indicate. For example, when we asked a question with citywide rather than neighborhood focus and which dealt with the general efficiency of the police rather than alleged violations of civil rights, we uncovered a different, and somewhat more negative, set of responses from both racial groups. In response to the question: "Some people say the police in Detroit do not do as good a job as they could in protecting people from burglars, purse snatchers, and other law breakers. Others disagree. What do you think?" whites display greater dissatisfaction than on previous questions. As revealed in Table 6, blacks are once again more critical of the police than whites, but now thirty-two percent of the whites, as opposed to eighteen percent using the earlier question, express discontent with the performance of the police force.

We followed up this question by asking our respondents: "It is sometimes said that the things we have just been talking about, such as unnecessary roughness and disrespect by the police, happen more to Negroes in Detroit than to white people. Do you think this is definitely so, probably so, probably not so, or definitely not so?" As the results reported in Table 7 indicate, there are still great differences of opinion between whites and blacks over the conduct of the police, but a surprisingly large percentage of our white respondents believe that the police do employ discriminatory standards of law enforcement. Although only eleven percent were definitely convinced, thirty-six percent of our white respondents felt it was probably so. In the black community almost half the respondents are definitely convinced that the police behave in a discriminatory fashion, and only nineteen percent believe they do not. These findings are surprising in light of the overwhelming support for the police expressed in the white community. Obviously, there are large numbers of people in both racial communities who believe the police to be discriminatory and abusive in their dealings with blacks, but are nonetheless still satisfied with their performance. These shortcomings of the police seem to be viewed by respondents either as inevitable or permissible aspects of vigorous law enforcement.

TABLE 7

ESTIMATES OF POLICE BIAS AGAINST NEGROES
BY RACE (in percentage)

Question: "It is sometimes said that the things we have just been talking about, such as unnecessary roughness and disrespect by the police, happen more to Negroes in Detroit than to white people. Do you think this is definitely so, probably so, probably not so, or definitely not so?"

	Whites	Blacks
Definitely so	11	49
Probably so	36	31
Probably not so	29	18
Definitely not so	21	1
Don't know, no answer	3	1
	100	100

THE STRUCTURE OF DISSATISFACTION
WITH PUBLIC SERVICES

We have asked our respondents about their satisfaction with a variety of public services ranging from garbage collection to the education provided by the public schools. Since many people's feelings about a particular service are determined by experiences which they share with only a few—a poor teacher in a child's class, a careless garbageman—and certain services like police protection seem to be connected to larger political issues which arouse intense passions in the community, we were not sure how the indicators of satisfaction with services would tie together. They are, however, strongly interrelated, with the dissatisfaction-with-police-protection item an integral part of the broader dimension we have labeled dissatisfaction with services.

The issue of the public's evaluation of police services, intimately tied to urban racial tensions, deserves extra attention. Our data indicate that whites are more satisfied than blacks with police and other services, but they also show that we must distinguish between discontent about police services as such, and anger about police mistreatment and abuse of citizens. The two are related (the gamma coefficients of correlation equal .34 for blacks and .46 for whites[2]), but the issue of police abuse is closely tied to racial attitudes, while feelings that the police do not provide adequate protection are more a part of the larger discontent with public services characteristic of elements of both racial communities. As part of our analysis, we will examine the correlates of both general dissatisfaction with services and perceptions of police mistreatment of neighborhood residents.

DISSATISFACTION INDICES

We constructed two indices in order to reduce the data and analysis to manageable proportions. The first, which we call the services dissatisfaction index, is a simple additive measure in which negative answers to the items on city services listed in Table 1 were cumulated. The second, labeled the police abuse index, does the same for the items presented in Table 5 on police mistreatment of neighborhood residents. Because of our interest in the distinction between perceived police mistreatment and abuse of citizens and discontent with police services, we also analyze the single item on Detroit police protection presented in Table 6. In all cases, a high score indicates a negative attitude.

SOURCES OF DISSATISFACTION

SOCIAL STATUS

One very plausible and simple explanation for dissatisfaction with services is that persons of lower status tend to receive poorer services and resent this. We already know that blacks are more dissatisfied than whites, which fits this simple model, and we would also expect that those who have lower incomes or levels of educational achievement would, regardless of race, live in poorer areas, receive inferior services and express more dissatisfaction than their higher status counterparts. However, the data do not confirm this proposition. Lower-status whites and blacks do live in poorer areas, and they may receive less adequate services, but they are no less (or more) satisfied than others in their respective racial communities.

DISCRIMINATION

While education and income do not help to explain dissatisfaction, racial discrimination does. In Table 8, we display the relationships for blacks between our measures of dissatisfaction and both reports of personal experiences of discrimination and a belief that the police are biased against Negroes.

TABLE 8

CORRELATIONS (GAMMA) FOR BLACKS BETWEEN THE
DISSATISFACTION INDICES AND (A) PERSONAL
EXPERIENCES OF DISCRIMINATION AND (B)
BELIEF THAT THE POLICE ARE BIASED AGAINST NEGROES

	Services Dissatisfaction Index	Police Don't Do a Good Job	Police Abuse Index
Personal experiences of discrimination index[a]	.27	.37	.47
Police bias against Negroes[b]	.32	.46	.61

[a] This is a simple additive index of reports of personal experiences of discrimination in the schools, housing or employment. A high score indicates many experiences of discrimination.

[b] The exact wording of this item and the distribution is found in Table 7. A high score indicates a definite perception of police discrimination.

Experiences of discrimination and the perception that the police are biased both relate to all measures of dissatisfaction. They are, however, more strongly associated with the police abuse index than with the services dissatisfaction index. When we take an independent indicator of dissatisfaction with police services, that is, one which focuses on strict police protection functions, not the mistreatment of citizens, and it is not asked in the same form as the dissatisfaction index questions, we see that reported experiences of discrimination or concern about police bias are less strongly related to it than to the police abuse index. The pattern established here holds throughout the rest of our data for both blacks and whites: factors and attitudes which are clearly connected in the public mind to the racial dilemma are more important in explaining beliefs that the police abuse citizens than they are in understanding dissatisfaction with the services cities are supposed to provide their residents, including ordinary police services. There is overlap between the two, but solving the problem of discrimination and police abuse of blacks will not totally eliminate dissatisfaction with police or other services.

RACIAL IDEOLOGY

If we turn from the realm of experience to that of ideas—the two are closely connected in the black community (Aberbach and Walker, 1968)—the data look very much like that presented above and we can examine the same phenomenon for whites. Table 9 plainly demonstrates that conservative racial ideas are strongly related to perceptions that police do not abuse citizens, but unimportant in predicting dissatisfaction with services.[3] Regular police services are a part of this pattern: dissatisfaction with them is totally unrelated to racial attitudes for whites and much less strongly related to militant racial attitudes than is the police abuse measure for blacks. Belief that the police abuse citizens is an integral part of the cycle of general discrimination and militant attitudes in the black community, and the only whites who even recognize it as a problem are the minority who hold very liberal racial attitudes.

TABLE 9

CORRELATIONS (GAMMA) BETWEEN THE DISSATISFACTION
INDICES AND RACIAL ATTITUDES, BY RACE

	Services Dissatisfaction Index	Police Don't Do a Good Job	Police Abuse Index
Whites			
Pro-integration[a]	-.05	-.03	.54
Blacks			
Favorable interpretation of "black power"[b]	.10	.26	.48

[a] The following question was employed to measure integrationist sentiments: "Speaking in general terms, do you favor racial integration, total separation of the races or something in between?"
A favorable attitude toward integration was given a high score.

[b] Respondents were asked a totally open-ended question: "What do the words black power mean to you?" Favorable interpretations of black power (given a high score on this index) consist almost exclusively of notions about a "fair share" for blacks or "racial unity" in the black community as a tactic in bettering conditions. See Aberbach and Walker (1968) for an extensive discussion of this.

RELATIVE QUALITY OF SERVICES AND RELATIONS WITH THE BUREAUCRACY

We have seen that racial experiences and attitudes are strong predictors of the recognition of police abuse, but we have not been too successful in understanding dissatisfaction with services. The public assesses services, in part, through considerations of whether they and their neighbors are receiving benefits equal to those given to others in the same jurisdiction. In addition, they often have actual contacts with public officials about problems they face and certainly hold beliefs about what officials would do for them if they sought a remedy from them. The relationships here are clearly part of a complex process. Belief that one's neighborhood receives unequal treatment, for example, leads to dissatisfaction with services which reinforces the perception of injustice. Table 10 displays the relationships between items designed to measure such feelings or reported behavior and dissatisfaction with public services.[4]

TABLE 10

CORRELATIONS (GAMMA) BETWEEN THE SERVICES DISSATISFACTION INDEX AND RELATIVE QUALITY OF SERVICES AND RELATIONS WITH THE BUREAUCRACY,[a] BY RACE

	Services Dissatisfaction Index	
	Whites	Blacks
Rank neighborhood services	.27	.27
Action if you complain	.15	.32
Have you complained?	-.20	-.49

[a] The exact wording of the items is found in Table 2 (rank neighborhood services), Table 3 (action if you complain) and Table 4 (have you complained?). High scores indicate worse neighborhood city services, a belief that city officials will not respond to complaints about poor service, and no actual complaints made to city officials.

All three variables are related to dissatisfaction with services in both the white and black communities. The correlation between the ranking of neighborhood services and the services dissatisfaction index is the same in both communities (.27), but the two complaint questions are more strongly associated with dissatisfaction in the black community. In each case, those satisfied with city services are more confident that city officials would respond to their complaints and are less likely to have complained.

There is a clear racial difference when we look at the inter-action between the two complaint items and dissatisfaction with services. As Table 11 indicates, if a white person has actually complained, success, or at least faith that the system will be responsive to his problem, predicts satisfaction. If he has not complained, such faith is much less important in predicting satis-faction. For blacks who have not actually complained, on the

TABLE 11

CORRELATIONS (GAMMA) BETWEEN THE SERVICES DISSATISFACTION INDEX AND THE BELIEF THAT CITY OFFICIALS WILL RESPOND TO COMPLAINTS FOR THOSE WHO HAVE AND HAVE NOT COMPLAINED, BY RACE

	Have Complained	Have Not Complained
Whites	.41	.10
Blacks	-.02	.46

other hand, a belief that city officials would do something about a complaint is associated with satisfaction. Blacks who have com-plained are so dissatisfied that even a belief that officials will respond to their own personal complaints is not enough to mod-erate their feelings.[5] Further analysis shows that blacks who have complained about services tend to be very distrustful of the gov-ernment and represent a militant group in the black community whose anger extends well beyond personal grievances. Large changes affecting the entire black community are necessary in order to satisfy them.

CONCLUSIONS

It is important for both the analyst and the policy maker to recognize that dissatisfaction with the services cities provide, including police services, and beliefs that the police abuse citizens have somewhat different roots. The issue of police abuse is closely tied to racial attitudes and experiences for both races. A feeling that the police do not provide adequate protection, however, is part of a general dissatisfaction with services which is less strongly tied to racial issues for blacks and independent of them for whites. A city like Detroit is plagued by crime, little of which is interracial, and many members of both races are deeply disturbed about the inability of the police to deal with the situation. Blacks are somewhat more likely than whites to believe that their neighborhoods receive worse services than other parts of the city, but about as many ascribe this to nonracial factors as to discrimination. Even on the specific question dealing with police protection only about twenty percent of the black respondents criticizing the police cite specifically racial reasons, while the rest stress general factors such as inadequate numbers of policemen and laxity in administration.

Vigorous public efforts to end discrimination, providing they were successful, should have an immediate impact on the perceptions of large numbers of blacks who believe that their neighborhood police are abusive. An end to discrimination would also have a noticeable, but smaller, impact on dissatisfaction with government services in general. Programs designed to end discrimination, especially those involving the police, however, cannot be carried out without political cost. At this point in time, highly publicized but ineffective programs will breed cynicism in the black community. At the same time, any such program will cause dismay among those whites who are very bitter about what they see as government's preferential treatment of blacks (Aberbach and Walker, 1969). When one adds to this the limited influence which elected officials often have on the police hierarchy and the fragile influence of the police hierarchy on the behavior of the man in the field, it is clear that even the best-intentioned municipal official may not be able to institute an effective program to end police abuses in black neighborhoods. Ending all discrimination in both the public and private sectors is a clear moral imperative. Increasingly, it is also a necessity if public order is to be preserved. Merely because it is right, however, does not make it any easier to do, politically or administratively.

It is clear that the ideal way to eliminate much of the dissatisfaction with services in both the black and white communities is simply to improve services. Unfortunately, this almost inevitably costs money which cities often do not have and cannot get. This is one area where massive federal aid could be put to reasonably good use, but the cost would be great because many cities have neglected for years needed maintenance and expansion of their facilities because they only had the revenue to cover operating expenses. The chances are that dissatisfaction with city services will increase drastically over the next few years as water systems, sewer systems and other basic facilities deteriorate further. This long overdue need for large capital expenditures in urban areas is another element which must be taken into account in establishing national priorities.

There are partial solutions to the public's dissatisfactions with services which some cities have attempted or are actively considering which are relatively inexpensive. These include decentralized administration with some form of public control, the establishment of neighborhood centers or local city halls and inaugurating a system of ombudsmen to aid citizens in processing their complaints. All of these devices might be successful in increasing the sense that the city government is responsive to complaints. Such a system would have to be effective, of course, or there is great danger that it would backfire. For blacks, the active citizens who now complain are unmoved by their perceptions of their own ability *as individuals* to successfully obtain actions on their complaints. Programs meant to improve governmental responsiveness would have to bring real results for the *entire* black community in order to increase their satisfaction with services and, consequently, the satisfaction of the increased number of people who would have contact with officials. Short-run symbols of attention (i.e., setting up the agencies or programs) will not be enough in an increasingly mobilized and action-oriented community. They will have to work and ultimately produce better results than the previous system. In the end, this must be expensive, no matter how improved the administrative structures.

Establishing more responsive administrative structure is clearly no easy task politically. One seemingly simple solution is to decentralize on racial grounds, giving blacks and whites control over their own communities. On the surface, this would appear to be a mechanism for satisfying both militant elements in the black community and angry whites. However, our data indicate that blacks are overwhelming in their desire for a form of equality which is inclusive rather than exclusive. They want to be a part of this society and they wish to share equally in its benefits. In addition, they do not want to exclude whites—even white policemen—from their communities.

In order to gauge the views of the two communities on these issues, we devised a series of questions to measure commitment to

TABLE 12

COMMITMENT TO AN INTERRACIAL SOCIETY, BY RACE (in percentage)

Question: "Speaking in general terms, do you favor racial integration, total separation of the races or something in between?"

	Whites	Blacks
Integration	42	84
In between	47	14
Separation	11	1
Don't know	——	1
	100	100

Question: "Some people say that white and Negro children should go to the same schools; others feel that they will learn more and be happier in separate schools. What do you think? [If same schools] Do you agree or disagree with this statement: Demonstrations to protest segregation of schools are a good idea, even if a few people get hurt."

	Whites	Blacks
Favor school segregation	29	3
Favor integration, but oppose demonstrations which risk violence	66	43
Favor integration and support demonstrations which risk violence	5	50
	100	100

Question: "Some people say that if Negro neighborhoods were patrolled only by Negro police, many problems of crime and community tensions would be solved. Others say that it is more important simply to have the best trained policemen no matter what their race. What do you think?"

	Whites	Blacks
Negro police only	22	12
Best-trained police, no matter what their race	78	88
	100	100

Question: "Some people believe that white merchants should be welcome in Negro neighborhoods as long as they hire Negro employees and charge reasonable prices; others believe that no matter what their policies white merchants should be relocated and replaced with Negro merchants. What do you think?"

	Whites	Blacks
White merchants welcome	83	86
Negro merchants only	16	13
Don't know	1	1
	100	100

Question: "Which do you think is *more* important now—to get more and better housing in and around areas where Negroes already live, or to open up integrated housing for Negroes in other parts of the city and in suburbs?"

	Whites	Blacks
More and better housing where Negroes already live	77	41
Integrated housing	20	54
Don't know	3	5
	100	100

an interracial society. On the broadest level, we simply asked people whether they favored general integration or separation, but we went beyond that in an attempt to understand their commitment to integrated schools and housing and to the presence of whites in various roles in black neighborhoods.

As Table 12 indicates, blacks are not only overwhelmingly committed to integration in the abstract, but many favor aggressive actions to bring it about in the schools. In addition, white merchants who are fair to their customers and white policemen who are trained to do their jobs in a professional manner are welcomed by most in the black community. Integrated housing is such an important symbol that more than half of the respondents questioned favor it over the immediate construction of better dwellings in the neighborhoods where blacks already live. A larger percentage of whites favor separationist policies in general and in housing, schools and police forces in particular.

There is a small, but significant, minority of blacks who do favor separationist policies and they are articulate. However, administrative or political division of the cities along racial lines would not be in tune with the desires of the vast majority of the black community and would certainly be a disaster if it did not ultimately result in better services and facilities for the residents. No matter what administrative devices are employed, cities need money to improve services and to better conditions for their citizens and no scheme which ignores this fact can be successful in the long run. Further separating whites from the problems of the black community may please them and those with similar opinions in the black community, but by itself, it is not an answer to the problems of the cities.

NOTES

1. The questions reported on in this table and in Table 2 are adapted from, and those in Tables 3 and 4 are directly borrowed from, Campbell and Schuman (1968: 39-46). The distributions we obtained are very similar to those reported by Campbell and Schuman.

2. The gamma coefficient is an ordinal measure of association proposed by Goodman and Kruskal (1954).

3. Separate measures were used for blacks and whites because virtually no blacks favor separation in any form and few whites interpret black power favorably (Aberbach and Walker, 1968).

4. In every case, the relationship between these items and the police abuse index is lower, and in one case, that involving "have you complained" in the white community, the sign is reversed. The magnitudes of the relationships with the item on police performance ("police don't do a good job") are closer to those with the services satisfaction index than to those with the police abuse index.

5. Actual complaints and beliefs about officials' responsiveness to complaints are virtually uncorrelated for blacks (Gamma = .05) and the correlation is actually negative (Gamma = -.20) in the white community. In other words, whites who have not complained are more likely to think that their complaints will prompt positive action. For them, contact seems to breed disillusion.

REFERENCES

ABERBACH, J. D. and J. L. WALKER (forthcoming, 1970) "The meanings of black power: a comparison of white and black interpretations of a political slogan." American Political Science Review (June).

––– (1969) "Political trust and racial ideology." Discussion Paper Number 8. Ann Arbor: Institute of Public Policy Studies, University of Michigan.

CAMPBELL, A. and H. SCHUMAN (1968) Racial Attitudes in Fifteen American Cities: A Preliminary Report Prepared for the National Advisory Commission on Civil Disorders. Ann Arbor: Institute for Social Research, University of Michigan.

GOODMAN, L. A. and W. H. KRUSKAL (1954) "Measures of association for cross classification." Journal of the American Statistical Association (December).

President's Commission on Law Enforcement and Administration of Justice (1967) Task Force Report: Crime and its Impact—An Assessment. Washington, D.C.: Government Printing Office.

19

The Impact of Anti-Poverty Expenditures in New York:

Determination of Basic Strategies at the Neighborhood Level

PETER K. EISINGER

☐ STUDENTS OF THE NATIONAL antipoverty effort have generally agreed that the publicly supported community action program was passed into law and then into practice before there was either widespread consensus on, or understanding of, the meaning and function of this social experiment. John C. Donovan, for example, has argued convincingly that the Congress probably had no clear notion of what the concept of community action entailed at the time of the passage of the Economic Opportunity Act of 1964 (Donovan, 1967: 35). At the same time, the planners of the program who subsequently came together to implement the act on the staff of the Office of Economic Opportunity were divided among themselves at the outset about the fundamental strategies of the community action program (Moynihan, 1966; 1969).

At least one thing about the program was clear in the early days, however; it was the explicit intent of the Congress that the administration of the community action program was to be lodged in the hands of the local communities, although community was defined to include a wide range of jurisdictions including states, counties, municipalities, and special districts (Economic Opportunity Act, 1964). Yet without consistent authoritative cues or directives from Congress or the federal antipoverty bureaucracy as to what, exactly, they were supposed to administer, antipoverty practitioners in various localities across the country proceeded to devise their own conceptions of the meaning and function of the community action program (Hallman, 1968; Clark, 1968). Even at the local level the staffs of some municipal community action agencies reflected, among themselves, the same diversity of conceptions that could be found at the federal level. The research reported here on the community action program in Brooklyn in 1967 suggests that such was the case in the New York City community action agency.

The effect in New York of both the federal and municipal dissensus was, to a great degree, to place ultimate determination of the meaning and function of the program in the hands of those organizations charged with the actual conduct of the program in the poverty neighborhoods, namely the community action groups, or, as they are called in New York, community corporations (CC's). Some of the CC's in Brooklyn chose to follow an intensely political strategy, conceiving the function of the community action program to be a means of providing local institutional support so that poor people could mobilize politically and articulate their interests in the municipal political arena. Other CC's chose to pursue an administrative-service strategy in which the public community action funds were seen as providing support for the participation of the poor in the planning and administration of social welfare and rehabilitation programs. It is the purpose of this paper to examine these two strategies in an attempt to determine one prime aspect of the nature of the impact of public community action expenditures at the neighborhood level.

THE RESEARCH SETTING

This report is based upon a comparative examination of the behavior of six antipoverty community corporations (CC's) in the borough of Brooklyn. The data for the study were gathered in 1967, primarily through participant-observer techniques, formal interviews with key staff members in the CC's and with officials in the city government's Community Development Agency, and a survey of the elected members of the CC boards of directors.

Each of the community corporations, whose establishment is authorized by the Economic Opportunity Act of 1964, serves its own geographical poverty area. The estimated populations of these areas range in size from Bedford-Stuyvesant's 285,000 to Fort Greene's 60,000. According to the best estimates of local community action participants and city antipoverty officials (Census figures are sadly out of date), nearly 965,000 people resided in the six poverty areas in 1967. Of the entire city's approximately two million poor, 37.8 percent (760,000) lived in those Brooklyn neighborhoods and nearly one-third of New York's welfare recipients could be found there (see New York City, Community Development Agency, 1966-1967; n.d.).

During the time span of this study, which extended over the entire calendar year of 1967, there were fifteen designated poverty areas in New York City, each of which had established, or was in the process of establishing, a community corporation. In the year after this research was completed, eleven more poverty areas were designated, bringing the current total to twenty-six. Each of the CC's is sponsored financially by both federal money authorized and appropriated for community action by Congress and by city funds. In the fiscal year (FY) of 1966-1967 the city supplied $15.4 millions for the community action program and the federal government contributed $23.8 millions. In FY 1967-1968 both sources decreased their contributions to $17.4 millions from the federal government and $14.1 millions from the city. Individual annual budgets of each of the six CC's averaged slightly over one million dollars.

The community corporations are governed by boards of directors, most of whose members are elected by the residents of the poverty areas. The boards themselves appoint the remaining members. The appointees, often drawn from the neighborhood, never comprise more than ten percent of the board. Nearly ninety percent of the board members reside in their respective poverty areas, and approximately one-half are poor. The boards, which average forty members, have the power to hire and fire the executive director and to make policy. Real control, however, tends to rest in the hands of the executive director, for the firing prerogative is seldom used or even brandished and the large, inexperienced boards are invariably manipulated easily.

MUNICIPAL AGENCY CONCEPTIONS OF
THE COMMUNITY CORPORATIONS

Every Community Development Agency (CDA) staff member interviewed expressed two different conceptions of the function of the community corporations. One was an official conception that seemed to provide the primary basis for making a host of relatively minor administrative decisions. The other was an informal conception that seemed to represent a crystallization of the predominant agency attitude toward the CC's. The informal conception was an expression of the long-range expectations of the program in New York. The two conceptions differ in important ways.

According to the official conception, set forth in the document relied upon by Mayor Lindsay while establishing the program, the function of the CC's is to plan and prepare an annual program budget and to coordinate all programs dealing with community action in their respective jurisdictions (Institute of Public Administration, 1966: 15-16). Every respondent in the CDA initially described the function of the CC's in these terms. While the CC's may engage to some limited extent in the actual operation of various service programs, the blueprint calls for operations to be delegated primarily to local organizations (or satellite groups, as they are called) such as church groups, block associations, tenant councils, and clubs.

To support programs devised in such areas as job training, consumer education, day care, remedial education, youth recreation, and so forth, the satellite groups submit applications to the CC program evaluation department, requesting a share of the CC budget. The CC boards of directors are vested with the authority to grant or withhold funds; the decision is based on the recommendations of the CC program evaluators.

Political mobilization and interest articulation activities are not included in the description of the functions of the CC's. The community action institution is a means of ensuring local participation in the control of public monies that flow to the poverty neighborhoods. Thus, to view the CC's and their satellite groups in this formal context is to place primary emphasis on the administrative and service functions of the community action program.

The informal conception of the CC's is expressly political in its intent. While this conception frequently provided a basis for making decisions in regard to the CC's, more importantly, it seemed to serve as the expression of the ultimate goals of the program. The informal conception, never expressed in writing, contained normative expectations in regard to the development of the CC's, and it often contained elements relating to personal motivations of individual staff members. One assistant commissioner spoke in the following way of the relationship between his personal motivations and his conception of the community corporations.

> We felt the community corporation could become a ghetto institution —a poor people's institution. We thought—some of us—that the CC's would become political instruments, a political power base. This implied working in opposition to the local political establishment. . . . I feel very strongly that powerlessness in the ghetto is a question of political power. . . . I'm in this whole thing because I'm an organizer, basically. This is what I hoped the CC's would do.

Political mobilization, interest articulation, and even the establishment of a basis for involvement by the poor in electoral politics were persistent themes expressed by the CDA officials in the interviews. One high official spoke of the community action program as a prerevolutionary technique in which the poor could learn the art of organization, and political leaders could learn and sharpen their political skills. He envisioned the politicization process of the community corporations and the populations they serve in the following terms:

> First the CC offers services and comes to understand that it can't compete with the professional service agencies. Then it organizes people to pressure other agencies in the city. . . . People will have to be shown through experience that political action is meaningful. . . .

A third stage, he continues, is engagement in electoral politics by a politicized poor.

The political (informal) and the administrative-service (official) conceptions of the community action program voiced simultaneously by officials in the CDA are not necessarily mutually exclusive. Political action and administration can and do exist side by side in organizations, as we shall see in the case of the Brooklyn community corporations. But the two conceptions do involve crucially different emphases on what the programmatic priorities of the community corporations ought to be. What is of interest to us here is that there is no evidence that either conception was consistently invoked by CDA staff members as the authoritative basis for decision-making in regard to the community action program as a whole.

The sum total of cues and directives emanating from the CDA were sufficiently diverse that CC's in Brooklyn, which approached the community in very dissimilar ways, could all report that their chosen approach appeared to be in basic agreement with the conception of the program held by the CDA. Thus, those CC's that chose a political approach and those devoted primarily to service and administration all saw their strategy as consistent with the cues supplied by the city agency. Clearly, the CDA of 1967 was a highly flexible institution in terms of communicating its own organizational aims.

An examination of the determinants of the strategic choices made by the CC's in Brooklyn suggests that the cues and directives derived from the CDA were not very important in determining the course of action that the CC's decided to take. The important choices were made at the neighborhood level—primarily by the individual executive directors—and, currently, CDA cues apparently serve to reinforce those early choices rather than to mold them.

One may conclude, then, that the municipal community action agency in New York in 1967 did not provide a single unambiguous and authoritative model which community action groups in the neighborhoods were obliged to follow. Hence the organizations at the neighborhood level, the community corporations, were remarkably free to devise their own courses of action. Their freedom may be seen analytically as involving the capacity to choose between the two general strategies mentioned above, one of which is clearly political in its predominant orientation while the other is service- and administratively oriented.

ELEMENTS IN THE STRATEGIC CHOICES

Regardless of the strategic choices they make, all of the CC's act to some extent both as political forces and administrative-service organizations. There are no pure types. Yet the making of a strategic choice between a political or administrative-service conception of the community corporation enables us to distinguish between categories of CC's. The strategic choice may be broken down into a series of choices. These choices emerge as: (a) the public announcement of distinctive major strategic commitments, and the consequent projection of characteristic organizational images; (b) the establishment of different programmatic, or active, priorities; and (c) the pursuit of distinctive tactical approaches to political mobilization. The actual choices, though theoretically independent of one another, tend empirically to be tightly clustered in two groups, creating a hypothetical but single dimension. Hence it was possible, initially, to describe the predominating tendencies of any given community corporation as leaning basically toward a political or administrative-service orientation.

ANNOUNCEMENT OF MAJOR STRATEGIC COMMITMENTS

All six of the Brooklyn CC's were organized prior to the establishment of the Community Development Agency by Mayor Lindsay's executive order of August 1966. Their respective origins are diverse. Three of the CC's—Williamsburg, Fort Greene, and South Brooklyn—were organized by the city's antipoverty agency established by Mayor Wagner in response to the funds made available by the Economic Opportunity Act of 1964. The announced intention of the city's Economic Opportunity Committee, the Wagner administration's predecessor of the Community Development Agency, was to decentralize the planning and administration of social welfare services. To accomplish this, the EOC organized city-controlled Community Progress Centers in impoverished areas. Local citizens were allowed an advisory role in the program planning process; they exercised no legal control over the paid executive staff hired by the city. Thus the initial, announced strategic commitment in the fight against poverty of the Community Progress Centers was to a decentralizing of the planning and provision of social welfare services.

Another CC, Youth-In-Action of Bedford-Stuyvesant, was organized by a federation of private Brooklyn welfare organizations at the behest of the New York City Youth Board. Modeled after the Lower East Side's Mobilization for Youth, Youth-In-Action was designed to combat juvenile delinquency by providing programs to increase social, economic, and educational opportunities for young people. The remaining two Brooklyn organizations, located in Brownsville and East New York, were originally established by private community organizers. Their announced intentions were to develop local leaders who would articulate community interests in the municipal political arena and to organize community interest groups such as tenant associations and welfare recipients leagues.

As the organizations matured, and as increased opportunities became available for neighborhood control through local hiring of the executive staff in the city-established groups after Lindsay's reorganization of the municipal antipoverty program, power struggles occurred in several of the groups. One consequence of these struggles was the announcement by the victorious new elites of new, primary strategic commitments in the neighborhood war against poverty. In Williamsburg, leaders among the executive staff emerged who were chiefly committed to community organizing for political ends rather than simply to the planning and provision of social welfare programs. In East New York the original community organizers were ousted by a conservative group of professional social workers, schooled in old-line public service agencies and committed to the development of service programs.

The initial commitments in Brownsville and Youth-In-Action to organizing and service, respectively, remained largely unchanged, for the configuration of power in these groups remained stable through 1967. But in Fort Greene and South Brooklyn, the early commitments to service were challenged, in the former case by two successive executive directors and in the latter by community organizers at the secondary staff level. However, the challenging forces in these two groups found scant support for their views within their respective organizations. Nevertheless, they wielded sufficient influence to make the executive staff uncertain as to the dominant nature of the primary strategic emphasis of their community corporations.

In 1967 the publicly announced strategic commitments of the CC's in Brooklyn were divided among those groups pledged to pursue the development of service programs, those that sought to use their resources to organize the community for political purposes, and those largely ambivalent about their commitment. In the latter category (see Chart I) there are indications that South Brooklyn leans more toward the administrative-service orientation than does Fort Greene.

CHART I

STRATEGIC COMMITMENTS, 1967

Political (Organizing) Commitment	Ambivalent Commitment	Administrative-Service Commitment
Brownsville	Fort Greene	Youth-In-Action
Williamsburg	South Brooklyn	East New York

The announced major strategic commitment of a community corporation is to be distinguished, theoretically, from the actual behavior of the organization. The commitment, announced and usually formulated by the executive director, is designed not simply to establish progam priorities, but, more importantly, to project an image of the group in the community. Conceivably, the image a group projects through its rhetoric of commitment may bear little relation to its actual behavior. The announced major strategic commitment, then, constitutes only one variable along the dimension of political and administrative-service orientations.

Let us explore briefly the implications, for the image of the CC's, of their announced commitment. Community organizing, on the one hand, and the planning and provision of social welfare services on the other, are activities which all the CC's pursue to one degree or another. But the announcement of a strategic commitment performs the function not of serving the community, as do the various types of activities, but of informing the community-at-large of how it is expected to act in relation to the community corporation.

An announced commitment to community organizing projects an image of the CC as an agent for the political mobilization of the community. An essential aspect of community organizing is the development of a network of voluntary associations capable of engaging independently in political conflict on behalf of members' collective interests. The image the CC wishes to project in this case is that of a force to stimulate self-help efforts in the political arena. The CC expects the members of the community to act, in great measure, independently. Hence the announcement of a commitment to a political strategy may be seen as the first step in the political education of the population the CC serves.

In contrast, the announcement of a commitment to planning and providing services, projects an image of an adjunct government agency. The poor have always been provided with services in greater or lesser measure. To choose which services they will receive, and to aid in the planning and administration of those services is, of course, a significant step toward humanizing the experience of dependence on the welfare system. But while local administration and planning of services may eventually pay dividends by increasing self-confidence, imparting a sense of dignity, and providing experience in management, it simply represents the decentralization and transfer of what has come to be regarded as a public social welfare function. The community is still expected to be generally dependent; people must still come to the CC and ask for help. The encouragement of civic self-sufficiency, especially through the development of collective skills, is not an explicit goal contained in the service-community announcement.

THE ESTABLISHMENT OF ACTIVE PRIORITIES

Because each of the CC's may allocate its resources to those activities which it feels to be most important, regardless of its announced commitment, the patterns of actual organizational behavior tend to differ among the six Brooklyn groups. Assuming that frequency of engagement in a certain type of activity is an indication of the real priority which that type of activity commands when resources are allocated, it may be said that some of the CC's have established distinct priorities in their modes of operation.

In order to determine the patterns of behavior of the CC's, a list was compiled of all the different activities in which each group engaged during the calendar year 1967. The various activities were classified according to their character and purpose primarily as "political" or "service" in nature. Political activities were defined as those that deal with mobilization (community organization, leadership development, or civic education), or interest articulation (defining issues on behalf of the interests of a collectivity and seeking to resolve those issues in the political arena). Service activities were defined as those that seek to provide welfare programs to help individuals. The main purpose of such programs, as Moynihan has observed, is to provide surrogate family services (Moynihan, 1965: 47). Political activities are designed to foster collective instincts. A residual category, classified as "other," refers to sponsorship of social and cultural activities, e.g., Senior Citizen's Picnic, Puerto Rican Literature Circle, and so forth.

The only systematic and comprehensive sources for the compilation of a list of activities are the monthly newsletters published by each of the CC's. An activity included any CC-sponsored or -administered endeavor that was either initiated during the calendar year or reported upon, either for the purpose of keeping a periodic record of achievement or making known the availability of a given service or opportunity. Thus the opening of a new day-care center, the organization of a new block association, the procurement of a contract to operate a Headstart program, or the organization of a demonstration at City Hall fell under the first category of those activities initiated during the year. The announcement that the CC had placed one hundred men in industrial jobs over the past three months or that a particular employment opportunity existed as a result of efforts by the CC job placement officer fell in the latter category, and each announcement was treated as a single activity. There is a bias in such accounting against ongoing, routine activities that are not reported as daily achievements. Thus, daily community organizing efforts are not included in the list except, for example, when a new association is formed. Also, the daily availability of family counseling services is not reported except when a new dimension is added to the service. Nevertheless, ongoing activities tend to be summarized each month in the newsletters and are thus reflected to some extent in the following table.

TABLE 1

TOTAL CC ACTIVITIES BY CATEGORY

Community Corporations	Political (percent)		Service (percent)		Other (percent)		N (percent)	
Brownsville	54.5	(36)	21.2	(14)	24.3	(16)	100	(66)
Williamsburg	51.4	(18)	40.0	(14)	8.6	(3)		(35)
Fort Greene	47.6	(20)	45.2	(19)	7.2	(3)		(42)
South Brooklyn	47.3	(9)	47.3	(9)	5.4	(1)		(19)
East New York	41.1	(14)	53.0	(18)	6.0	(2)		(34)
Youth-In-Action	23.4	(11)	68.0	(32)	8.6	(4)		(47)

It is clear from this table that certain CC's have established distinct priorities in terms of their behavior. The two CC's which, we may recall, made a major announced commitment to community organizing—Brownsville and Williamsburg—are also the two groups that have engaged most frequently in a whole range of political activities, of which community organizing is only one part. The two CC's that announced a major commitment to an administrative-service strategy—Youth-In-Action and East New York—are the two groups that have engaged most frequently in service activities. The two middle CC's—South Brooklyn and Fort Greene—exhibited ambivalent strategic commitments, and their behavior reflects this graphically. We may conclude, then, that among the Brooklyn CC's there is a strong relationship between the announced commitment to a strategy for combating poverty and the actual behavior of the organization.

Research not reported here indicates that the executive directors are not only usually responsible for formulating and announcing the strategic commitment, but are also key figures in the determination of active priorities (Eisinger, 1969). However, the finding that among the Brooklyn groups the announced commitment and actual behavior coincide, limited as its scope may be, is worthy of note, for it is somewhat inconsistent with the broader findings of Kenneth Clark (1968: 234). In his study of the community action programs in twelve cities, Clark concluded that while practitioners normally voiced commitments to political action and power strategies for solving the problems of the poor, "in the actual operation of community action programs, the traditional social services approach, not the action approach, dominated."

TACTICAL DIFFERENCES IN THE PURSUIT OF MOBILIZATION

Using the Organizing Staff

A central aspect of political mobilization efforts is community organizing, the attempt to form local para-political voluntary associations. All of the CC's maintain sizable staffs of organizers and block workers. Yet while there is apparent agreement among the CC's as to what, precisely, community organizing seeks to accomplish, the two types of CC's differ in the way that they actually employ their organizing staffs. One approach is to use the staff to organize a constituency, and this is the tactic followed by the politically oriented community corporations. The other way to employ the organizing staff is to use it to amass a clientele, a tactic that is mainly the preserve of the administrative-service CC's.

A constituency may be defined simply as a population both desirous and capable of making political demands in the bargaining arena for the purpose of satisfying collective needs or grievances. A constituency is endowed with the organizational resources to make such demands. The particular tactic designed to create a constituency, using the organizing staff, is characterized by the effort to establish an associational complex with the following properties: (a) the individual organizations in the complex are capable of engaging in political action in concert with the CC, in coalition with other local associations, or alone; (b) the associational complex is capable of serving as an organizational base and a communications network for communitywide political activities; and (c) individuals who belong to these associations are conscious of their membership in a potentially political collectivity, and they are willing to use political action as a means of solving their collective grievances.

A clientele, on the other hand, is a population whose members satisfy their needs or demands individually by choosing among solutions offered on the market. Because the CC's offer a range of social services, as do many public and private service agencies, from which the poor client may choose, we may say that these service solutions are on the market to the extent that they constitute a set of offerings subject to acceptance or rejection by the client. In other words, the members of a clientele do not make explicit collective demands but take or reject what is available on an individual basis. The tactic that employs the organizing staff, mainly to create a clientele, involves the effort to recruit users of the services offered by the community corporation.

The two tactical approaches to the use of the organizing staff are not mutually exclusive. All six CC's conduct service programs, and all of them rely to some extent on their organizers to recruit a clientele. But the recruitment task is generally performed at the expense of organizing. Thus in some of the CC's, explicit priorities are established governing the use of the organizing staff. The two most striking cases in which priorities have been set are in Williamsburg, where organizers are used to organize, and in South Brooklyn, where organizers are used to recruit.

The Williamsburg community organizing efforts spring from what Bloomberg and Rosenstock (1968) have called the "social action-social protest school"; the organizers are deeply embedded in the Alinsky tradition. Their primary goal has been to organize associations, especially by relying upon the the catalytic and polarizing effects of conflict. The resolution of issues has been secondary. One organizer explained:

> Our goal is the complete involvement of the community in a systematic coordinated effort. The [organizers] have found that the issues concerning housing problems create a fertile foundation for organization, and therefore they are now being used as an organizational approach.

When a welfare mothers' association was formed after the women had unsuccessfully sought extra allowances for winter clothing at the local city welfare center, the community organizer said:

> The primary intention on our part was to help the welfare client see the effectiveness and the importance of group pressure, and that issues could be resolved by means of this pressure. Our secondary intention was to obtain winter clothing.

It is expected that once an association is organized it will eventually act as an independent political force, without the aid and prompting of the CC organizers. Indeed, a number of block and tenant associations in Williamsburg have initiated rent strikes, and some have succeeded in obtaining court orders to force landlords to make repairs. One Williamsburg group concerned with better education organized a successful ten-day boycott of a grade school to protest certain administration practices.

The use to which the organizing staff in South Brooklyn has been put has been quite different. In South Brooklyn, as in all the CC's, the organizers operate out of storefront neighborhood action centers. These storefronts are designed to diffuse the influence of the CC in the community and to bring its physical presence closer to the residents. Generally, a community organizer is in charge of the neighborhood action center, and he is responsible to the head of the CC community action department. In South Brooklyn, however, the centers are headed by job placement specialists who report directly to the executive director of the CC. The organizing staff, which is subordinate only in South Brooklyn to the job specialists, is used chiefly to recruit people for the massive manpower, or job placement and training program, that the South Brooklyn CC offers. The organizers uniformly oppose this approach. "The organization set-up here," one organizer remarked, "is strictly for manpower. They're even going to change the name of the neighborhood action centers to Neighborhood Manpower Service Centers. . . . The community organizers are just going to be recruiting people for jobs."

The criterion of success for the South Brooklyn manpower program is the number of people it serves. In other words, the CC must develop a substantial clientele. Success in manpower programs can, of course, be measured quantitatively, using such indices as number trained, number placed, number still working after one year, and so on. It is considerably more difficult to quantify the success of a community organizing effort, and the effects which do appear may be expected to emerge only after a long period of time. The South Brooklyn CC has chosen to use its organizers to produce fast, measurable results, and such a course is necessarily at the expense of efforts to establish local voluntary associations. Thus, the organizing staff members, as recruiters, have contributed little to the politicization of the population of the area.

Of the remaining four groups, Brownsville organizers are committed primarily to organizing tasks, Fort Greene and Youth-In-Action organizers are used for both functions with no explicit priorities in evidence, and East New York organizers are employed mainly for recruitment.

Approaches to Activating the Poor

Whether or not a CC had made a public commitment to a political or an administrative-service strategy, CC staff members invariably claimed in the interviews that they had but one ultimate goal in mind: the involvement of people in civic action directed toward self-determination, i.e., "so they can control their own destiny," by supplying them, on the one hand, with prerequisite collective or organizational resources or, on the other, with pre-requisite personal resources such as a job or education. Both types of CC, then, could be said to be working ostensibly toward the attainment of a kind of existential freedom for those they serve, an essentially political goal. The CC participants agree that the main obstacle to achieving this goal is the apathy with which they are generally met in their poverty areas.

To overcome this apathy, the political and administrative-service CC's employ quite different tactics. The political CC's seek normally to form a constituency at the outset by using the experience, both actual and vicarious, of mass protest action to destroy communal apathy. This tactic is thought to depend for its success on the active participation of only a relatively small segment of the population, for it is geared to penetrate the apathy of observers as well as participants. The tactic seeks to accomplish this widespread penetration by demonstrating in the first instance to the residents of the poverty area that the community is capable of acting as a collectivity on its own behalf.

The virtue of mass protest is that it is easily comprehended. As Piven (1966: 78) has observed: "Social protest actions, because they offer simple and dramatic definitions of problems, may penetrate apathy and override the puzzled disengagement bred of lack of information." What is important about this approach is that the CC is perceived by area residents primarily as a political force.

In contrast, the administrative-service CC's appear first in the role of providers of social services. Their tactic for achieving subsequent civic involvement is to attempt to politicize their clientele. This is what Brager and Specht (1967: 145) have called the "social brokerage technique." The tactic involves attracting individuals by the offer of social services to meet their immediate needs. Thus the CC assembles a captive audience to whom it can then proceed to teach the values of political action.

At first glance, the social brokerage technique might seem to have certain advantages over the purely political conflict approach. The offer of tangible services probably penetrates the basic apathy about the prospects for change of more people than does the vicarious excitement of mass protest. Through its services the CC can reach an extremely wide and diverse population, including those who are least likely to establish and sustain membership in a local association or allow themselves to be recruited for a political activity by a CC organizer. Occasionally the CC can subsequently organize the natural constituencies that comprise its clientele, such as the mothers of children who attend Headstart, or the young people enrolled in the Neighborhood Youth Corps program. More frequently the CC can seek to politicize the members of its clientele by asking them if they are registered to vote when they come into the neighborhood action center for help, or by having them sign up for a seat on a bus going to the state capital for a demonstration. These are standard procedures in all of the CC's. The clientele may be continually exposed to the rudiments of a political education, and its members may be pressured to be active.

All of the CC's use their services to a greater or lesser degree to build a reputation in the community and to develop a following. In terms of the goal of mobilizing the poor, reliance solely on the experience of political conflict to overcome apathy is a risky tactic, for the degree to which the apathy of the observer population—the nonparticipants—is penetrated is, in fact, unknown.

Given the objectives of politically mobilizing the poor, there are dangers too in the exclusive use of the social brokerage approach. Frequently the services and their administration became an end in themselves. To maintain the claim that the services are used to attract people for other purposes is simply to provide the basis for self-delusion, as is the case in several of the Brooklyn CC's. Little more than fragmentary efforts are made in East New York and Youth-In-Action to politicize the clientele, despite assurances by most of the CC staff members in those groups that organizing and political assertion are the ultimate keys to the solution of ghetto poverty. But the administration of the services in those areas has come to have its own fascination. A kind of routinization of political expectations occurs in which the magnification and embellishment of the service programs becomes all-engrossing, and political goals are forgotten in all but their most superficial form. The resources allocated to the more difficult, often abrasive, tasks of politicization grow smaller. If efforts at preparing the clientele for political action do occur, they are likely to be fragmentary and elementary in the absence of a major commitment to political action. Thus the man who is told to register to vote when he comes into the neighborhood action center, or to "learn the value of organization," as a flyer on one CC bulletin board urged, is likely to have no collectively sponsored outlet for whatever political impulses may be stirred within him.

In Brownsville and Williamsburg the CC's have managed to integrate the two tactical approaches for the conquest of apathy. In Youth-In-Action, especially in East New York, the administration and provision of services for their own sake is in the ascendancy. In Fort Greene and South Brooklyn, opposing forces struggle with one another over whether to engage the CC more actively in the politicization of the clientele.

THE IMPACT OF COMMUNITY ACTION
EXPENDITURES AND THE IMPLICATIONS
OF PUBLIC SPONSORSHIP

It is apparent that one significant consequence of public spending for community action in New York City has been to confer sufficient resources on neighborhood organizations to enable them to devise strategies of their own choosing by which to combat poverty. Some of these organizations, as we have seen, choose a strategy that brings them into frequent political conflict with their government sponsors. Not only are their mobilization efforts pursued with the end in mind of building a base from which to confront established municipal political actors with organized strength, but the demands they actually do make in the political arena are more likely than not to be directed at some city agency or official. Government initiation and sponsorship of the participation of groups pursuing private interests in pluralistic group bargaining are a striking notion in a society that purports to base its political life on private initiative. There is in fact no precedent in recent American history which readily comes to mind. The only similar, but certainly not analogous, instance of government participation in regard to the right of such groups to bargain was the provision in the National Industrial Recovery Act of 1933 which declared that workers had the right to bargain collectively with their employers. When this act was declared unconstitutional by the Supreme Court, Congress passed the Wagner-Connery Act in 1935, incorporating the same labor guarantees. But this provision assured labor the right to bargain through the threat of sanctions with private, not government, institutions. Furthermore, labor unions did not receive financial support from the government in order to enable them to carry out their bargaining activities. The community action group, however, as a publicly sponsored institution that participates in political bargaining, often directs its energies against government itself. Paradoxically, government (using the term in its generic sense) has borne and nurtured an antagonist.

How can we begin to explain this unusual, if not unprecedented, situation? Can we safely argue, for example, that the community action program, insofar as it enables organizations to mobilize the poor, represents a belated attempt to realize some long-suppressed democratic ideal by sponsoring the entry of hitherto powerless groups into the political process? In a limited sense, I suspect that this is not far from the truth. If this is the case, then we must be prepared to say that the government has admitted, at least implicitly, that organized political activity in a system based on pluralistic bargaining among organized power holders is a right so important that it must be sustained, if necessary, by the state.

This guarantee may not be so radical as it sounds, for indeed the motives for such sponsorship are undoubtedly mixed. By supporting what are, essentially, interest groups in the pluralistic bargaining arena, government may hope to forestall attempts to employ alternative means by which such groups might hope to achieve the goals they now must bargain for. In other words, public sponsorship is designed in large measure to perpetuate the viability of pluralistic politics by assuming that participants admitted to the bargaining process will develop a commitment to the maintenance of the system in the hopes of winning what they seek through this peaceful means. By supporting these groups, even as they engage in political bargaining, government makes the prospect of extrasystem efforts at political gain, such as violence or Black separatism, less attractive.

There are a number of implications of this government sponsorship in particular which may serve to finally vitiate and destroy this curious and paradoxical democratic experiment. For example, the primary sponsor of the community action program is the federal government. Not only does the legal basis for community action originate in congressional action, but the main source of funds is the national government. Indeed, the chief overseer of the community action program is a federal agency, the Office of Economic Opportunity. The primary targets of community action groups, however, are local governments and their agencies. Thus the New York community corporations, supported by a flexible framework of federal laws, bring pressure to bear on the New York City government. The possibilities for tension between levels of government under such an arrangement are manifest. The CC can, it would seem, either exploit the situation by playing one level of government against the other, or, and this is more likely, the CC can be caught in the middle of such tension. In the latter instance the powers of the CC face curtailment. Indeed, this is exactly what happened when Congress passed the amendments to the Economic Opportunity Act in the fall of 1967, a move in response to pressure from big-city mayors afraid of the political challenge posed by the federally sponsored community action program (Donovan, 1967: 56-57). The entire pattern of challenge and response between the community action groups and their sponsors poses the question of whether such a relationship can survive in the long run. The prognosis is not optimistic.

Another implication of government sponsorship is that community action groups like the CC's can become the cause of tensions between different branches of the same level of government. When the mayor of a city is committed to hearing demands from the community action groups and promising attempts to satisfy those demands, as Mayor Lindsay has generally been committed, but at the same time elements in the municipal bureaucracy are loath to take such action, the community action groups can, on occasion, drive a wedge between the executive and his agencies. Certainly the community corporation in Brownsville may take part of the credit for the estrangement between Lindsay and the city school bureaucracy, for the CC was strident and energetic in its involvement in the school decentralization controversy of 1967-1968. A similar situation arises when a city council refuses to augment municipal financial support for community action after the mayor or the city antipoverty agency, responding to CC demands, requests such an increase. The repercussions of such intragovernmental tensions would, one might expect, largely affect the power and credibility of people or agencies in the government. The mayor who makes promises and cannot deliver, or the council that refuses to act on promises made by others, is likely to suffer politically in the city's poverty areas. For all government parties concerned, the community action groups can, in any given instance, become a liability.

A third implication is that, even in the act of helping to establish structures through which demands are made on certain levels of government, both federal and local governments have contributed to the stimulation of expectations that neither is prepared to satisfy. Cloaking its sponsorship in the rhetoric of self-help, government is engaged in an attempt, whether explicit or not, to persuade the ghettos that they can achieve what they want by resorting to pluralistic bargaining. In one sense this serves to exacerbate the tensions that I have described above. In another more serious sense, however, government sponsors, both on the federal and local levels, are bound to be accused of bad faith. Government has helped to establish channels through which demands can be made, but cannot be, or are not met in many instances.

It is conceivable, then, that such sponsorship, the implicit function of which is to perpetuate a pluralistic system, will do just the opposite. By failing to respond in adequate measure to demands made through channels which it has helped to establish, government contributes directly to the growth of alienation and bitterness in the ghettos. Here, then, is still another stimulus alongside those of poverty and discrimination which may periodically serve to shatter the tenuous stability of the cities.

A final implication of government sponsorship of community action structures is that the course of the relationship described above makes it probable that increasingly restrictive controls will be imposed on the political behavior of those neighborhood groups as government becomes fully conscious of what it has done. Eventually the program as we know it may be ended. Such a course of action represents the final logical response to the tensions provoked by the community groups at all governmental levels. If government is unwilling to satisfy the demands of the very groups it supports and fosters, then it will no longer be able to rationalize the contradiction of its sponsorship of political antagonists against itself. By ending the program, the government may hope to gain a moment of what is likely to be a highly illusory peace.

REFERENCES

BLOOMBERG, W., Jr. and F. W. ROSENSTOCK (1968) "Who can activate the poor? one assessment of 'maximum feasible participation.'" Pp. 313-354 in W. Bloomberg, Jr. and H. J. Schmandt (eds.) Power, Poverty, and Urban Policy. Beverly Hills: Sage Publications.

BRAGER, G. A. and H. SPECHT (1967) "Social action by the poor: prospects, problems, and strategies." In G. A. Brager and F. Purcell (eds.) Community Action Against Poverty. New Haven: College and University Press.

CLARK, K. (1968) A Relevant War Against Poverty. New York: Metropolitan Applied Research Center.

DONOVAN, J. C. (1967) The Politics of Poverty. New York: Pegasus.

Economic Opportunity Act of 1964.

EISINGER, P. K. (1969) The Anti-Poverty Community Action Group as a Political Force in the Ghetto. Ph.D. dissertation. New Haven: Yale University.

HALLMAN, H. (1968) "The Community Action Program: an interpretative analysis." Pp. 285-311 in W. Bloomberg, Jr. and H. J. Schmandt (eds.) Power, Poverty, and Urban Policy. Beverly Hills: Sage Publications.

Institute of Public Administration (1966) Developing New York City's Human Resources. A report of a study group of the Institute of Public Administration to Mayor John V. Lindsay, Volume I (June).

MOYNIHAN, D. P. (1969) Maximum Feasible Misunderstanding. New York: Free Press.

——— (1966) "What is 'Community Action'?" Public Interest (Fall).

——— (1965) "Three problems in combatting poverty." In M. S. Gordon (ed.) Poverty in America. San Francisco: Chandler Publishing.

New York City, Community Development Agency (1966-1967) Community Action Allocations, 1966-1967 (mimeo).

——— (n.d.) Poverty Indices Chart (mimeo).

PIVEN, F. F. (1966) "Participation of residents in neighborhood community action programs." Social Work (January).

20

Servers and Served in Service

ALBERT J. REISS, JR.

☐ SINCE 1932, debate in American society about domestic policy and programs has focused on big issues such as the development of a welfare state and civil rights, and on big programs, whether those of social security and unemployment compensation of the thirties of of economic opportunity and Medicare of the sixties. Indeed, the big issues merge and submerge in slogans: "Full Employment" and "Fair Employment" or "Economic Opportunity" and "Equal Opportunity." The great Depression of the thirties seemed to convince one and all that our local communities were organizationally, as well as fiscally, bankrupt in dealing with social problems and that federal solutions were required since the problems were common to most if not all communities.

A CORE ISSUE OF THE SEVENTIES

Beginning with the mid-sixties a new domestic issue has emerged that crosscuts programs: the form, control, and quality of the organization and administration of Great Society programs. There have been growing doubts about the responsiveness of federal programs to local needs and requirements, about local as well as federal bureaucratic administration of such programs, about citizen as opposed to bureaucratic control of programs. This new issue undoubtedly will emerge as a big issue of the seventies. Briefly, the issue is who shall serve whom, how, and for what in which ways. How, for example, shall those who serve meet demands to the satisfaction of those served?

Sociologists contributed to the rise of this new issue. They spawned the belief that the community had declined, even disappeared, as a viable organization in American society. For it, they substituted the mass society with its mass media, its mass middle class, and its mass opinion. At the same time, they undermined the very institutions and organizations of the Great Society by documenting the failures rather than successes of bureaucracy and by focusing on what they regarded as dysfunctional consequences of bureaucratic processing, such as labeling clients as deviant. In its vulgar form, sociologists held that both the organized political and administrative community and large-scale bureaucracy were bankrupt. The only remaining and saving grace was informal social organization where "people" solved problems together, a return to "community" with a small "c."

THE DECLINE OF COMMUNITY?

Squaring off the mass society and the community, sociologists neglected much that characterizes local organization in American society. Perhaps in this period of social unrest, it has become less necessary to be reminded that political and administrative communities bear much of the burden of social unrest for a very important reason. Though the domestic issues that surround administrators beset the total society and derive in part from its mass character, when they involve the common man, they focus as issues in our local communities (Reiss, 1968). The local community becomes the main arena for conflict. For it is in our communities where men live and work that protest is most evident; mass institutions and organizations are locally organized and situated. Even though the Poor People's March on Washington, set in the tradition of such marches, was to focus attention on national problems, it inevitably became an issue for the District of Columbia's community as well. The administrative community faced not only the issues of how to police the march and its aftermath, but also issues such as its effect on the tourist trade.

More importantly, in the American system of local government, national issues must come to focus in the local community. Actually, what is often at issue today is both the scope and scale of government and its capacity to deal with domestic issues. While local governments must contend with social unrest as it focuses on *local* schools, *local* industry, *local* business, or the *local* police and courts, their capacity to resolve these issues often may lie beyond their jurisdiction. The community has not declined as a focal point for social issues and unrest, but its capacity to resolve at least some aspects of such issues apparently has changed.

Discussion of these issues has been retarded by a mindless view of "the community" as a more or less homogeneous entity. Proposals for "local" control of schools or the police, thus, often rest on simple notions of a homogeneous population and community consensus—for example, that all Blacks are alike. Simple recall of our failures to achieve consensus and control over vice should be sufficient to draw attention to the fact that municipal communities are mosaics of local areas, and that any local area, generally, is a mosaic of groups and interests. The policing required or wanted by businessmen or the aged may not be that wanted by the youth in an area. As a matter of fact, community not uncommonly is made up of conflicts among interests, as well as like or common interests. A less mindless view of community must take account of the fact that interdependence in creating community does not necessarily create like-mindedness or common agreement. Solutions that recognize conflicting as well as common interests may prove more viable.

SOME ISSUES OF CONCERN

Citizens and their organizations increasingly show concern about how their demands for service are recognized, what resources are allocated to them, and the form and quality of the service they receive. A number of their demands are closely related to problems of *discretionary justice*—the way that administrators and officials exercise discretion in deciding matters for the citizens they serve (Davis, 1969). Whether it be the discretion of the housing official in deciding matters relating to tenants, or the discretion of the police to arrest, citizens are concerned about delegated powers to decide matters that affect them. Two of the ways that this has become symbolized is in "citizen participation" in the decision-making and running of organizations, and in "civic accountability" of organizations. The demand for local control of the police or for a citizen's review board are examples. In truth, there is virtually a social movement to expand client participation in service organizations, be they universities or welfare offices. The rights of citizens before the bureaucracy is at issue here.

Other symbolic demands are closely related to a recognition of differences in demand and the allocation of resources to meet them. At issue here are matters of where to locate program offices, how to equalize opportunity in employment, how to provide defense counsel or to afford a divorce, and similar matters. Whether such interests masquerade under labels of poverty programs, rights of the poor, or other symbolic forms, they highlight a client revolt in the system. The recipients of service no longer are willing to accept agent definitions of client needs and services. They seek to influence them actively. No longer, for instance, are they willing to have the board of the United Fund or its component agencies decide matters of resource allocation. They, the recipients, want a part of the action as well.

The remainder of this paper is given over to an examination of some of these issues and a statement of some requirements for their "resolution."

CITIZEN DEMANDS FOR SERVICE

Each day, a substantial number of citizens in any large community face a problem they regard as a crisis demanding immediate attention. Such problems include domestic crises, arguments between landlords and tenants or among neighbors, and crises of health or accident, to list a few. What shall they do? Generally, they call the police.

Calling the police in most cities is not surprising, since, apart from the limited emergency service provided by fire departments or ambulances, the police generally are the only service available twenty-four hours a day to serve citizens in need of help. Some years ago a police official in Chicago wryly commented:

> Where else in the City of Chicago can you put a dime in the telephone, call for help and it comes, and get your dime back? Any welfare agency is on a nine to five basis and even if you call them, they'll give you an appointment two weeks hence. But if you call the police, they go.

Setting aside the question of how and in what way the police should be involved in providing such emergency services, it is clear that communities face an enormous problem of linking citizens to emergency services. A major urban problem is to provide an appropriate range of services (undoubtedly in cooperation with the police) to deal with the range of problems citizens present twenty-four hours a day. This is no simple matter.

Let us consider an example. If a citizen gets into a family argument or a dispute with a landlord, and it escalates to the point where someone defines it as a law enforcement problem, the police will be called. On hearing the complaint, it helps neither the citizen nor the police very much for the officer to define it as a "civil" rather than a "criminal" matter and then to take leave. Though danger of altercation may have passed, the source of their dispute—the problem—remains. Suppose the officer is more helpful than that and suggests ways the citizens may proceed to deal with the problem—for example, by consulting a lawyer. If the citizen has no lawyer, cannot afford one, or doesn't understand how to go about handling the matter, such advice fails to meet the problem. Suppose we carry it a step further; suppose the police officer advises a citizen to go to a particular legal aid service if he is unable to afford legal help. How well is the legal service organized to meet the demand that would ensue were the police to do such a "good" job and the citizen to follow through? The answer, one suspects, is that legal organizations are unprepared to meet such demands through referral. Yet many of these demands appear to be legitimate ones and certainly ones that, in one way or another, the government is trying to meet through some new programs for the underclass.

Our consideration of this simple case suggests several requirements for any urban service model that would meet many citizen demands for service: (1) that there be provision for emergency response to crises as well as organized response to more routine matters, (2) that the citizen be linked to an agency that can deal with his problem, (3) that agencies be available to meet effectively the demand for service. It should not be assumed that either the police or an emergency service or crisis intervention unit is sufficient, for that is not the case. Such a unit is a necessary but not a sufficient condition for an urban service that meets client needs.

A CLIENT-CENTERED SYSTEM

Let us further examine current models for offering and providing services to clients. Most services to clients are offered on a model of the "walk-in," a client who presents himself at a place where diagnosis and treatment can take place—a hospital, a clinic, a welfare office, even a Community Action Center. Ideally, the client should be preprocessed before much is done. What this means is that, ideally, the professional person or decision maker wants information about the client before contact is made with him.

What is more, service agencies usually are structured on what used to be called "banker's hours"—a nine a.m. to five p.m. schedule of work. Over and beyond that, their high specialization of services often means that clients must move episodically to different specialists in different agencies. Such a system seems designed to cool clients out of their demands for service rather than to meet client demands. The client is required to adapt to the agency personnel's schedule rather than his own schedule.

There are other contingencies that arise in such a system. The client soon discovers that he loses contacts with previous agencies. The referral has been incorrect and he moves to another that refers to yet another. Often there is no agency to which one returns. Even where, as in some of the Community Action Centers, attempts are made to accommodate the clock and calendar of the client and to make some follow-up, all too often the referral for a client is to an agency that operates from nine to five. Employment offices, social welfare agencies, even parole agents generally are on a nine-to-five basis.

It must be recognized that life is more comfortable for the professionals that way and these agencies are and must be regarded as being professional rather than client-centered.

The capacity of professionals to subvert the goals of clients is enormous. Some years ago, in New York City, when the city attempted to speed up the processing of offenders in the system of criminal justice, courts were set up to operate "around the clock and calendar." However, someone had forgotten there were vacation schedules and the swing shift was eliminated when the judges wanted their vacation! Apart from the emergency services of police, fire, and hospital organizations, no provision generally is made for services where demand exists and where there are needs to be met around the clock and calendar.

One suspects that much of the success of voluntary organizations like Alcoholics Anonymous or Synanon depends not only upon the fact that they develop informal relationships among social deviants, but also because a call for service or help is readily met by those who are members. Indeed, movement to a client is a relatively rare event today apart from police, fire, and sanitation services—the latter being one of the few regularly programmed services. A visiting nurse association service is relatively uncommon in the system and, like most such services, is based on notions of client inability to become a walk-in.

Questions about professional criteria for client preparation for service abound. Do their distinctions about what constitutes emergency make sense? Is social work or correctional work outside the closed institution really a nine-to-five job? May not a parolee need attention more at one a.m. than at three p.m.? If the Traveler's Insurance Company can advertise that no matter where and when you are involved in an automobile accident, a simple call to a number brings the Traveler's agent to work on your problem, why can't many public urban services do so?

Should one think further about what clients mean when they say, "the system doesn't meet our needs"? They seem to say more than that it doesn't meet needs for equal rewards. Such needs are not to be slighted, but they point up the fact that "systems" do not treat clients as "persons," as people with integrity, dignity, and worth. Part of the clients' reaction to not being regarded as "people" arises from the way they are treated when professionals are in their presence; but much of it also arises from the way that we have organized the bureaucratic processing of people as *cases* (Scheff, 1966).

Recognizing this dilemma, some social work and clinical agencies have come to speak of developing "client-centered" agencies. What they emphasize is doing something *with* people rather than *to* people. Important as such a distinction may be, it also may miss the point people are making. What may be more important to many people is what it is like to be processed in any system. All too rarely do we look at systems from the standpoint of people who are processed within them. How many agencies apprise citizens of their rights? (The police may do a better job of this now than any welfare or correctional agency!) What attempt is made to communicate with citizens about what is happening *to* them, and why. Part of this failure to communicate lies in the fact that the system is built more for the professionals than for the clients. All too often a continuance in court, for example, is to suit the convenience of the lawyers or the judge—not to suit the convenience of the client. Recently I listened to a man describe the process of being arrested and processed in the criminal courts of Detroit. He saw it as a process of going up and down in elevators in the same building, and in his mind, there were no separate elements in the system of justice. The police, the prosecutor, the judge, and the agency were all one, not several. While one could argue whether a system of criminal justice should be designed so that people like it, one should hardly argue that ignorance of how clients view it is rational.

It is not clear that clients are fully aware of what they mean when they say that the "system doesn't work or meet my needs." Yet if one listens not only to what people say about when the system doesn't work but about when it does work well, one secures information that is highly relevant to changing conditions of systems. Some years ago, I had occasion to review a study of probationer evaluations of their probation officers (Honnard, 1961). It was an unusual study, not only because of its novel approach of having probationers tell what they liked and didn't like about their probation officers, but also in its major findings. What emerged from the study was that many probationers didn't like officers who wanted to change *them* even when the probation officer viewed it as doing something *with* the person. They resented almost all therapeutic efforts and particularly disliked being labeled "sick." What probationers did like was an officer who did something *for* them, particularly *changing their environment in a significant way* so that they then could do something for themselves. There was the girl who liked her probation officer because she managed to get the girl into a school where her reputation was left behind. There was the boy who was helped to quit school and get a job that held a future. And so on. Even more striking was the fact that when probation officers did something for probationers, their chances of success on probation were far greater than when probation officers were doing something to change probationers as persons.

Perhaps the single most important failure of many service organizations is the failure to manipulate the environment in the interest of the client. All too often the burden of changing the environment is shifted to the client, the service organization frequently justifying doing so on grounds that it is in the client's "best interest" or that it is "therapeutic." It is argued that the client must be able to do things on his own and display independence in solving problems. But just what is "doing things on one's own"? It seems strange that where agencies often fail, clients are expected to succeed.

What administrators and staff in agencies need to keep in mind is that often it is easier to change the constraints in the environment to which a client can adapt than to change adaptive mechanisms in persons. Indeed, often changing the constraints is a condition for adaptation on the part of clients. The important lesson one can learn, if one listens to what is said about the system when it works, is that a significant change is required in the environment. A significant change is one that *permits the client to adapt in the environment.*

These additional considerations suggest additional requirements for any organization of urban services: (4) that the organization of a system of services so that clients are not only referred but followed up and aided in having their demands met, (5) that the system be organized to meet client needs when they are presented and, if possible, at client-designated rather than professional-designated time and place, and (6) that agencies be organized to effect significant changes in the environment so that clients may cope with it.

CIVIC ACCOUNTABILITY OF PUBLIC EMPLOYEES

Some of the fiercest debate about urban services today rages around the issue of how citizen complaints about service are to be handled. In many communities, there are sharp debates about citizen review boards for the police or grievance commissions to deal with citizen complaints of discrimination in housing or employment. There should be no doubt that, in a democratic society, citizens have both a right and an obligation to complain about treatment by public officials, whether or not their complaints prove warranted on proper and appropriate investigation. A central issue lost sight of is: how are we, on the one hand, to be regulated by any public service and yet make public servants accountable to us for that regulation?

Perhaps, we need to remind ourselves that the problem is a central one for all democratic societies. It only takes different forms at different times. The dilemma took a somewhat different form for citizens in our nineteenth century cities where service and accountability rested in a "spoils system." Good government rebelled against the spoils system and introduced the "merit system" of appointment to insure quality and responsibility for service by tenure rather than by political fealty. We also removed many offices from direct election and made them appointive on the theory that accountability to the electorate could be exercised more wisely if fewer persons are held responsible to the electorate. We are still engaged in this process of "streamlining" government. Finally, we developed vast bureaucratic, publicly sponsored welfare services that replaced the services of the precinct and ward committeemen in local parties.

In the cause of "good government," however, we have lost sight of some important consequences of "machine politics." When public servants were enmeshed in the local political system of cities, they had to be fairly responsive to the demands of citizens because their continuation depended upon a local electorate. This was particularly true for the underclass and for minority groups in our cities; an underclass or minority group vote was as good as any other. Indeed, the successive waves of immigrant domination of sectors of public service (for example, that of the German and Irish in police departments) rested less on merit than on the responsiveness of local government organization to the organized political pressures of each successive wave of immigrants. There is little doubt that local politics in the nineteenth century often made for the rapid movement of the underclass into positions in the urban polity, thereby representing their interests. Furthermore, the local political organization often served quite effectively as a channel for grievance. The decline of machine politics carried with it some measure of loss of accountability of public officials and servants to citizens (Cornwell, 1964).

Our current system of civic accountability, then, is built primarily upon three assumptions: (a) that we can make elected officials responsible for the conduct of officials and servants in the bureaucracies under them, (b) that each bureaucracy establishes effective machinery for handling citizen complaints and insuring justice, and (c) that civil servants are responsive to the public. We are satisfied that over and above this, when a criminal violation is involved, the citizen can seek redress through our legal system.

All of these are questionable assumptions about how to make public agencies and civil servants responsive to citizens. A major task facing citizens and government in cities at all levels of government is how one can make bureaucratic officials and servants responsive to citizen demands and complaints. Note that mention is made not only of responsiveness to *complaints about* service but to *demands for* service that the agency should render.

Clearly, elected officials should be held responsible for the bureaucracies under them. Citizens should insist that their mayors, city managers, and common councils be responsible for all municipal services. They must also insist that each agency have formal machinery to investigate and process complaints, and that a merit system continue to be the main basis for public service. Finally, citizens want assurance that where there are clear criminal violations, prosecution will follow. But that is not enough. Some other things are necessary.

EXTERNAL ACCOUNTABILITY OF PUBLIC AGENCIES

Any system must be considered from the standpoint of the people processed by it. Consider the above description of a complaint system from the standpoint of a citizen making a complaint. Consider first, the requirement that the citizen shall lodge and pursue his complaint through the very agency against which the complaint is lodged. If I have a complaint about a police officer, I should go to a police official; if I have a complaint about a school teacher, I should go to a school official; if I have a complaint about health service, I should go to the hospital or clinic staff.

No citizen should be required to make and pursue his complaint through the very agency against which the complaint is lodged. Such a system fosters reluctance on the part of citizens to make complaints because of fear the agency may be punitive. Furthermore, if one complains to the offending agency, one is open to organizational manipulation by personnel in the agency, often to "cool out" the complainant. In short, agency personnel are not

"disinterested" parties; justice lies partly in disinterest in the outcome. A major requirement, then, for any complaint system that is
accountable to the citizens it serves is that the registration of
complaints lie outside the agency against which the complaint is
lodged.

There are grounds for arguing that the responsibility for inves-
tigating complaints and sanctioning offending staff should lie with
the agency against which the complaint is lodged, rather than with
the agency with which it is registered. The main reasons offered
for separating investigation-sanction and complaint-receiving functions is the argument that professional responsibility requires
investigation and sanctioning by peers and that, generally, if change
is to be effective, it must be undertaken by the administration and
staff of the agency affected. There are also arguments against
lodging responsibility with officials of the offending agency on the
grounds that it opens the way for any agency to manipulate
matters in its own interest.

Regardless of the stance taken on this matter, it is clear that the
agency which registers complaints should maintain control over
these inputs, at least to the degree that it receives a full report of
any investigation, and of any actions taken by the alleged offending agency. Acquisition of the basic input and output information
probably is the most powerful monitoring device in any system
since the organization that controls this information can assess
where organizational problems lie. The patterning of complaints
whether by tape, by staff, by program, or by any other organizational component, provides useful information for changing the
organization. Characteristically legalists and moralists want to
change organizations on the basis of "making a case." Usually, that
means they want to bring action against an individual or group as
adversaries or, at most, to adjudicate the particular case to the
satisfaction of the offended party. While that may be good public
relations, it hardly is good organizational practice since all too
often it leaves the agency itself unchanged. Indeed, the aggregative
effects of complaints are more readily discernible and demonstrable
than are the merits of an individual case.

There are problems in insuring that citizens will register their
complaints in such a way as to affect the offending agency. Part of
the problem lies in the fact already mentioned—that citizens are
reluctant to lodge complaints when the agency is an interested
party. We found in a survey of the problems of Detroit citizens,

however, that there are other problems as well. Often citizens lack information on *where* to take their complaints. When they wish to complain, most don't complain. When citizens do complain, the majority turn to the offending agency, usually to a supervisor of the offending party. Others either seek legal advice or seek to arouse an elected official's interest in their complaint. Almost all citizens lack specific information on where to lodge a complaint even when many such specific agencies exist. Most people, for example, don't know there is a Better Business Bureau in Detroit that will pay attention to problems of fraudulent practices in business. More people know more about Action Line, a column of a daily newspaper, than about the Michigan Civil Rights Commission or the Detroit Human Relations Commission. Briefly, our system fails to educate citizens about where they can take their complaints.

To make a complaint system responsive to citizens, ways of making them aware of where and how they can lodge complaints are needed. Since it is difficult to communicate a large amount of specific information, it seems reasonable to conclude that there should be but one agency (or at most a very few) to which complaints are taken if we are to link citizens effectively to complaint systems. The problem is an organizational one. Citizens should not be required to link their complaints to a whole variety of complaint-receiving and processing agencies. In a bureaucratically organized society, that task should be the province of an agency whose responsibility it should be to aid the citizen in preparing the complaint and forwarding it for action and response to the appropriate agency.

INTERNAL ACCOUNTABILITY

All too often proposals for making public agencies accountable to citizens focus on how external agencies may audit the public agencies and make them accountable. There are demands for "civilian review," "citizen control," or an "ombudsman." In these days of quality control, it is surprising that agency administrators give so little attention to how they can monitor the quality of service their agency gives clients. A few examples of agency monitoring may make apparent how such means can be developed and foster accountability.

Many agencies could readily develop an internal audit unit that regularly and systematically samples the client population and assesses the adequacy of service from both the administrative and the client point of view. If that agency is the police department, for example, it is possible to make an immediate follow-up after the police have engaged in a transaction with a citizen. Since the computer can select such cases, it is difficult to "beat" the system by controlling selection. Where the agency does not wish to conduct the audit itself, it can secure the services of an independent agency that reports to it. Such audits have become an essential part of many private industrial establishments. They need to become part of management for any public agency as well.

Much of the emphasis on internal accountability rests on presumptions that one investigates the behavior of offending persons and punishes them in some way when they are adjudged guilty of violations. Mention has been made of the fact that aggregative control may be superior to such case control. Apart from the merits of that argument, a case can be made for handling investigations in other ways. Much research from modern psychology suggests that punishment is a poor way to get people to conform— that it is more likely to generate deviation from expectations than conformity. Punishments that follow upon agency investigation tend to be self-defeating, giving rise to employee subcultures that subvert investigation. The "blue curtain" of the police department that descends to protect a fellow police officer under investigation is no different from the protective silence that falls in the welfare temple when mistakes are made.

Apart from the developing of ways to provide professional incentives for service, rewards for changing the system when mistakes are made should be more productive of organizational change than any form of punishment. What is more, failure to handle a position properly need not be grounds for dismissal when often transfer is possible. Indeed, transfers need not be punitive and involve a loss in prestige or income. The possibilities for a civil service bureaucracy to handle transfers is considerable and should not be scanted as a means for handling complaints against persons. Above all, however, the major focus of any complaint system should be on how organizations generate "mistakes" or "errors" by personnel and how changes in the *organization* can control their occurrence rather than on changing or punishing people. It is axiomatic in sociology that changing organizations usually changes persons in roles within organizations.

One final method of developing internal accountability should be mentioned. Any system of internal accountability should provide for continual review, as already noted. While such systems can be efficient by monitoring small samples of cases, in other cases, gains can be made if the accountability function is built into each and every transaction. Ideally any system of accountability has both a regular-accounting and an audit function. One of the ways that an organization can regularly account for itself is in providing citizens with information on the transactions that serve as a "receipt" of the encounter. Apart from the fact that such a receipt is evidence of the encounter, it provides factual information for lodging any complaint—such as who was the agency representative, what they regarded as the purpose of the transaction, what action was taken, and so on.

Though arguments can be lodged against the cost of such systems, they also can bring benefits that reduce costs in other areas. Just for an example, if a police department provided citizens with information on investigation of routine events such as burglary, that receipt automatically provides the information usually required by insurance companies to settle complaints, thereby reducing such inquiries to the department and simplifying the citizen's need to file for the information.

EXTERNAL AND INTERNAL ACCOUNTABILITY

It has been suggested that any system with a goal to make public organizations accountable to the citizenry must build in certain elements that now are generally absent. A number of these elements have been reviewed: (1) that the registration of any complaint and its final review must lie with an agency, other than the one against which the complaint is lodged; (2) that a single agency best meets the requirement of informing citizens where they may go to lodge a complaint; (3) that an organization, rather than persons, should be responsible for formulating the complaint and transmitting it for appropriate action; (4) that both a regular system of accounting and an auditing service for clients must be built into the agency; (5) that punishment systems of sanctions against offending agency personnel militate against changing the organization; (6) that the focus of action based on complaints should be on how the agency, rather than on how its staff, generates the complaints.

The closest approximation we have to such an agency today anywhere in the Western world is the *office* of ombudsman. What many Americans fail to understand about the operation of that office is the fact that the ombudsman functions primarily to alter organizational structures rather than to adjudicate cases. Having considerable informal as well as some formal powers to alter bureaucratic structures and practices, the ombudsman is able to alter environmental conditions and constraints so that the system works in the sense that it is less likely to produce complaints.

REFERENCES

CORNWELL, E. E., Jr. (1964) "Bosses, machines, and ethnic groups." Annals of the American Academy of Political and Social Science (May).

DAVIS, K. C. (1969) Discretionary Justice. Baton Rouge: Louisiana State University Press.

HONNARD, R. et al. (1961) Views of Authority: Probationers and Probation Officers. Research Paper 1. Los Angeles: Youth Studies Center, University of Southern California.

REISS, A. J., Jr. (1968) "Some sociological issues about American communities." Pp. 66-74 in T. Parsons (ed.) American Sociology. New York: Basic Books.

SCHEFF, T. (1966) "Typification of the diagnostic practices of rehabilitation agencies." In M. Sussman (ed.) Sociology and Rehabilitation. Washington, D.C.: American Sociological Association.

BIBLIOGRAPHY

Bibliography

I. THE URBAN ECONOMY: An Overview

ABRAMS, CHARLES. The City is the Frontier (New York: Harper & Row, 1965).

BANFIELD, EDWARD C. and MARTIN MEYERSON. Boston: The Job Ahead (Cambridge: Harvard University Press, 1966).

BIRKHEAD, G. S. (ed.) Metropolitan Issues: Social, Governmental, Fiscal (Syracuse: Syracuse University Press, 1962).

BOLLENS, JOHN C. (ed.) Exploring the Metropolitan Community (Berkeley-Los Angeles: University of California Press, 1961).

BOLLENS, JOHN C. and HENRY J. SCHMANDT. The Metropolis: Its People, Politics and Economic Life (New York: Harper & Row, 1965).

BUCHANAN, JAMES M. Academia in Anarchy: An Economic Analysis (New York: Basic Books, 1970).

CALLOW, ALEXANDER B., Jr. (ed.) American Urban History (New York: Oxford University Press, 1969).

CHINITZ, BENJAMIN (ed.) City and Suburbs: The Economics of Metropolitan Growth (Englewood Cliffs, N.J.: Prentice-Hall, 1964).

CLARK, COLLIN. "The Economics of a City in Relation to Its Size," Econometrica (April 1965).

CLARK, W. A. V. and ROBERT O. HARVEY. "The Nature and Economics of Urban Sprawl," Land Economics, Volume 41, Number 1, February 1965.

CLEAVELAND, FREDERIC N. and Associates. Congress and Urban Problems: A Casebook on the Legislative Process (Washington, D.C.: Brookings Institution, 1969).

EDITOR'S NOTE: *Compiled by Mr. Robert Cline, graduate student, Department of Economics, The University of Michigan.*

DUHL, LEONARD J. (ed.) The Urban Condition (New York: Basic Books, Inc., 1963).

DYCKMAN, JOHN W. "The Public and Private Rationale for a National Urban Policy," in Bass, S. B., Jr. (ed.) Planning for a Nation of Cities (Cambridge: MIT Press, 1966).

DYE, THOMAS. Politics, Economics and the Public (Chicago: Rand McNally, 1966).

ELAZAR, DANIEL J. "Megalopolis and the New Sectionalism," The Public Interest, Number 11 (Spring 1968) pp. 67-85.

FALTERMAYER, EDMUND K. Redoing America (New York: Harper and Row, 1968).

FRIEDEN, BERNARD. Metropolitan America: Challenge to Federalism, Advisory Commission on Intergovernmental Relations, Committee Print of Intergovernmental Relations Subcommittee of the House Committee on Government Operations, Eighty-ninth Congress, Second Session, October, 1966.

FUSFELD, DANIEL R. The Age of the Economist (New York: Scott, Foresman, 1966; Morrow, 1968).

HANDLIN, OSCAR and JOHN BURCHARD (eds.) The Historian and the City (Cambridge: MIT Press and Harvard University, 1963).

HARRIS, BRITTON, "The Uses of Theory in the Simulation of Urban Phenomena," Journal of the American Institute of Planners, Volume 32, Number 5 (September 1966).

HATT, PAUL K. and ALBERT J. REISS, Jr. Cities and Society (Glencoe: The Free Press, 1957).

HAUSER, PHILIP M. and LEO F. SCHNORE (eds.) The Study of Urbanization (New York: John Wiley, 1965).

HOOVER, E. M. The Location of Economic Activity (New York: McGraw-Hill, 1948).

HOOVER, E. M. and R. VERNON. Anatomy of a Metropolis. (Cambridge: Harvard University Press, 1959).

ISARD, W. Location and Space Economy (New York: MIT Press and John Wiley, 1956).

JACOBS, JANE. The Economy of Cities (New York: Random House, 1969).

LOGUE, E. "Prospects for the Metropolis," in A. H. Pascal (ed.), Contributions to the Analysis of Urban Problems (P-3868) (Santa Monica: RAND Corporation, 1968).

MAHOOD, H. R. and EDWARD L. ANGUS (eds.) Urban Politics and Problems (New York: Scribner's, 1969).

MARGOLIS, JULIUS (ed.). The Public Economy of Urban Communities (Washington: Resources for the Future, 1965).

MARTIN, ROSCOE C. The Cities and the Federal System. (New York: Atherton Press, 1965).

MILLER, H. "Urban Crisis: Prospects Not All Black," Los Angeles Times (July 13, 1969), p. 11, columns 4-8.

PARK, R. Human Communities: The City and Human Ecology (Glencoe: Free Press, 1962).

PASCAL, A. H. (ed.). Cities in Trouble: An Agenda for Urban Research (RM-5603-RC) (Santa Monica: RAND Corporation, 1968).

PASCAL, A. H. Contributions to the Analysis of Urban Problems (P-3868) (Santa Monica: RAND Corporation, 1968).

PASCAL, A. H. Thinking About Cities: New Perspectives on Urban Problems (Belmont, Calif.: Dickenson, 1970).

PELL, CLAIBORNE. Megalopolis Unbound: The Super-City and the Transportation of Tomorrow (Frederick A. Praeger, 1966).

PERLOFF, HARVEY S. (ed.). The Quality of the Urban Environment (Washington: Resources for the Future, distributed by Johns Hopkins Press, 1969).

PERLOFF, HARVEY S. and LOWDON WINGO, Jr. (eds.) Issues in Urban Economics (Baltimore: Johns Hopkins Press for Resources for the Future, 1968).

PERLOFF, HARVEY S., EDGAR S. DUNN, ERIC E. LAMPARD, and RICHARD F. MUTH. Regions, Resources, and Economic Growth (Baltimore: The Johns Hopkins Press for Resources for the Future, Inc., 1960).

SCHNORE, LEO, and HENRY FAGIN. Urban Research and Policy Planning (Beverly Hills: Sage Publications, 1967).

SYELL, ANWAR. The Political Theory of American Local Government (New York: Random House, 1966).

THOMPSON, WILBUR R. A Preface to Urban Economics (Baltimore: The Johns Hopkins Press for Resources for the Future, Inc., 1965).

U.S. Advisory Commission on Intergovernmental Relations. Metropolitan Social and Economic Disparities. Implications for Intergovernmental Relations in Central Cities and Suburbs (Washington 1965).

——— Urban and Rural America: Policies for the Future (Washington: U.S. Government Printing Office, 1968).

U.S. National Commission on Urban Problems. Building the American City (Washington: U.S. Government Printing Office, 1968).

U.S. National Committee on Urban Growth Policy. The New City (New York: Praeger for Urban America, Inc., 1969).

U.S. Senate, Committee on Government Operations. Government in Metropolitan Areas (New York Metropolitan Region), (Washington: Government Printing Office, 1963).

"Urban Problems and Prospects." Law and Contemporary Problems, Volume 30, Winter 1965.

WEAVER, ROBERT C. The Urban Complex (Garden City, N.Y.: Doubleday, 1964).

WEBBER, MELVIN M., et al. Explorations into Urban Structure (Philadelphia: University of Pennsylvania Press, 1964).

WILDAVSKY, AARON. "The Private Markets and Public Arenas." The American Behavioral Scientist, Vol. 9, No. 7 (Sept. 1965).

WILLIAMS, OLIVER P., HAROLD HERMAN, CHARLES S. LIEBMAN, and THOMAS R. DYE. Suburban Differences and Metropolitan Policies: A Philadelphia Story (Philadelphia: University of Pennsylvania Press, 1965).

WILLMAN, JOHN B., The Department of Housing and Urban Development (New York: Praeger, 1967).

WILSON, JAMES Q. (ed.) The Metropolitan Enigma (New York: Chamber of Commerce of the United States, 1967).

ZIMMERMAN, JOSEPH F. (ed.) Government of the Metropolis (New York: Holt, Rinehart and Winston, Inc., 1968).

II.   URBAN FISCAL POLICY

A. GENERAL

ANDO, ALBERT, E. CAREY BROWN, and EDWARD W. ADAMS, Jr. "Government Revenues and Expenditures," in James S. Duesenberry et al., The Brookings Quarterly Econometric Model of the United States (Chicago: Rand McNally and Company, 1965).

BECKER, GARY S. Human Capital (New York: National Bureau of Economic Research, 1964).

BRAZER, HARVEY E. (ed.) Essays in State and Local Finance. (Ann Arbor: Institute of Public Administration, The University of Michigan, 1967).

BRAZER, HARVEY E. "The Federal Government and State-Local Finances," National Tax Journal, XX (June, 1967), pp. 155-164.

BRAZER, HARVEY E. "Some Fiscal Implications of Metropolitanism," in Benjamin Chinitz (ed.), City and Suburb: The Economics of Metropolitan Growth (Englewood Cliffs, N.J.: Prentice-Hall, 1964).

BREAK, GEORGE F. Intergovernmental Fiscal Relations in the United States (Washington: The Brookings Institution, 1967).

BUCHANAN, JAMES M. Cost and Choice: An Inquiry in Economic Theory (Chicago: Markham, 1969).

BUCHANAN, JAMES M. Public Finance in Democratic Process: Fiscal Institutions and Individual Choice. (Chapel Hill: The University of North Carolina Press, 1967).

BUCHANAN, J. (ed.) Public Finances: Needs, Sources and Utilization, National Bureau of Economic Research (Princeton: Princeton University Press, 1961).

BUCHANAN, JAMES M. and GORDON TULLOCK. "Public and Private Interaction under Reciprocal Externality," in Julius Margolis (ed.), The Public Economy of Urban Communities (Baltimore: Johns Hopkins Press for Resources for the Future, 1965).

CAMPBELL, ALAN K. and SEYMOUR SACKS. Metropolitan America: Fiscal Patterns and Governmental Systems (New York: The Free Press, 1967).

CAMPBELL, JACK M. "Are the States Here to Stay?," Public Administration Review, Vol. 28 (January/February 1968), pp. 26-39.

CAREY, WILLIAM D. "Intergovernmental Relations: Guides to Development", Public Administration Review, Vol. 28 (January/February 1968), pp. 22-25.

CLEVELAND, HARLAN. "Are the Cities Broke?," National Civic Review, 50 (March, 1961).

Committee for Economic Development. Modernizing Local Government: to Secure a Balanced Federalism (New York: CED, 1966).

CRECINE, JOHN P. A Dynamic Model of Urban Structure (Santa Monica: RAND Corporation, 1968).

DAVIS, OTTO A. and GEORGE H. HAINES, Jr. "A Political Approach to a Theory of Public Expenditure: The Case of Municipalities," National Tax Journal, XIX (September 1966).

DORFMAN, ROBERT (ed.). Measuring Benefits of Government Investments (Washington: Brookings Institution, 1965).

DUE, JOHN F. Government Finance: Economics of the Public Sector, 4th ed. (Homewood, Illinois: Richard W. Irwin, 1968).

EPPS, R. W. "The Metropolitan Money Gap," Business Review, Federal Reserve Bank of Philadelphia, V. (June, 1968).

FABRICANT, S. The Trend of Government Activity in the United States Since 1900 (New York: National Bureau of Economic Research, Inc., 1952).

FISHER, GLENN W., and FAIRBANKS, ROBERT P. Illinois Municipal Finance: A Political and Economic Analysis. (Urbana: University of Illinois Press, 1968).

FITCH, LYLE C. "Metropolitan Financial Problems," in Benjamin Chinitz (ed.), City and Suburb: The Economics of Metropolitan Growth (Englewood Cliffs, New Jersey: Prentice-Hall, 1964).

GANS, HERBERT J. "Some Proposals for Government Policy in an Automating Society," The Correspondent, No. 30 (January-February, 1964) pp. 74-82.

HAWLEY, A. H. "Metropolitan Population and Municipal Government Expenditures in Central Cities," Journal of Social Issues, Vol. 7, pp. 100-108.

HEAD, J. "Public Goods and Public Policy," Public Finance, 17, No. 3 (1962), pp. 197-219.

KAYSEN, CARL. "Model-makers and Decision-makers: Economists and the Policy Process," The Public Interest, No. 12 (Summer, 1968) pp. 80-95.

KEE, W. S. "City-Suburban Differentials in Local Government Fiscal Effort," National Tax Journal, Vol. 18 (June, 1968), pp. 183-189.

KEY, V. O., Jr. Public Opinion and American Democracy (New York: Alfred A. Knopf, 1961).

KRISTOL, IRVING. "Decentralization for What?" The Public Interest No. 11 (Spring, 1968) pp. 17-25.

LICHFIELD, NATHANIEL, "Spatial Externalities: A Case Study", in Julius Margolis (ed.), The Public Economy of Urban Communities (Baltimore: Johns Hopkins Press for Resources for the Future, 1965).

LINDBLOM, C. "Decision-Making in Taxation and Expenditures," in J. Buchanan (ed.) Public Finances: Needs, Sources and Utilization, National

Bureau of Economic Research (Princeton: Princeton University Press, 1961).

LINEBERRY, ROBERT L. and EDMUND P. FOWLER, "Reformism and Public Policies in American Cities," The American Political Science Review, LXI (September 1967).

MARGOLIS, JULIUS. "Metropolitan Finance Problems: Territories, Functions, and Growth," in J. Buchanan (ed.) Public Finances: Needs, Sources, and Utilization, National Bureau of Economic Research, (Princeton: Princeton University Press, 1961).

MARGOLIS, JULIUS, "Municipal Fiscal Structure in a Metropolitan Region", Journal of Political Economy (June 1951).

MAXWELL, JAMES A. Financing State and Local Governments (Washington: The Brookings Institution, 1965).

Municipal Finance Officers Association, "National Conference on Local Government Fiscal Policy," Municipal Finance, XXXIX (February, 1967).

MUSGRAVE, RICHARD A. (ed.) Essays in Fiscal Federalism. (Washington: The Brookings Institution, 1965).

MUSGRAVE, RICHARD A. The Theory of Public Finance: A Study in Public Economy (New York: McGraw-Hill Book Company, 1959).

MUSGRAVE, R. A. and A. T. PEACOCK (eds.) Classics in the Theory of Public Finance, International Economic Association (London: Macmillan and Company, Ltd., 1958).

MUSHKIN, S. J. and G. RUPO. "Project '70: Projecting the State-Local Sector," The Review of Economics and Statistics, 49, No. 2 (May, 1967).

NETZER, DICK. "Financial Needs and Resources over the Next Decade: State and Local Governments," in National Bureau of Economic Research, Public Finances: Needs, Sources and Utilization (Princeton: Princeton University Press, 1961).

NETZER, D. State-Local Finance and Intergovernment Fiscal Relations (Washington, D.C.: Brookings Institution, 1969).

PEACOCK, ALAN T. and WISEMAN, JACK. The Growth of Public Expenditure in the United Kingdom (Princeton: Princeton University Press, 1961).

PECKMAN, JOSEPH A. Federal Tax Policy (Washington, D.C.: The Brookings Institution, 1966), pp. 201-231.

PIDOT, GEORGE B., Jr. "The Public Finances of Local Government in the Metropolitan United States." (Unpublished Ph.D. dissertation, Harvard University, 1966).

Research Advisory Committee (Francis Boddy, Chairman). Twin Cities Metropolitan Tax Study: Recommendations of the Research Advisory Committee, December, 1966.

ROSENTHAL, ALBERT H. The Current Scene: Approaches and Reproaches, Public Administration Review, Vol. 28 (January/February 1968). pp. 3-9.

SACHS, S. and W. F. HELLMUTH, Jr. Financing Government in a Metropolitan Area: The Cleveland Experience (New York: Free Press of Glencoe, Inc., 1961).

SCHERER, JOSEPH and JAMES A. PAPKE. Public Finance and Fiscal Policy (Boston: Houghton Mifflin Company, 1966).

SHAPIRO, H. "Economies of Scale and Local Government Finance," Land Economics, 39 (May, 1963), pp. 175-186.

SHARKANSKY, I. "Regionalism, Economic Status and the Public Policies of American States," Southwestern Social Science Quarterly (June, 1968), pp. 9-26.

SHOUP, CARL S. Public Finance (Chicago: Aldine Publishing Company, 1969).

Tax Foundation, Inc. Fiscal Outlook for State and Local Government to 1975 (Research Publication No. 6) (New York: Tax Foundation, Inc. 1966).

THOMPSON, WILBUR. "The City as a Distorted Price System," Psychology Today, II (August, 1968).

TURVEY, R. "On Divergences Between Social Cost and Private Cost," Economics, 30 (1963), pp. 309-313.

U.S. Advisory Commission on Intergovernmental Relations. State and Local Finances: Significant Features, 1966 to 1969, 1968.

U.S. Department of Commerce, Bureau of the Census. City Government Finances in 1967-68 (Washington: U.S. Government Printing Office, 1968).

U.S. Senate, Committee on Government Operations. Federal Role in Urban Affairs, Hearings before the Subcommittee on Executive Reorganization 89th Congress, 2nd Session, August-December 1966.

VICKREY, WILLIAM W. "General and Specific Financing of Urban Services," in Howard G. Schaller (ed.), Public Expenditure Decisions in the Urban Community (Washington: Resources for the Future, Inc., 1963).

VIEG, J. A. (ed.) California Local Finance (Palo Alto: Stanford University, 1960).

WOOD, ROBERT C. 1400 Governments: The Political Economy of the New York Metropolitan Region (Cambridge, Massachusetts: Harvard University Press, 1961).

II.   URBAN FISCAL POLICY

B. REVENUES

The Academy of Political Science. "Municipal Income Taxes," Proceedings, ed. Robert H. Connery XXVIII, No. 4. (New York: Columbia University, 1968).

ALDERFER, H. F. "Is Authority Financing the Answer?", in Financing Metropolitan Government (Princeton: Tax Institute, Inc., 1955).

ANDERSON, WILLIAM. "The Myths of Tax Sharing", Public Administration Review, Vol. 28 (January/February 1968). pp. 10-14.

BOSKIN, MICHAEL JAY. "The Negative Income Tax and the Supply of Work Effort." National Tax Journal, Vol. XX, No. 4 (December 1967).

——— "Reply". National Tax Journal, Vol. XXII, No. 3 (September 1969). p. 417.

BROWNLEE, O. H. "The Effects of the 1965 Federal Excise Tax Reduction Prices," National (September, 1967). pp. 235-249.

BROWNLEE, O. H. "User Prices vs. Taxes," in Public Finances: Needs, Sources and Utilization, National Bureau of Economic Research (Princeton: Princeton University Press, 1961), pp. 421-32.

BRUNO, JAMES E. "Achieving Property Tax Relief with a Minimum Disruption of State Programs," National Tax Journal, Vol. XXII, No. 3 (September 1969), pp. 379-389.

BUCHANAN, J. M. "The Economics of Earmarked Taxes," Journal of Political Economy, Vol. 71, October 1963.

BUCHANAN, J. M. (ed.) Public Finances: Needs, Sources and Utilization, National Bureau of Economic Research (Princeton: Princeton University Press, 1961).

BURKHEAD, JESSE. State and Local Taxes for Public Education (Syracuse, Syracuse University Press, 1963).

DAVID, ELIZABETH LIKERT. "A Comparative Study of Tax Preferences," National Tax Journal, XXI, No. (March, 1968).

DAVID, ELIZABETH LIKERT. "Public Preferences and State-Local Taxes," in H. Brazer, (ed.) Essays in State and Local Finance. (Ann Arbor: Institute of Public Administration, The University of Michigan, 1967).

DAVID, ELIZABETH LIKERT, and ROGER B. SKURSKI. "Property Tax Assessment and Absentee Owners." National Tax Journal, XIX, No. 4 (December 1966).

DAVIES, DAVID. "Financing Urban Functions and Services," Law and Contemporary Problems, Vol. 30, Winter 1965.

DERAN, ELIZABETH Y. "Earmark and Expenditures: A Survey and A New Test," National Tax Journal, Vol. 18 (December 1965).

DERAN, ELIZABETH Y. Financing Capital Improvements: The "Pay-As-You-Go" Approach (Berkeley: Bureau of Public Administration, University of California, 1961).

DERAN, ELIZABETH Y. "An Overview of the Municipal Income Tax" in Robert H. Connery, (ed.) Municipal Income Taxes, Proceedings of the Academy of Political Science, Columbia University, Vol. XXVIII, No. 4 (1968), pp. 441-448.

DERAN, ELIZABETH Y. "Tax Structure in Cities Using the Income Tax," National Tax Journal, Vol. XXI, No. 2 (June, 1968), pp. 147-152.

DONHEISER, A. D. "The Incidence of the New York City Tax System," Financing Government in New York City (New York: Graduate School of Public Administration, New York University, 1966).

FISHER, GLENN W. and ROBERT P. FAIRBANKS. "The Politics of Property Taxation," Administrative Science Quarterly, XII (June 1967).

GENSEMER, BRUCE, JANE A. LEAN, and WILLIAM B. NEENAN. "Awareness of Marginal Income Tax Rates Among High-Income Taxpayers," National Tax Journal, 18 (September, 1965), pp. 258-267.

GRAVES, L. F. "State and Local Tax Burdens in California," in California Assembly Interim Committee on Revenue and Taxation, 1964.

GROVES, H. "Innovation in Tax-Sharing—The Wisconsin Experience" in Revenue Sharing and Its Alternatives: What Future for Fiscal Federalism?, Joint Economic Committee, 90th Congress, 1st Session, Vol. 1 (Washington, 1967).

HAMOVITCH, WILLIAM. "Sales Taxation: An Analysis of the Effects of Rate Increases in Two Contrasting Cases," National Tax Journal, XIX, No. 4 (December 1966).

HARBERGER, A. C. "Taxation, Resource Allocation, and Welfare," in J. F. Due (ed.), The Role of Direct and Indirect Taxes in the Federal Revenue System.

HELLER, WALTER W. New Dimensions of Political Economy (Cambridge: Harvard University Press, 1966).

HELLER, W. W. "A Sympathetic Reappraisal of Revenue Sharing," in Harvey S. Perloff and Richard P. Nathan (eds.), Revenue Sharing and the City (Baltimore: Johns Hopkins Press, 1968).

HELLER, WALTER W. and J. A. PECHMAN. "Questions and Answers on Revenue Sharing," in Revenue Sharing and Its Alternatives: What Future for Fiscal Federalism? Joint Economic Committee, 90th Congress, 1st Session, Vol. 1 (Washington, D.C., 1967).

HELLER, WALTER W., RICHARD RUGGLERS et al., Revenue Sharing and the City (Baltimore: The Johns Hopkins Press for Resources for the Future, 1968).

HILLHOUSE, A. M., and S. KENNETH HOWARD. Revenue Estimating By Cities (Chicago: Municipal Finance Officers Association of the U.S. and Canada, 1965).

KEE, WOO SIK. "City-Suburban Differentials in Local Government Fiscal Effort," National Tax Journal, XXI (June, 1968), pp. 183-89.

League of California Cities. Statement of Principles and Alternatives Regarding Tax Reform. October, 1966.

LEWIS, WILLIAM C. "Tax Concessions and Industrial Location: A Review," Reviews in Urban Economics, (September 1968).

LYNN, ARTHUR D., Jr. (ed.) The Property Tax and Its Administration (Madison: The University of Wisconsin Press, 1969).

Maryland University of Maryland Tax Study (College Park, Maryland: University of Maryland, 1965).

MORGAN, WILLIAM. "The Effect of State and Local Tax Inducements on Industrial Location." unpublished Ph.D. dissertation, University of Colorado, 1964.

MUSGRAVE, R. A. and D. W. DAICOFF. "Who Pays the Michigan Taxes?" Michigan Tax Study Staff Papers (Lansing, Michigan: Michigan Legislative Tax Study Committee, 1958).

NETZER, D. Economics of the Property Tax (Washington, D.C.: The Brookings Institution, 1966).

NETZER, D. "Federal, State and Local Finance in a Metropolitan Context," in Perloff and Wingo (eds.) Issues in Urban Economics (Baltimore: Johns Hopkins University Press, 1968).

New York University, Graduate School of Public Administration. Financing Government in New York City. Final Research Report to the Temporary Commission on City Finances, City of New York (New York: 1966).

PECHMAN, JOSEPH A. Financing State and Local Government, Reprint 103 (Washington: The Brookings Institution, 1965).

PENNIMAN, CLARA. "The Politics of Taxation," in Herbert Jacob and Kenneth N. Vines (eds.), Politics in the American States. (Boston: Little, Brown and Company, 1965).

QUINDRY, KENNETH E. and BILLY D. COOK. "Humanization of the Property Tax for Low Income Households," National Tax Journal, Vol. XXII, No. 3 (September 1969). pp. 357-367.

ROSTVOLD, GERHARD N. "Distribution of Property, Retail Sales, and Personal Income Tax Burdens in California: an Empirical Analysis of Inequity in Taxation." National Tax Journal, XIX, No. 1 (March, 1966).

SCHEFFER, WALTER F. "Problems in Municipal Finance," The Western Political Quarterly, 15 (September, 1962).

STEPHENS, G. ROSS. "The Suburban Impact of Earnings Tax Policies," National Tax Journal, Vol. XXII, No. 3 (September 1969). pp. 313-333.

STEPHENS, G. ROSS and HENRY J. SCHMANDT. "Revenue Patterns of Local Government," National Tax Journal, Vol. XV, No. 4 (December 1962), pp. 432-436.

STOCKFISCH, J. A. "The Outlook for Fees and Service Charges as a Source of Revenue for State and Local Governments." 1967 Proceedings of the Sixtieth Annual Conference on Taxation (Columbus: National Tax Association, 1968).

Tax Foundation, Inc. City Income Taxes (New York: Tax Foundation, Inc. December, 1967).

U.S. Advisory Commission on Intergovernmental Relations. Federal-State Coordination of Personal Income Taxes (Washington, 1965).

U.S. Advisory Commission on Intergovernmental Relations. Fiscal Balance in the American Federal System, Vol. 1 (Washington, D.C., 1967).

U.S. Advisory Commission on Intergovernmental Relations. Sources of Increased State Tax Collections: Economic Growth vs. Political Choice. (Washington, 1968).

U.S. Advisory Commission on Intergovernmental Relations. "Tax Overlapping in the United States." in Scherer (ed.), Public Finance and Fiscal Policy (Boston: Houghton Mifflin, 1966).

U.S. Joint Economic Committee. Revenue Sharing and Its Alternatives: What Future for Fiscal Federalism?, 90th Congress, 1st Session, Vol. 1 (Washington, D.C., 1967).

U.S. National Commission on Urban Problems. Impact of the Property Tax: Its Implication for Urban Problems (Washington, D.C.: Joint Economic Committee Print, 90th Congress, 2nd Session).

U.S. Subcommittee on Fiscal Policy of the Joint Economic Committee, Revenue Sharing and its Alternatives: What Future for Fiscal Federalism?, 90th Congress, 1st Session. Vol. 1-3 (Washington: U.S. Government Printing Office, 1968).

VANDERMEULEN, ALICE JOHN. "Reform of a State Fee Structure: Principles, Pitfalls, and Proposals for Increasing Revenue," National Tax Journal, 17 (December, 1964).

WILLIAMSON, ROBERT B. "Some Evidence in Support of State Industrial Financing Programs. The Southwestern Case," Land Economics, (August, 1968), pp. 388-393.

Wisconsin University of Tax Study Committee. Wisconsin's State and Local Tax Burden (Madison, Wisconsin: University of Wisconsin, 1959).

II. URBAN FISCAL POLICY

C. EXPENDITURES

BAHL, R. W. Metropolitan City Expenditures: A Comparative Analysis (Louisville: University of Kentucky, 1968).

BAHL, R. W. "Studies on Determinants of Expenditures: a Review," in S. Mushkin and J. Cotten, Functional Federalism: Grants in Aid and PPB Systems (Washington: George Washington University, 1968).

BAHL, R. W., and SAUNDERS, R. J. "Determinants of Changes in State and Local Government Expenditures." National Tax Journal, Vol. 18, No. 1. (March 1965).

BAHL, R. W., and SANDERS, R. J. "Fabricant's Determinants After Twenty Years: A Critical Reappraisal", paper presented at the 78th Annual Meeting of the A.E.A., New York. December 1965, published in the American Economist, Spring 1966.

BAHL, ROY W. and ROBERT J. SAUNDERS. "Factors Associated with Variations in State and Local Government Spending," Journal of Finance, Vol. 21 (September 1966).

BARLOW, R. "Comment on Alternative Federal Policies for Stimulating State and Local Expenditures," National Tax Journal, Vol. 22 (June, 1969), pp. 282-285.

BARR, JAMES L. and OTTO A. DAVIS, "An Elementary Political and Economic Theory of the Expenditures of Local Governments," Southern Economic Journal, Vol. 32 (October 1966).

BIRDSALL, WILLIAM C. "A Study of the Demand for Public Goods," in Richard A. Musgrave (ed.), Essays in Fiscal Federalism (Washington: The Brookings Institution, 1965).

BOOMS, B. H. "City Governmental Form and Public Expenditure Levels," National Tax Journal, 19 (June, 1966), pp. 187-199.

BRAZER, HARVEY E. City Expenditures in the United States, Occasional Paper 66 (New York: National Bureau of Economic Research, 1959).

BROWN, W. H., Jr. and C. E. GILBERT. Planning Municipal Investment: A Case Study of Philadelphia (Philadelphia: University of Pennsylvania, 1961).

DAVIS, O. A. and G. H. HARRIS, Jr. "A Political Approach to a Theory of Public Expenditure: The Case of Municipalities," National Tax Journal, 19 (September, 1966) pp. 259-275.

DORFMAN, ROBERT (ed.). Measuring Benefits of Government Investments (Washington: Brookings Institution, 1965).

ECKSTEIN, OTTO. Trends in Public Expenditures in the Next Decade (New York: Committee for Economic Development, 1959).

FISHER, G. W. "Determinants of State and Local Governments Expenditures: A Preliminary Analysis," National Tax Journal, Vol. 14 (December, 1961), pp. 349-355.

FISHER, G. W. "Interstate Variation in State and Local Government Expenditure," National Tax Journal, 17 (March, 1964), pp. 57-74.

FREDLAND, J. E., S. HYMANS and E. L. MORSS. "Fluctuations in State Expenditures: An Econometric Analysis," Southern Economic Journal, Vol. 33 (April, 1967), pp. 496-517.

GILLESPIE, W. I. "Effects of Public Expenditures on the Distribution of Income," in R. A. Musgrave (ed.), Essays in Fiscal Federalism (Washington, D.C.: Brookings Institution, 1966).

GINSBURG, A. L., G. SCHRAMM and G. R. WILENSKY. "Determinant Studies: the Need for New Directions." Unpublished paper

HANSEN, NILES M. "The Structure and Determinants of Local Public Expenditures." Review of Economics and Statistics, Vol. 47, No. 2 (May 1965).

HEILBRONER, ROBERT L. and PETER L. BERNSTEIN. Primer on Government Spending (New York: Vintage, 1963).

HELLER, W. W. "Reflections on Public Expenditure Theory," in Phelps (ed.), Private Wants and Public Needs, 1962.

HIRSCH, W. Z., "Determinants of Public Education Expenditures", National Tax Journal, Vol. 13 (March 1960).

HIRSCH, W. Z. "Factors Affecting Local Governmental Expenditure Levels," in John C. Bollens (ed.), Exploring the Metropolitan Community (Berkeley -Los Angeles: University of California Press, 1961).

KATZMAN, MARTIN T. "Distribution and Production in a Big City Elementary School System," Yale Economic Essays (Spring 1968).

KEE, WOO SIK. "City Expenditures and Metropolitan Areas: Analysis of Intergovernmental Fiscal Systems," National Tax Journal (December 1965).

KURNOW, E. "Determinants of State and Local Expenditure Reexamined," National Tax Journal, 16 (September, 1963), pp. 252-255.

McKEAN, ROLAND N. Public Spending. (New York: McGraw-Hill, Inc., 1966).

MEYER, DONALD GORDAN. "The Impact of Credit Conditions on the Timing of State Capital Expenditures," unpublished Ph.D. dissertation, Northwestern University, 1965).

MILLIMAN, J. W. "Beneficiary Charges and Efficient Public Expenditure Decisions," in Joint Economic Committee, The Analysis and Evaluation of

Public Expenditures: The PPB System (Washington, D.C.: Government Printing Office, 1966).

MILLS, EDWIN S. "An Aggregative Model of Resource Allocation in a Metropolitan Area, AER, Vol. 57, May 1967.

MORSS, ELLIOTT R. "Some Thoughts on the Determinants of State and Local Expenditures," National Tax Journal, XIX, No. 1 (March, 1966).

MUNSE, ALBERT R. State Programs for Public School Support, U.S. Department of HEW (Washington U.S. Government Printing Office, 1965).

MUSHKIN, SELMA. "Intergovernmental Aspects of Local Expenditure Decisions," in Howard G. Schaller (ed.), Public Expenditure Decisions in the Urban Community (Baltimore: Johns Hopkins Press for Resources for the Future, 1963).

OSMAN, J. "The Dual Impact of Federal Aid on State and Local Government Expenditure," National Tax Journal, 19 (December, 1966), pp. 362-372.

OSMAN, J. "On the Use of Intergovernmental Aid as an Expenditure Determinant," National Tax Journal, 21 (December, 1968), pp. 437-447.

PHELPS, C. "The Impact of Tightening Credit on Municipal Capital Expenditures in the United States," Yale Economic Essays, Vol. 1, No. 2 (Fall 1961).

PIDOT, G., Jr. "A Principal Components Analysis of the Determinants of Local Government Fiscal Patterns," Review of Economics and Statistics, Vol. 51 (May, 1969), pp. 176-188.

POGUE, T. F. and L. G. SGONTZ. "The Effect of Grants-in-Aid on State-Local Spending," National Tax Journal, 21 (June, 1968), pp. 190-199.

RENSHAW, E. F. "A Note on the Expenditure Effect of State Aid to Education," Journal of Political Economy, Vol. 68 (April, 1960), pp. 170-174.

RIEW, JOHN. "Economies of Scale in High School Operation," The Review of Economics and Statistics, Vol. 48 (August 1966).

SACKS, SEYMOUR. "Spatial and Locational Aspects of Local Government Expenditures," in Howard G. Schaller (ed.) Public Expenditure Decisions in the Urban Community (Baltimore: Johns Hopkins Press for Resources for the Future, 1963).

SACKS, SEYMOUR and ROBERT HARRIS. "Determinants of State-Local Government Expenditures and Intergovernmental Flows of Funds," National Tax Journal, Vol. 17 (March, 1963).

SAMUELSON, PAUL A. "Aspects of Public Expenditure Theories," Review of Economics and Statistics, Vol. 24 (November 1958).

SAMUELSON, PAUL A. "Diagrammatic Exposition of a Theory of Public Expenditure," Review of Economics and Statistics, Vol. 37 (November, 1955), pp. 350-56.

SAMUELSON, P. A. "A Pure Theory of Public Expenditures." Review of Economics and Statistics (November 1954).

SAMUELSON, P. A. "Pure Theory of Public Expenditure and Taxation" (unpublished paper, M.I.T., 1966).

SCHALLER, HOWARD G. (ed.). Public Expenditure Decisions in the Urban Community. (Washington: Resources for the Future, 1963).

SCHMANDT, HENRY J. and G. ROSS STEPHENS, "Local Government Expenditure Patterns in the United States," Land Economics, Vol. XXXIX, No. 4, (November, 1963) pp. 397-406.

SHAPIRO, S. "Some Socioeconomic Determinants of Expenditures for Education: Southern and Other States Compared," Comparative Education Review, Vol. 6 (October, 1962), pp. 160-166.

SHARKANSKY, IRA. "Some More Thoughts About the Determinants of Government Expenditures." National Tax Journal, Vol. 20, No. 2 (June 1967).

SHARKANSKY, I. Spending in the American States (Chicago: Rand McNally and Company, 1968).

SHOUP, CARL S. "Standards for Distributing a Free Governmental Service: Crime Prevention," Public Finance, Vol. 19 (1964).

SUNLEY, E. M., Jr. The Determinants of Government Expenditures Within Metropolitan Regions. (Unpublished doctoral dissertation, The University of Michigan, 1968).

TIEBOUT, CHARLES M. "A Pure Theory of Local Expenditures," Journal of Political Economy, Vol. 64 (October 1956).

WILL, ROBERT E. "Scalar Economies and Urban Service Requirements," Yale Economie Essays, Vol. 5 (Spring 1965).

WRIGHT, D. S. Trends and Variations in Local Finances: The Case of Iowa Counties (Iowa City: Institute of Public Affairs, University of Iowa, 1965).

II.   URBAN FISCAL POLICY

D. THE BUDGETARY PROCESS

ANTON, THOMAS. Budgeting in Three Illinois Cities (Urbana: Institute of Government and Public Affairs, University of Illinois, 1964).

ANTON, THOMAS. Politics of State Expenditure in Illinois (Urbana: University of Illinois Press, 1966).

BANFIELD, E. C. Political Influence. (Glencoe: The Free Press, 1961).

BANFIELD, E. C. and WILSON J. Q. City Politics (Cambridge: Harvard University Press, 1965).

BARBER, JAMES D. Power in Committees: An Experiment in the Governmental Process (Chicago: Rand McNally, 1966).

BATOR, F. M. "Budgetary Reform: Notes on Principles and Strategy," Public Finance and Fiscal Policy," Scherer, J. (ed.). (Boston: Houghton Mifflin, 1966).

BAUMOL, WILLIAM L. "Urban Services. Interactions of Public and Private Decisions," in Howard G. Schaller, Public Expenditure Decisions in the Urban Community (Baltimore: Johns Hopkins Press for Resources for the Future, 1963).

BURKHEAD, J. Government Budgeting (New York: Wiley, 1956).

CARLSON, JACK W. "Federal Support for State and Local Government Planning, Programming, and Budgeting," in U.S. Congress, Joint Economic Committee, Innovations in Planning, Programming, and Budgeting in State and Local Governments, 91st Congress, 1st session (Washington, D.C.: U.S. Government Printing Office, 1969).

CHINITZ, BENJAMIN and CHARLES M. TIEBOUT. "The Role of Cost-Benefit Analysis in the Public Sector of Metropolitan Areas," in Julius Margolis (ed.), The Public Economy of Urban Communities (Baltimore: Johns Hopkins Press for Resources for the Future, 1965).

COLM, GERHARD and PETER WAGNER. "Some Observations of the Budget Concept," in Scherer, J. (ed.), Public Finance and Fiscal Policy (Boston: Houghton Mifflin, 1966).

COTTON, J. F. and HARRY P. HATRY. Program Planning for State, County, and City. State-Local Finances Project of the George Washington University (Washington, D.C.: 1966).

CRECINE, JOHN P. Government Problem-Solving: A Computer Simulation of Municipal Budgeting (Chicago: Rand McNally and Company, 1969).

CRECINE, JOHN P. PPBS Implementation: Some Considerations Based on the Federal Experience. Paper presented for American Management Association Seminar on Financial Planning and Control for the Governmental Unit,

Chicago, Illinois, April 28, 1969. (University of Michigan, Institute of Public Policy Studies Discussion Paper No. 5.)

DAVIS, OTTO A. "Empirical Evidence of Political Influences Upon the Expenditure Policies of Public Schools," in Julius Margolis, (ed.), The Public Economy of Urban Communities (Washington: Resources for the Future, 1965).

DAVIS, OTTO A., M. DEMPSTER and A. WILDAVSKY. "A Theory of the Budgetary Process," American Political Science Review. (September 1966).

DAVIS, OTTO A., M. DEMPSTER and A. WILDAVSKY. "On the Process of Budgeting: An Empirical Study of Congressional Appropriations," in Tullock, G. (ed.), Papers on Non-Market Decision Making, Thomas Jefferson Center, University of Virginia, 1967.

DOWNS, ANTHONY. Inside Bureaucracy (Boston: Little, Brown and Company, 1967).

DREW, ELIZABETH B. "PPBS: Its Scope and Limits, HEW Grapples with PPBS," The Public Interest, No. 8 (Summer, 1967) pp. 9-29.

FENNO, RICHARD F., Jr. The Power of the Purse: Appropriations Politics in Congress. (Boston: Little, Brown and Company, 1966).

FREDERICKSON, H. GEORGE. "Exploring Urban Priorities, The Case of Syracuse," Urban Affairs Quarterly, Vol. 5, No. 1 (September 1969), pp. 31-43.

GERWIN, DONALD. Budgeting Public Funds: The Decision Process in an Urban School District (Madison: The University of Wisconsin Press, 1969).

GORHAM, WILLIAM. "PPBS: Its Scope and Limits, Notes of a Practitioner," The Public Interest, No. 8 (Summer, 1967), pp. 4-8.

GRIZZLE, GLORIA. "PPBS in Dade County: Status of Development and Implementation," in U.S. Congress, Joint Economic Committee, Innovations in Planning, Programming and Budgeting in State and Local Governments, 91st Congress, 1st Session (Washington, D.C.: U.S. Government Printing Office, 1969).

GRODZINS, MORTON. "Centralization and Decentralization in the American Federal System," in A Nation of States: Essays on the American Federal System (New York: Rand McNally and Company, 1963), pp. 1-23.

HATRY, HARRY P. and JOHN F. COTTON. Program Planning for State, County, City (Washington, D.C.: The George Washington University, 1968).

HIRSCH, WERNER Z. "Input-Output Techniques for Urban Government Decisions," Papers and Proceedings, American Economic Review, Vol. LVIII, No. 2 (May, 1968) pp. 162-181.

HIRSCH, WERNER Z. (ed.). Regional Accounts for Policy Decisions (Baltimore: The Johns Hopkins Press, for Resources for the Future, Inc. 1966).

HITCH, CHARLES. "The Uses of Economics," in Research for Public Policy (Washington, D.C.: The Brookings Institution, 1961) p. 104.
1961) p. 104.

HOLLINGER, L. S. "Changing Rules of the Budget Game: The Development of a Planning-Programming-Budgeting System for Los Angeles County," in U.S. Congress, Joint Economic Committee, Innovations in Planning, Programming, and Budgeting in State and Local Government, 91st Congress 1st session. (Washington, D.C.: U.S. Government Printing Office, 1969).

HORTON, ROBERT A. "Planning, Programming, and Budgeting in Metropolitan Nashville-Davidson County, Tennessee," in U.S. Congress, Joint Economic Committee, Innovations in Planning, Programming, and Budgeting in State and Local Governments, 91st Congress, 1st session, (Washington, D.C.: U.S. Government Printing Office, 1969).

JONES, MARTIN V. A Regional Project Evaluation Model (Bedford, Massachusetts: The Mitre Corporation, 1968).

LECHT, LEONARD A. Goals, Priorities and Dollars (New York: Free Press, 1966).

LICHFIELD, NATHANIEL and JULIUS MARGOLIS. "Benefit-Cost Analysis as a Tool in Urban Government Decision Making", in Howard G. Schaller (ed.), Public Expenditure Decisions in the Urban Community (Baltimore: Johns Hopkins Press for Resources for the Future, 1963).

LYDEN, TREMONT J. and ERNEST G. MILLER. (eds.). Planning, Programming, Budgeting: A Systems Approach to Management (Chicago: Markham, 1967).

McKEAN, R. N. "Evaluating Alternative Expenditure Programs," in J. M. Buchanan (ed.), Public Finances: Needs, Sources, and Utilization (Princeton: Princeton University Press, 1961).

MAO, JAMES C. T. "Efficiency in Public Urban Renewal Expenditures Through Benefit-Cost Analysis," Journal of American Institute of Planners, Vol. 32 (March 1966).

MARVIN, K. E. and A. M. ROUSE. "The Status of PPB in Federal Agencies: A Comparative Perspective," in U.S. Congress, Joint Economic Committee, The Analysis and Evaluation of Public Expenditures: The PPB System, 91st Congress, 1st session, Vol. 3 (Washington, D.C.: Government Printing Office, 1969), pp. 801-814.

MUSHKIN, SELMA J. "PPB in Cities," Public Administration Review, 29 (March/April, 1969) pp. 167-178.

MUSHKIN, SELMA J. and ROBERT F. ADAMS. "Emerging Patterns of Federalism," National Tax Journal, XIX (September 1966).

MUSHKIN, SELMA J. and BRIAN HERMAN. The Search For Alternatives (Washington, D.C.: The George Washington University).

MUSHKIN, SELMA J., HARRY P. HATRY, JOHN F. COTTON, et al. Implementing PPB in State, City, and County (Washington, D.C.: The George Washington University, 1969).

NOVICK, DAVID (ed.). Program Budgeting, Program Analysis, and the Federal Budget. (Cambridge: Harvard University Press, 1965).

PREST, A. and R. TURVEY, "Cost-Benefit Analysis: A Survey," Economic Journal, Vol. 75 (1965), pp. 683-735.

RIVLIN, ALICE M. "The Planning, Programming and Budgeting System in the Department of Health, Education and Welfare: Some Lessons From Experience," in U.S. Congress, Joint Economic Committee, The Analysis and Evaluation of Public Expenditures: The PPB System, Vol. 3, 91st Congress, 1st session (Washington, D.C.: U.S. Government Printing Office) p. 917.

ROTHENBERG, JEROME. Economic Evaluation of Urban Renewal: Conceptual Foundation of Benefit-Cost Analysis (Washington: Brookings Institution, 1967).

ROTHENBERG, JEROME. "A Model of Economic and Political Decision Making," in Julius Margolis (ed.), The Public Economy of Urban Communities (Baltimore: Johns Hopkins Press for Resources for the Future, 1965).

SCHALLER, HOWARD G. (ed.). Public Expenditure Decisions in the Urban Community (Washington: Resources for the Future, Inc., 1963).

State-Local Finances Project. PPB Note II (Washington, D.C.: The George Washington University, 1968).

TIEBOUT, C. M. "An Economic Theory of Fiscal Decentralization" in Universities–National Bureau of Economic Research, Public Finances: Needs, Sources and Utilization (New York: 1961).

U.S. Joint Economic Committee. The Analysis and Evaluation of Public Expenditures: The PPB System (Washington, D.C.: U.S. Government Printing Office, 1966).

U.S. Joint Economic Committee. Innovations in Planning, Programming, and Budgeting in State and Local Governments, 91st Congress, 1st session (Washington, D.C.: U.S. Government Printing Office, 1969).

WEIDENBAUM, MURRAY L. "Program Budgeting–Applying Economic Analysis to Government Expenditure Decisions", Business and Government Review, Vol. 7, No. 4 (July/August, 1966), pp. 22-31.

WHITELAW, W. An Econometric Analysis of a Municipal Budgetary Process Based on Time Series Data, Doctoral Dissertation at Massachusetts Institute of Technology, (discussion paper under the Harvard Program on Regional and Urban Economics, September 1968).

WILDAVSKY, AARON. "PPBS. Its Scope and Limits, The Political Economy of Efficiency," The Public Interest, No. 8 Summer, 1967. pp. 30-48.

WILDAVSKY, AARON. "The Political Economy of Efficiency: Cost-Benefit Analysis, Systems Analysis, and Program Budgeting," Public Administration Review, XXVI (December, 1966).

WILDAVSKY, AARON. The Politics of the Budgetary Process (Boston: Little, Brown, 1964).

WILDAVSKY, A. "Rescuing Policy Analysis From PPBS," Public Administration Review, 29 (March/April, 1969), pp. 194-195.

WILDAVSKY, AARON and ARTHUR HAMMAN. "Comprehensive Versus Incremental Budgeting in the Department of Agriculture", Administrative Science Quarterly, Vol. 10, No. 3 (December 1965), pp. 321-346.

III. SUB-ECONOMIES

A. GENERAL

ABERBACH, JOEL D. and JACK L. WALKER. Race and the Urban Community (Little, Brown, 1970).

ALLPORT, G. W. The Nature of Prejudice (Garden City: Doubleday, 1958).

BAUMOL, WILLIAM J. Welfare Economics and the Theory of the State (Cambridge: Harvard University Press, 1952, Revised Second edition, 1965).

BECKER, GARY S. The Economics of Discrimination (Chicago: University of Chicago Press, 1957).

BLOOMBERG, WARNER Jr. and HENRY J. SCHMANDT (eds.). Power, Poverty, and Urban Policy (Beverly Hills: California: Sage Publications, 1968).

BOWEN, HOWARD. Toward Social Economy (New York: Rinehart, 1948).

BRAGER, G. A. and F. PURCELL. Community Action Against Poverty (New Haven, Connecticut: College and University Press, 1967).

BRIMMER, ANDREW J. "The Negro in the National Economy," in John Davis (ed.), The American Negro Reference Book (Englewood Cliffs, New Jersey: Prentice-Hall, 1966).

BROOKS, MICHAEL P. and MICHAEL A. STEGMAN. "Urban Social Policy, Race, and the Education of Planners," Journal of the American Institute of Planners, 34 (July, 1967), pp. 275-86.

BROWN, CLAUDE. Manchild in the Promised Land (New York: Macmillan, 1965).

CAMPBELL, A. and H. SCHUMAN. Racial Attitudes in Fifteen American Cities: A Preliminary Report Prepared for the National Advisory Commission on Civil Disorders, (Institute for Social Research, The University of Michigan, Ann Arbor, Michigan, 1968).

CAMPBELL, ALAN K., et al., The Negro in Syracuse (University College of Syracuse University, 1964).

CAPLOVITZ, DAVID. The Poor Pay More (New York: The Free Press of Glencoe, 1963).

CLARK, KENNETH B. "Sex, Status and Underemployment of the Negro Male," in Arthur M. Ross and Herbert Hill, Employment, Race and Poverty (New York: Harcourt, Brace and World, 1967).

DRAKE, S. and H. R. CLAYTON. Black Metropolis: A Study of Negro Life in a Northern Ghetto (New York: Harcourt-Brace, 1945).

DuBOIS, W. E. B. The Philadelphia Negro (Philadelphia: Publications of the University of Pennsylvania, 1899).

DUNCAN, B. and O. D. DUNCAN. The Negro Population of Chicago (Chicago: University of Chicago Press, 1957).

DUNCAN, B. and P. M. HAUSER. Housing a Metropolis–Chicago (Glencoe: The Free Press, 1960).

DUNCAN, O. D. and S. LIEBERSON. "Ethnic Segregation and Assimilation," American Journal of Sociology, Vol. 64 (January, 1959), pp. 364-374.

FRANKLIN, JOHN H. From Slavery to Freedom, A History of American Negroes (New York: Alfred A. Knopf, 1956).

FLEISHER, B. M. The Economics of Delinquency (Chicago: Quadrangle Books, 1966).

GLAZER, NATHAN and DANIEL MOYNIHAN. Beyond the Melting Pot (Cambridge: M.I.T. Press, 1965).

GOODMAN, L. H. (ed.). Economic Progress and Social Welfare (New York, Columbia University Press, 1966).

GORDON, D. M. "Income and Welfare in New York City," The Public Interest (Summer, 1969).

GORDON, DANIEL N. The Social Bases of Municipal Government (Ph.D. dissertation, University of Wisconsin, 1967).

GORDON, KERMIT (ed.). Agenda For The Nation (Washington, D.C.: Brookings Institution, 1968).

GORDON, MARGARET S. (ed.). Poverty in America (San Francisco: Chandler, 1965).

GREEN, CONSTANCE McL. The Secret City: A History of Race Relations in the Nation's Capital. (Princeton, New Jersey: Princeton University Press, 1967).

GRODZINS, MORTON. The Metropolitan Area as a Racial Problem (Pittsburgh: University of Pittsburgh Press, 1959).

Harlem Youth Opportunities Unlimited, Inc. Youth in the Ghetto: A Study of the Consequences of Powerlessness (Harlem Youth Opportunities Unlimited, Inc., 1964).

HARNS, B. "Some Problems in the Theory of Intra-Urban Location," Operations Research, Vol. 9 (September-October 1961), pp. 695-721.

HUBBARD, HOWARD. "Five Long Hot Summers and How They Grew," The Public Interest, No. 12 (Summer, 1968), pp. 3-24.

HUNTER, DAVID R. The Slums, Challenge and Response (New York: Free Press of Glencoe, 1964).

HYMAN, H. H. and P. B. SHEATSLEY. "Attitudes toward Desegregation," Scientific American, Vol. 211 (July 1964), pp. 16-23.

JOHNSON, HARRY G. "The Economic Approach to Social Questions," The Public Interest, No. 112 (Summer, 1968) pp. 68-79.

KAIN, JOHN F. "Housing Segregation, Negro Employment and Metropolitan Decentralization," Quarterly Vol. 82 Journal of Economics, (May 1968), pp. 175-198.

KAIN, JOHN F. and JOSEPH J. PERSKY. "Alternatives to the Gilded Ghetto," paper prepared for the Economic Development Administration Research Conference in Washington D.C., February, 1968. Also available as Discussion Paper No. 21, Program on Regional and Urban Economics, Harvard University (September, 1967).

KOLKO, GABRIEL. Wealth and Power in America: An Analysis of Social Class and Income Distribution. (New York: Praeger, 1962).

LEVITAN, SAR A. The Great Society's Poor Law: A New Approach to Poverty (Baltimore: Johns Hopkins Press, 1969).

LEWIS, OSCAR. La Vida (New York: Random House, 1965).

LIEBERSON, S. Ethnic Patterns in American Cities (Glencoe: The Free Press, 1963).

LIEBOW, ELLIOTT. Tally's Corner, A Study of Street-Corner Men (Boston: Little, Macmillan, 1965).

LOEWENSTEIN, LOUIS K. The Location of Work Places in Urban Areas (New York: The Scarecrow Press, 1965).

MALCOLM X. Autobiography (New York: Grove Press, 1965).

MOONEY, JOSEPH D. "Housing Segregation, Negro Employment and Metropolitan Decentralization," Quarterly Journal of Economics, Vol. 83, (May 1969), pp. 299-311.

MOYNIHAN, DANIEL P. (ed.). On Understanding Poverty (New York: Basic Books, 1969).

MYRDAL, GUNNAR. An American Dilemma (New York: Harper and Row, 1944).

Newsweek. "Report from Black America," Vol. 83 (June 30, 1969), pp. 17-39.

OLSON, MANCUR, Jr. "Economics, Sociology and the Best of All Possible Worlds," The Public Interest, No. 12 (Summer, 1968), pp. 96-118.

PARSONS, TALCOTT (ed.). The Negro American (Boston: Houghton Mifflin and Cambridge: Riverside Press, 1966).

PASCAL, ANTHONY H. (ed.) The American Economy in Black and White: Essays in Racial Discrimination in Economic Life (forthcoming: 1970).

PINKNEY, A. Black Americans (Englewood Cliffs: Prentice-Hall, 1969).

RAINE, WALTER J. "Los Angeles Riot Study: The Ghetto Merchant Survey" (Los Angeles: Institute of Government and Public Affairs, University of California, 1967).

RAINWATER, LEE and WILLIAM L. YANCEY. The Moynihan Report and the Politics of Controversy (Cambridge, Massachusetts: M.I.T. Press, 1967).

SANDALOW, TERRANCE, et al. Government in Urban Areas (forthcoming: 1970).

SCHELLING, T. C. Models of Segregation, RM-6014-RC (Santa Monica: The RAND Corporation, 1969).

SCHORR, ALVIN L. Slums and Social Insecurity (Washington: Social Security Administration, 1963).

SCHULTZ, T. W. "Investing in Poor People: An Economist's View," American Economic Review, Vol. 55 (May 1965).

SELIGMAN, BEN B. (ed.). Poverty as a Public Issue. (New York: The Free Press, 1965).

SHEATSLEY, P. "White Attitudes toward the Negro," Daedalus, Vol. 95 (1968), pp. 217-38.

THOMAS, PERI. Down These Mean Streets (New York: Knopf, 1967).

U.S. Bureau of the Census. "Characteristics of Selected Neighborhoods in Cleveland, Ohio: April 1965," Current Population Reports Series P-23, No. 21 (Washington: U.S. Government Printing Office, January 23, 1967).

U.S. Bureau of Labor Statistics and Bureau of the Census. Social and Economic Conditions of Negroes in the United States (B.L.S. Report No. 32, CPS Report No. 24) (Washington: U.S. Government Printing Office, 1967).

U.S. National Advisory Commission on Civil Disorders. Report (New York: Bantam Books, 1968).

U.S. Office of Policy Planning and Research. The Negro Family (The "Moynihan Report") (Washington: U.S. Department of Labor, 1965).

U.S. President's Commission on Law Enforcement and Administration of Justice. Task Force Report: Crime and Its Impact—An Assessment (Washington: U.S. Government Printing Office, 1967).

VALENTINE, C. A. Culture and Poverty: Critique and Counter Proposals (Chicago: University of Chicago Press, 1968).

WEAVER, R. The Negro Ghetto (New York: Harcourt, Brace, 1948).

WEISSBOURD, BERNARD. Segregation, Subsidies and Megapolis. Occasional paper No. 1 on The City, (Center for the Study of Democratic Institutions, 1964).

WILSON, JAMES Q. "The Urban Unease: Community vs. City," The Public Interest, No.72 (Summer, 1968) pp. 25-39.

III. SUBECONOMIES

B. WELFARE

COLEMAN, J. S. "The Possibility of a Social Welfare Function." American Economic Review (December 1966).

GALLAWAY, LOWELL E. "Negative Income Tax Rates and The Elimination of Poverty," National Tax Journal, Vol. 19, No. 3. (September, 1966), pp. 298-307.

GREEN, C. Negative Income Taxes and the Poverty Program (Washington, D.C.: Brookings Institution, 1967).

KESSELMAN, JONATHAN. "The Negative Income Tax and the Supply of Work Effort: Comment," National Tax Journal, Vol. 22, No. 3, (September, 1969), pp. 411-416.

LEIBENSTEIN, HARVEY. "Long-run Welfare Criteria," in Julius Margolis (ed.), The Public Economy of Urban Communities (Baltimore: Johns Hopkins Press for Resources for the Future, 1965).

MORGAN, JAMES N. et al., Income and Welfare in the United States, Survey Research Center, University of Michigan (New York: McGraw Hill, 1962).

MOYNIHAN, DANIEL P. "The Crises in Welfare," The Public Interest, No. 10 (Winter, 1968), pp. 3-29.

TOBIN, J., J. A. PECHMAN, and P. MIESZKOWSKI, "Is a Negative Income Tax Practical?" Yale Law Journal, (1967), pp. 1-27.

VADAKIN, JAMES C. "A Critique of the Guaranteed Annual Income," The Public Interest No. 11 (Spring, 1968), pp. 53-66.

III. SUBECONOMIES

C. HOUSING

ABRAMS, C. Forbidden Neighbors: A Study of Prejudice in Housing (New York: Harper and Brothers, 1955).

ANDERSON, M. The Federal Bulldozer (Cambridge: MIT Press, 1964).

BAILEY, MARTIN J. "Note on the Economics of Residential Zoning and Urban Renewal," Land Economics Vol. 35 (August 1959), pp. 288-90.

——— "Effects of Race and Other Demographic Factors on the Values of Single-Family Houses," Land Economics, Vol. 42 (May 1966), pp. 215-20.

BEYER, GLENN. Housing and Society. Revised Edition (New York: Macmillan, 1965).

CARROLL, J. D. "Some Aspects of the Home-work Relationship of Industrial Workers," Land Economics, Vol. 25 (November 1949), pp. 414-422.

DOWNS, A. "Moving Toward Realistic Housing Goals," in K. Gordon (ed.), Agenda for the Nation (Washington, D.C.: Brookings Institution, 1968).

FARLEY, REYNOLDS and KARL E. TAEUBER. "Population Trends and Residential Segregation Since 1960," Science, Vol. 159, No. 3818 (March 1, 1968), pp. 953-956.

FRIEDEN, BERNARD J. "Housing and National Urban Goals: Old Policies and New Realities," in James Q. Wilson (ed.), The Metropolitan Enigma, (New York: Chamber of Commerce of the United States, 1967).

GLAZER, NATHAN. "Housing Problems and Housing Policies," The Public Interest, No. 7 (Spring, 1967) pp. 21-51.

GLAZER, N. and D. McENTIRE. Studies in Housing and Minority Groups (Berkeley: University of California Press, 1960).

HARTMAN, CHESTER. "The Limitations of Public Housing: Relocation Choices in a Working-Class Community," Journal of the American Institute of Planners, Vol. 29 (November 1963), pp. 283-296.

HERBERT, J. D. and B. H. STEVENS. "A Model for the Distribution of Residential Activities in Urban Areas," Journal of Regional Science, Vol. 2 (Fall 1960), pp. 21-36.

KAIN, J. "The Effect of Housing Segregation." Savings and Residential Financing Conference Proceedings. (Chicago: U.S. Savings and Loan League, 1969).

LANSING, JOHN B. New Homes and Poor People (Ann Arbor, Michigan: Institute for Social Research, Survey Research Center, 1969).

LAURENTI, L. Property Values and Race: Studies in Seven Cities (Berkeley: University of California Press, 1960).

LOWRY, I. S. "Housing" in A. H. Pascal (ed.), Cities in Trouble: An Agenda for Urban Research (RM-5603-RC) (Santa Monica: The RAND Corporation, 1968).

McENTIRE, D. Residence and Race: Final and Comprehensive Report to the Commission on Race and Housing (Berkeley: University of California Press, 1960).

MUTH, R. F. Cities and Housing (Chicago: The University of Chicago Press, 1969).

MUTH, RICHARD F. "The Distribution of Population within Urban Areas," in Robert Ferber (ed.), Determinants of Investment Behavior. (New York: National Bureau of Economic Research, 1967).

MUTH, R. F. "Urban Residential Land and Housing Markets," in Harvey S. Perloff and Lowdon Wingo, Jr. (eds.), Issues in Urban Economics (Baltimore: The Johns Hopkins Press for Resources for the Future, Inc., 1968).

NEVITT, ADELA ADAM, (ed.), The Economic Problems of Housing (London: Macmillan, 1967).

OLSEN, EDGAR O. A. Welfare Economic Evaluation of Public Housing, doctoral dissertation, Rice University, Houston, Texas (May 1968).

PASCAL, A. H. The Economics of Housing Segregation, RM-5510-RC (Santa Monica: The RAND Corporation, 1967).

STERNLIEB, GEORGE. The Tenement Landlord, 2nd Edition (New Brunswick, New Jersey: Rutgers University Press, 1969).

SUDMAN, S., N. M. BRADBURN and G. GOCKEL. "The Extent and Characteristics of Racially Integrated Housing in the United States," Journal of Business, Vol. 42 (January, 1969), pp. 50-87.

TAEUBER, KARL E. "The Effect of Income Distribution on Racial Residential Segregation," Urban Affairs Quarterly, Vol. 4 (September 1968), pp. 7-14.

TAEUBER, KARL E. and ALMA F. TAEUBER. Negroes in Cities, Residential Segregation and Neighborhood Change (Aldine Publishing Company, 1965).

U.S. Bureau of the Census. U.S. Census of Population and Housing: 1960 Final Report, PHC(1) (Washington: U.S. Government Printing Office, 1962).

U.S. Joint Economic Committee. Industrialized Housing (Washington: U.S. Government Printing Office, 1969).

U.S. President's Committee on Urban Housing. A Decent Home (Washington: U.S. Government Printing Office, 1969).

III. SUBECONOMIES

D. EMPLOYMENT

BATCHELDER, ALAN. "Decline in the Relative Income of Negro Men," Quarterly Journal of Economics, Vol. 78 (November 1964).

BLUESTONE, BARRY. "Low Wage Industries and the Working Poor," Poverty and Human Resources Abstracts, Vol. 3, No. 2 (March-April, 1968), pp. 1-13.

COHEN, BENJAMIN I., and ROGER G. NOLL. "Employment Trends in Central Cities," Social Science Discussion Paper 69-1 (Pasadena: California Institute of Technology, 1969).

DOERINGER, PETER. "Manpower Programs for Ghetto Labor Markets." Proceedings of the 21st Annual Winter Meeting of the Industrial Relations Research Association, 1969, pp. 257-267.

FLAIM, PAUL O. "Jobless Trends in Twenty Large Metropolitan Areas," Monthly Labor Review (May 1968).

GINZBERG, ELI and Associates. Manpower Strategy for the Metropolis (New York: Columbia University Press, 1968).

KAIN, JOHN F. "The Distribution and Movement of Jobs and Industry," in James Q. Wilson (ed.), The Metropolitan Enigma (Washington: Chamber of Commerce of the United States, 1967).

KAIN, J. F. "The Effect of the Ghetto on the Distribution and Level of Non-White Employment." Paper presented at the Annual Meetings of the American Statistical Association, Chicago, December 1964. (Published as P-3059 by The RAND Corporation, Santa Monica, California, January, 1965).

McCALL, J. Economics of Information and Job Search, RM-5745-OEO, (Santa Monica: The RAND Corporation, 1968).

NOLL, ROGER GORDON. "Central City–Suburban Employment Distribution," in President's Task Force on Urban Growth, Report (Washington: U.S. Government Printing Office, 1969).

PIORE, MICHAEL J. "Public and Private Responsibilities in On-the-Job-Training of Disadvantaged Workers," Department of Economics Working Paper No. 23, Department of Economics, M.I.T. (Mimeo, 1968).

REISS, ALBERT J., Jr. Occupations and Social Status (New York: Free Press, 1961).

SCHNORE, L. F. "The Separation of Home and Work," Social Forces, Vol. 32 (May, 1954), pp. 336-343.

SHEPPARD, HAROLD L. "The Nature of the Job Problem and the Role of New Public Service Employment" (Kalamazoo, Michigan: Upjohn Institute for Employment Research, 1969), pp. 2-11.

SHEPPARD, HAROLD L. and A. HARVEY BELITSKY. The Job Hunt (Baltimore: Johns Hopkins Press, 1966).

U.S. Bureau of Labor Statistics, Handbook of Labor Statistics—1968 (Washington: U.S. Government Printing Office, 1968).

U.S. Department of Commerce, Bureau of the Census. "Recent Trends in Social and Economic Conditions of Negroes in the United States," Current Population Reports, Series P-23, No. 26 (July 1968).

U.S. Department of Labor. "Employment Surveyed in Slum Areas of Six Large Cities," U.S. Department of Labor news release, February 20, 1969.

U.S. Joint Economic Committee. "Employment and Manpower Problems in the Cities," in Hearings, May 28-June 6, 1968 (Washington: U.S. Government Printing Office, 1968).

WHEELER, JAMES O. "Residential Location by Occupational Status," Urban Studies, Vol. 5 (February 1968), pp. 24-32.

III. SUBECONOMIES

E. EDUCATION

BISHOP, G. A. "Stimulative versus Substitutive Effects of School Aid in New England," National Tax Journal, Vol. 17 (June, 1964), pp. 133-143.

BOWLES, SAMUEL and HENRY LEVIN. "The Determinants of Scholastic Achievement," Journal of Human Resources (Winter, 1968).

BURKHEAD, JESSE, THOMAS G. FOX, and JOHN W. HOLLAND. Inputs and Outputs in Large City Education (Syracuse: Syracuse University Press, 1967).

COLEMAN, JAMES S. Equality of Educational Opportunity. U.S. Department of HEW, Office of Education (Washington: U.S. Government Printing Office, 1966).

GARMS, WALTER I. "The Financial Characteristics and Problems of Large City School Districts," Educational Administration Quarterly Vol. 3 (1967), pp. 14-27.

JAMES, H. THOMAS, JAMES A. KELLY and WALTER I. GARMS in Determinants of Educational Expenditures in Large Cities of the United States (Stanford, California: Stanford University, School of Education, 1966).

MINER, JERRY. Social and Economic Factors in Spending for Public Education. The Economics and Politics of Public Education Series, No. 11 (Syracuse, New York: Syracuse University Press, 1965).

REISS, ALBERT J., Jr. Schools in a Changing Society (New York: Free Press, 1965).

RIBICH, T. I., Education and Poverty (Washington, D.C.: Brookings Institution, 1968).

WEISBROD, BURTON A. "Geographic Spillover Effects and the Allocation of Resources to Education", in Julius Margolis (ed.), The Public Economy of Urban Communities (Baltimore: Johns Hopkins Press for Resources for the Future, 1965).

WILENSKY, GAIL R. State Aid and Educational Opportunity (Beverly Hills: Sage Publications, 1970).

III. SUBECONOMIES

F. TRANSPORTATION

FAUX, GEOFFREY. "Demands and Trends in Prices for Transit," Monthly Labor Review (March, 1964).

FITCH, LYLE C. and Associates. Urban Transportation and Public Policy (San Francisco: Chandler Publishing Company, 1964).

Franklin Institute, Philadelphia. Journal. New Concepts in Urban Transportation Systems (Lancaster, Pennsylvania, 1968).

HARRIS, BRITTON. Urban Transportation Planning: Philosophy of Approach (Philadelphia: University of Pennsylvania, Institute for Environmental Studies, 1966).

HEYMANN, H., Jr. Transportation Technolocy and the Real World (Santa Monica, California: The RAND Corporation, P-2755), 1963.

HOFFMAN, GEORGE A. Automobiles—Today and Tomorrow (Santa Monica, California: The RAND Corporation, 1963).

HOOVER, EDGAR M. "Motor Metropolis: Some Observations on Urban Transportation in America," (Center for Regional Economic Studies, University of Pittsburgh, 1965).

KAIN, JOHN F. "The Big Cities' Big Problem," Challenge, (September-October 1966).

KAIN, JOHN F. "A Contribution to the Urban Transportation Debate: An Econometric Model of Urban Residential and Travel Behavior," Review of Economics and Statistics, Vol. 46 (February, 1964), pp. 55-64.

KAIN, JOHN F. "The Journey-to-Work as a Determinant of Residential Location." Papers and Proceedings of the Regional Science Association, Vol. 9 (1962).

KUHN, TILLO E. "The Economics of Transportation Planning in Urban Areas," in Transportation Economics (New York: National Bureau of Economic Research, 1965).

KUHN, TILLO E. Public Enterprise Economic and Transportation Problems (Berkeley: University of California, 1962).

LANSING, JOHN B. Transportation and Economic Policy. (New York: Free Press, 1966).

LOEWENSTEIN, LOUIS K. The Location of Residences and Work Places in Urban Areas (New York: The Scarecrow Press, Inc., 1965).

MEYER, J. R. "Transportation in the Program Budget," in D. Novick (ed.), Program Budgeting, Program Analysis, and the Federal Budget (Cambridge, Harvard University Press, 1965).

MEYER, J. R., J. F. KAIN, and M. WOHL. The Urban Transportation Problem (Cambridge: Harvard University Press, 1965).

MOHRING, HERBERT. "Urban Highway Investments", in Robert Dorfman (ed.), Measuring Benefits of Government Investments (Washington: Brookings Institution, 1965).

MOSES, L. N. and H. F. WILLIAMSON, Jr. "Value of Time, Choice of Mode, and the Subsidy Issue in Urban Transportation," Journal of Political Economy, 71 June, 1963), pp. 247-64.

National Bureau of Economic Research. Transportation Economics (New York: National Bureau of Economic Research, 1965).

OLSSON, GUNNAR. Distance and Human Interaction: A Review and Bibliography (Bibliography Series, No. 2, Philadelphia: Regional Science Research Institute, 1965).

OWEN, WILFRED. The Metropolitan Transportation Problem (Washington: Brookings, 1966).

ROBERTS, PAUL O., DAVID KIESGE, and JOHN R. MEYER (eds.). Techniques of Transport Planning, Vol. II, Models for Transport System Simulation (Washington: Brookings Institute, forthcoming).

SCHMANDT, HENRY J. and G. ROSS STEPHENS. "Public Transportation and the Worker," Traffic Quarterly, Vol. XVII, No. 4. (October, 1963), pp. 573-583.

SHERMAN, ROGER. "A Private Ownership Bias in Transit Choice." American Economic Review, Vol. LVII, No. 5 (December, 1967).

SMERK, GEORGE M. (ed.). Readings in Urban Transportation (Bloomington: Indiana University Press, 1968).

SMITH, WILBUR and Associates. Future Highways and Urban Growth (New Haven, 1966).

SMITH, WILBUR and Associates. Transportation and Parking for Tomorrow's Cities (New Haven, 1966).

STROTZ, ROBERT H. Urban Transportation Parables, in Julius Margolis (ed.), The Public Economy of Urban Communities (Baltimore: Johns Hopkins Press for Resources for the Future, 1965).

WINGO, LOWDON, Jr. Transportation and Urban Land (Washington: Resources for the Future, 1961).

ZETTEL, RICHARD M. and RICHARD R. CAILL. Summary Review of Major Metropolitan Area Transportation Studies in the United States, Institute of Transportation and Traffic Engineering, (Berkeley, 1962).

ZWICK, C. J. "The Demand for Transportation Services in a Growing Economy." 42nd Annual Meeting of the Highway Research Board (January 1963). Also available as The RAND Corporation Publication No. P-2682.

III. SUBECONOMIES

G. URBAN RENEWAL, PLANNING, AND LAND USE

ALONSO, WILLIAM. Location and Land Use—Toward a General Theory of Land Rent (Cambridge: Harvard University Press, 1964).

ALONSO, WILLIAM and JOHN FRIEDMANN (eds.). Regional Development and Planning (Cambridge: M.I.T. Press, 1964).

ALTSHULER, ALAN. The City Planning Process: A Political Analysis (Ithaca, New York: Cornell University Press, 1965).

ANDERSON, MARTIN, The Federal Bulldozer: A Critical Analysis of Urban Renewal (Cambridge: M.I.T. Press, 1964).

BANFIELD, EDWARD C. and MARTIN MEYERSON. Politics, Planning and the Public Interest: The Case of Public Housing in Chicago (Glencoe: The Free Press, 1955).

BRIDGES, BENJAMIN. "State and Local Inducements for Industry," National Tax Journal, Vol. XVIII, No. 1 (March, 1965), pp. 1-14, and (June, 1968) pp. 175-192.

CHAPIN, F. STUART, Jr. "Existing Techniques of Shaping Urban Growth," in H. Wentworth Eldredge (ed.), Taming Megalopolis, Vol. II, (New York: Frederick A. Praeger, Inc., 1967).

CHAPIN, F. STUART, Jr. Urban Land Use Planning 2nd Edition (Urbana: University of Illinois, 1965).

CRECINE, JOHN P., OTTO A. DAVIS, and JOHN E. JACKSON. "Externalities in Urban Property Markets: Some Empirical Results and Their Implications for the Phenomenon of Municipal Zoning." (paper read at the annual meetings of the Southern Economies Association, 1966), Journal of Law and Economics (October, 1967).

DAHL, ROBERT A. "The Politics of Planning", International Social Science Journal, Vol. 11, No. 3 (1959).

DAVIES, J. CLARENCE, III. Neighborhood Groups and Urban Renewal (New York: Columbia University Press, 1966). Metropolitan Series No. 5.

DAVIS, OTTO A. and ANDREW B. WHINSTON. "The Economics of Urban Renewal," Law and Contemporary Problems, Vol. 26, No. 1 (Winter 1961), pp. 105-117.

DENTLER, ROBERT and PETER ROSSI. The Politics of Urban Renewal (New York: Free Press of Glencoe, Inc., 1961).

DYCKMAN, JOHN W. "State Development Planning in a Federal Economic System," in H. W. Eldredge, Taming the Megalopolis, Vol. II (New York: Praeger, 1967).

EICHLER, EDWARD P. The Community Builders (Berkeley: University of California Press, 1967).

FAGIN, HENRY. "The Policies Plan: Instrumentality for a Community Dialogue," Eleventh Annual Wherrett Lecture on Local Government at the University of Pittsburgh, 1965.

FRIEDEN, B. J. The Future of Old Neighborhoods: Rebuilding for a Changing Population (Cambridge: M.I.T. Press, 1964).

GLAZER, NATHAN. "Problems of Poverty and Race Confront the Planner," in American Institute of Planners, Proceedings of the 1963 Annual Conference (Washington: AIP, February, 1964) pp. 144-9.

GODSCHALK, DAVID and WILLIAM MILLS. "A Collaborative Approach to Planning through Urban Activities," Journal of the American Institute of Planners. Vol. XXXI (March, 1966).

GOODMAN, WILLIAM I. (ed.) Principles and Practice of Urban Planning (Washington: International City Managers' Association, 1968).

GREER, S. Urban Renewal and American Cities: The Dilemma of Democratic Intervention (Indianapolis: Bobbs-Merrill, 1965).

GRIER, EUNICE and GEORGE. "Human Needs and Public Policy in Urban Development," a paper delivered before the annual meeting of the American Orthopsychiatric Association, March 9, 1963.

GROBERG, ROBERT P. Centralized Relocation (Washington: National Association of Housing and Redevelopment Officials, 1969).

GROBERG, ROBERT P. "Urban Renewal Realistically Reappraised," Law and Contemporary Problems, Vol. 30, No. 1 (Winter 1965).

GROSS, BERTRAM M. Action Under Planning (New York: McGraw-Hill, 1967).

GROSSMAN, DAVID A. "The Community Renewal Program: Policy, Development, Progress, and Problems," Journal of the American Institute of Planners, Vol. 29, No. 4 (November 1963), pp. 259-269.

HAAR, CHARLES (ed.) Law and Land: Anglo-American Planning Practice (Cambridge: Harvard University Press and the M.I.T. Press, 1964).

HAMILTON, RANDY. "The Regional Commissions: A Restrained View", Public Administration Review, Vol. 28 (January/February 1968), pp. 19-21.

HARRIS, BRITTON. "Urban Development Models: New Tools for Planner." in Eldredge, H. Wentworth, (ed.), Taming Megalopolis, Vol. II. (New York: Frederick A. Praeger, 1967). Also in Journal of the American Institute of Planners, Special Issue, Vol. XXXI, No. 2 (May, 1965).

HEARLE, EDWARD F. R. "Regional Commissions: Approach to Economic Development", Public Administration Review, Vol. 28 (January/February 1968), pp. 15-18.

KAPLAN, MARSHALL (ed.). The Model Cities Program: a History and Analysis of the Planning Process in Three Cities (Washington: U.S. Government Printing Office, 1969).

LINDBLOOM, CARL G. and MORTON FARRAH. The Citizen's Guide to Urban Renewal (West Trenton, New Jersey: Chandler-Davis Publishing Company, 1968).

LOEWENSTEIN, LOUIS K. and CYRIL C. HERRMANN. "The San Francisco Community Renewal Program: A Summary," in H. W. Eldredge, (ed.), Taming the Megalopolis, Vol. II (New York: Frederick A. Praeger, 1967).

LOWRY, IRA S. "A Model of Metropolis." RAND Research Memorandum, RM-4035-RC, August, 1964.

LOWRY, IRA S. Seven Models of Urban Development: A Structural Comparison, P3673 (Santa Monica, California: The RAND Corporation, September 1967).

MORRIS, ROBERT (ed.), Centrally Planned Change: Prospects and Concepts (New York: National Association of Social Workers, 1964).

MOSES, LEON. "Location and the Theory of Production," Quarterly Journal of Economics, Vol. 73 (May, 1958) pp. 259-272.

MOSES, LEON and H. F. WILLIAMSON, Jr. "The Location of Economic Activity in Cities". Papers and Proceedings of the American Economic Association (May 1967).

MURPHY, RAYMOND E. The American City: An Urban Geography (New York: McGraw-Hill, 1966).

Oakland, City of. Application for Planning Grant Model Cities Program, 1967.

PARSONS, K. C. "The Role of Universities in City Renewal," in Eldredge H. W. Taming the Megalopolis, Vol. II (New York: Praeger, 1967). pp. 979-1001.

PERLOFF, HARVEY S. "New Directions in Social Planning." Journal of the American Institute of Planners, Vol. 31, No. 4 (November, 1965).

PRESCOTT, JAMES R. and WILLIAM C. LEWIS. "State and Municipal Locational Incentives: A Discriminant Analysis," National Tax Journal, Vol. 22, No. 3 (September, 1969), pp. 399-407.

REPS, JOHN W. "Requiem for Zoning," Planning, 1964 (Chicago: American Society of Planning Officials, 1964).

ROTHENBERG, JEROME. Economic Evaluation of Urban Renewal (Washington: Brookings Institution, 1967).

ROTHENBERG, JEROME. "Urban Renewal Programs," in Robert Dorfman, (ed.), Measuring Benefits of Government Investments. (Washington: The Brookings Institution, 1965).

SLAYTON, WILLIAM L. "The Operation and Achievements of the Urban Renewal Program," in James Q. Wilson (ed.), Urban Renewal: The Record and the Controversy (Cambridge: The M.I.T. Press, 1966), pp. 189-229.

SLAYTON, WILLIAM L. "Rehabilitation Potential Probed for Urban Renewal, Public Housing," Journal of Housing, Vol. 22, No. 11 (December 1965).

STEGER, WILBUR A. "Review of Analytic Techniques for the CRP," Journal of the American Institute of Planners, Special Issue, Vol. 31, No. 2 (May, 1965).

STOKES, CHARLES J., PHILIP MINTZ, and HANS VON GELDER, "Economic Criteria for Urban Redevelopment." American Journal of Economics and Sociology, Vol. 24, No. 3 (July 1965).

U.S. Joint Center for Urban Studies of M.I.T. and Harvard University. The Effectiveness of Metropolitan Planning. Prepared for the Senate Subcommittee on Intergovernmental Operations of the Committee on Governmental Operations (Washington: U.S. Government Printing Office, June 30, 1964).

VAN HUYCK, ALFRED and JACK HORNUNG. The Citizen's Guide to Urban Renewal (West Trenton, New Jersey: Chandler-Davis, 1962).

WEBBER, MELVIN M. "Prospect for Policies Planning," in Leonard J. Duhl (ed.), The Urban Condition (New York: Basic Books, Inc., 1963). pp. 319-330.

WILSON, JAMES Q. (ed.). Urban Renewal: The Record and the Controversy (Cambridge: M.I.T. Press, 1966).

WILSON, JAMES Q. "The War on Cities," The Public Interest, Vol. 1, No. 3 (Spring 1966).

WINGO, LOWDON, Jr. (ed.). Cities and Space: The Future Use of Urban Land. (Baltimore: The Johns Hopkins Press, for Resources for the Future, Inc., 1963).

III. SUBECONOMIES

H. POLLUTION

American Association for the Advancement of Science. Air Conservation Commission. Air Conservation (Washington: American Association for the Advancement of Science, Publication No. 80, 1968).

BATTAN, LOUIS J. The Unclean Sky: A Meteorologist Looks at Air Pollution (Garden City, New York: Anchor Books, 1966).

BEUCHE, ARTHUR M. Technology and the Pollution Problem (Schenectady, New York: General Electric Company, 1966).

BLUM, E. H. "Engineering and Public Affairs: Some Directions for Education and Research," Transactions of the New York Academy of Sciences, April, 1967. (The RAND Corporation No. P-3628).

BRICE, ROBERT M. and JOSEPH F. ROSELER. "The Exposure to Carbon Monoxide of Occupants of Vehicles Moving in Heavy Traffic," Journal of the Air Pollution Control Association, (November, 1966), pp. 597-600.

CLARKE, A. C. Profiles of the Future (New York: Harper and Row, 1963).

DAVIDSON, B. and R. W. BRADSHAW. "Thermal Pollution of Water Systems," Environmental Science and Technology, Vol. 1 (1967), pp. 618-630.

EDELSON, EDWARD and FRED WARSHOFSKY. Poisons in the Air (New York: Pocket Books, Inc., 1966).

ELDIB, I. ANDREW (ed.). "Technical and Social Problems of Air Pollution," Symposium Sponsored by the Metropolitan Engineers Council on Air Resources and Engineering Foundation, February 24, 1966.

GOLDMAN, MARSHALL I. (ed.). Controlling Pollution: the Economics of a Cleaner America (Englewood Cliffs, New Jersey: Prentice-Hall, Inc., 1967).

HALEY, THOMAS J. "Chronic Lead Intoxication from Environmental Contamination: Myth or Fact?," Archives of Environmental Health, Vol. 12 (June 1966), pp. 781-785.

HEAD, ROY. "Technological Change as it Relates to Air Pollution," in National Commission on Technology, Automation, and Economic Progress, "Technology and the American Economy," Vol. V (Washington: U.S. Government Printing Office, 1966).

"Health Hazards of Community Air Pollution," in Air Pollution—1966; hearings Before Subcommittee on Air and Water Pollution of the Committee on Public Works, U.S. Senate, 89th Congress, 2nd Session (Washington: U.S. Government Printing Office, 1966).

KNEESE, A. V. "Rationalizing Decisions in the Quality Management of Water Supply in Urban-Industrial Areas," in J. Margolis (ed.), The Public

Economy of Urban Communities (Washington, D.C.: Resources for the Future, 1965).

LOWRY, WIELEAN P. "The Climate of Cities," Scientific American, Vol. 217, No. 2 (August 1967), pp. 15-23.

National Academy of Sciences–National Research Council Committee on Pollution. "Waste Management and Control." Report to the Federal Council for Science and Technology (Washington: NAS-NRC, 1966).

RIDKER, R. G. and J. A. HENNING. "The Determinants of Residential Property Values with Special Reference to Air Pollution," Review of Economics and Statistics, Vol. 49 (1967), pp. 246-257.

SMITH, I. "Cost of Conventional and Advanced Treatment of Waste Water," Journal Water Pollution Control Federation, Vol. 40 (September, 1968), pp. 1546-1574.

SPORN, PHILIP. "Development of National Policies with Respect to Natural Gas, Coal, and Oil, and Nuclear and other New Sources of Power." National Conference on Air Pollution, December 12-14, 1966, Washington, D.C.

STARR, ROGER and JAMES CARLSON. "Pollution and Poverty. The Strategy of Cross-commitment," The Public Interest, No. 10 (Winter, 1968) pp. 104-131.

STERN, ARTHUR C. (ed.). Air Pollution (New York: Academic Press, 1962).

U.S. Congress, House of Representative, Committee on Science and Astronautics, Sub-Committee on Science, Research and Development. The Adequacy of Technology for Pollution Abatement, 89th Congress, 2nd Session, (Washington: U.S. Government Printing Office, 1966).

U.S. Congress, House of Representatives, Committee on Science and Astronautics, Sub-Committee on Science, Research and Development. Environmental Pollution: A Challenge to Science and Technology, 89th Congress, 2nd Session (Washington: U.S. Government Printing Office, 1966).

U.S. Congress, Senate Committee on Public Works. Air Pollution–1967 (Air Quality Act). Hearings before the Subcommittee on Air and Water Pollution, 90th Congress, 1st Session. (Washington: U.S. Government Printing Office, 1967).

U.S. Department of Health, Education and Welfare. Automotive Air Pollution, Reports to the U.S. Congress (Washington: U.S. Government Printing Office, 1965-1967).

U.S. Department of Health, Education and Welfare, Public Health Service, Division of Air Pollution. Air Pollution Publications; A Selected Bibliography, 1963-66. (Washington: U.S. Government Printing Office, 1966).

U.S. Department of Health, Education and Welfare, Public Health Service, Division of Air Pollution. The Federal Air Pollution Program (Washington: U.S. Government Printing Office, 1966).

U.S. Department of Health, Education and Welfare, Public Health Service, Division of Air Pollution. A Digest of State Air Pollution Laws (Washington: U.S. Government Printing Office, 1966).

U.S. Department of Health, Education and Welfare, Public Health Service, Division of Air Pollution. State and Local Programs in Air Pollution Control (Washington: U.S. Government Printing Office, 1966).

U.S. Department of Health, Education and Welfare, Task Force on Environmental Health and Related Problems. A Strategy for a Livable Environment (Washington: U.S. Government Printing Office, 1967).

U.S. National Commission on Technology, Automation and Economic Progress. Technology and the American Economy; A Report, Vol. 1 (Washington: U.S. Government Printing Office, 1966).

U.S. President's Science Advisory Committee, Environmental Pollution Panel. Restoring the Quality of Our Environment (Washington: U.S. Government Printing Office, 1965).

WOLOZIN, HAROLD (ed.). The Economics of Air Pollution (New York: W. W. Norton and Company, 1966).

IV. URBAN SERVICES

A. THE DECISION MAKING PROCESS

ABERBACH, J. D. and J. L. WALKER. "The Meanings of Black Power: A Comparison of White and Black Interpretations of a Political Slogan," American Political Science Review (forthcoming, June, 1970).

ABERBACH, J. D. and J. L. WALKER. "Political Trust and Racial Ideology." Institute of Public Policy Studies, The University of Michigan, Discussion Paper Number 8, 1969.

AGGER, ROBERT E., DANIEL GOLDRICH and BERT E. SWANSON. The Rulers and the Ruled (New York: John Wiley, 1964).

ALFORD, ROBERT L. "The Comparative Study of Urban Politics," in Henry Fagin and Leo F. Schnore (eds.). Urban Research and Policy Planning (Beverly Hills, California: Sage Publications, Inc., 1967).

ALFORD, ROBERT R. and HARRY M. SCOBLE. Bureaucracy and Participation: Political Cultures in Four Wisconsin Cities (Chicago: Rand McNally, forthcoming).

BANFIELD, EDWARD C. (ed.). Urban Government: A Reader in Administration and Politics (New York: Free Press, 1961).

BANFIELD, EDWARD C. and JAMES WILSON. City Politics (Cambridge: Harvard University and M.I.T. Press, 1963).

BANFIELD, EDWARD C. and JAMES WILSON, "Voting Behavior on Municipal Public Expenditures: A Study in Rationality and Self-Interest," in J. Margolis (ed.), The Public Economy of Urban Communities (Washington: Resources for the Future, 1965).

BOSKOFF, ALVIN and HARMON FEIGLER. Voting Patterns in a Local Election (Philadelphia: J. B. Lippincott Company, 1964).

BRAYBROOKE, DAVID and CHARLES E. LINDBLOM. A Strategy of Decision (New York: The Free Press of Glencoe, 1963).

BRETON, A. "A Theory of the Demand for Public Goods," The Canadian Journal of Economics and Political Science, Vol. 32 (1966), pp. 455-467.

BUCHANAN, JAMES M. Demand and Supply of Public Goods (Chicago: Rand McNally, 1968).

BUCHANAN, JAMES M. and GORDON TULLOCK. The Calculus of Consent (Ann Arbor: The University of Michigan Press, 1962).

CARTER, RICHARD F. and WILLIAM G. SAVARD. Influence of Voter Turnout on School Bond and Tax Elections, U.S. Department of Health, Education, and Welfare, Cooperative Research Monograph, No. 5 (Washington: Government Printing Office, 1961).

CLARKE, TERRY W. (ed.). Community Structure and Decision-Making: Comparative Analysis (San Francisco: Chandler Publishing Company, 1968).

CYERT, RICHARD M. and J. G. MARCH. A Behavioral Theory of the Firm (Englewood Cliffs, New Jersey: Prentice-Hall, 1963).

DAHL, ROBERT A. Who Governs? (New Haven: Yale University Press, 1961).

DOWNS, ANTHONY. An Economic Theory of Democracy (New York: Harper and Row, 1957).

DROR, YEHEZKEL. Public Policymaking Reexamined. (San Francisco: Chandler Publishing Company, 1968).

DYE, THOMAS R. Politics, Economics, and the Public: Policy Outcomes in the American States. (Chicago: Rand McNally and Company, 1966).

DYE, THOMAS R. "Urban Political Integration: Conditions Associated with Annexation in American Cities," Midwest Journal of Political Science, Vol. 8, (November, 1964).

FIELD, JOHN O. and RAYMOND E. WOLFINGER, "Political Ethos and the Structure of City Government," American Political Science Review, Vol. 60, No. 2 (June 1966).

FRIEDMAN, ROBERT, et al. "Administrative Agencies and the Publics They Serve", Public Administration Review, Vol. 26 (1966), pp. 192-204.

GREENSTONE, J. O. and P. PETERSON, "Reformers, Machines and the War on Poverty," Proceedings of 1966 American Political Science Association.

GREER, S. Metropolitics (New York: John Wiley, 1963).

HIRSCH, WERNER Z. "Local Versus Areawide Urban Government Services," National Tax Journal, Vol. 17 (December 1964).

HORTON, JOHN E., and WAYNE E. THOMPSON. "Powerlessness and Political Negativism: A Study of Defeated Local Referendums," The American Journal of Sociology, Vol. 57 (March, 1962), pp. 485-93.

LINDBLOM, C. E. "The Science of 'Muddling Through', Public Administration Review (Spring, 1959), pp. 79-88.

LONG, NORTON. "Political Science and the City," in Fagin and Schnore (eds.), Urban Research and Policy Planning (Beverly Hills, California: Sage Publications, Inc., 1967).

MANN, LAWRENCE D. "Studies in Community Decision-Making," Journal of the American Institute of Planners, 30 (February, 1964), pp. 58-65.

MARCH, JAMES G. and HERBERT A. SIMON, Organizations (New York: John Wiley and Sons, Inc., 1958).

MARTIN, ROSCOE C., et al. Decisions in Syracuse (New York: Anchor Books, 1965).

MERANTO, PHILIP. The Politics of Federal Aid to Education in 1965 (Syracuse: Syracuse University Press, 1967).

MITAU, G. THEODORE. State and Local Government: Politics and Processes. (New York: Charles Scribner's Sons, 1966).

MOYNIHAN, DANIEL P. "A Crisis of Confidence?," The Public Interest No. 7 (Spring, 1967) pp. 3-10.

MYRDAL, GUNNAR. The Political Element in the Development of Economic Theory (Cambridge, Mass.: Harvard University Press, 1954).

OLSON, MANCUR. The Logic of Collective Action (Cambridge: Harvard University Press, 1965).

POLSBY, NELSON. Community Power and Political Theory (New Haven: Yale University Press, 1963).

RANNEY, AUSTIN. Political Science and Public Policy (Chicago: Markham Publishing Company, 1968).

ROTHENBERG, J. "A Model of Economic and Political Decision Making," in J. Margolis (ed.), The Public Economy of Urban Communities (Washington: Resources for the Future, 1965).

SIMON, HERBERT. Administrative Behavior, 2nd Edition (New York: Macmillan Company, 1961).

STONE, CLARENCE N. "Local Referendums: An Alternative to the Alienated-Voter Model," The Public Opinion Quarterly, Vol. 29, No. 2 (1965).

UYEKI, E. S. "Patterns of Voting in a Metropolitan Area, 1938-62," Urban Affairs Quarterly, Vol. 1, No. 4 (June 1966), pp. 65-77.

WARD, BENJAMIN. "Majority Voting and Alternative Forms of Public Enterprise," in Julius Margolis (ed.), The Public Economy of Urban Communities (Baltimore: Johns Hopkins Press for Resources for the Future, 1965).

WILLIAMS, OLIVER P., HAROLD HERMAN, CHARLES S. LIEBMAN, and THOMAS R. DYE. Suburban Differences and Metropolitan Policies: A Philadelphia Story. (Philadelphia: University of Pennsylvania Press, 1965).

WILSON, JAMES Q. and EDWARD C. BANFIELD. "Public-Regardiness as a Value Premise in Voting Behavior," The American Political Science Review, 63 (December, 1964), pp. 876-87.

IV. URBAN SERVICES

B. PROVISION AND DELIVERY OF SERVICES

ADAMS, ROBERT F. "On the Variation in the Consumption of Public Services," in Harvey E. Brazer (ed.), Essays in State and Local Finance (Ann Arbor, Michigan: Institute of Public Administration, The University of Michigan, 1967), pp. 9-45.

CUROW, KENNETH J. Social Choice and Individual Values (New York: John Wiley and Sons, 1951; Revised Edition, 1963).

BLOOMBERG, WARNER, Jr. and FLORENCE W. ROSENSTOCK. "Who Can Activate the Poor? One Assessment of 'Maximum Feasible Participation'," in Warner Bloomberg, Jr. and Henry J. Schmandt (eds.), Power, Poverty and Urban Policy (Beverly Hills, California: Sage Publications, 1968).

BRAGER, GEORGE A. and HARRY SPECHT. "Social Action by the Poor: Prospects, Problems, and Strategies," in G. A. Brager and F. Purcell (eds.), Community Action Against Poverty (New Haven, Conn.: College and University Press, 1967).

CLARK, KENNETH. A Relevant War Against Poverty (New York: Metropolitan Applied Research Center, 1968).

CLEGG, R. K. The Administrator in Public Welfare (Springfield: Thomas, 1966).

COASE, R. H. "The Problem of Social Cost," Journal of Law and Economics, Vol. 3 (October, 1960), pp. 1-44.

DAVIES, BLEDDYN. Social Needs and Resources in Local Services (London: Michael Joseph Ltd., 1968).

DONOVAN, JOHN C. The Politics of Poverty (New York: Pegasus, 1967).

EISINGER, PETER K. "The Anti-Poverty Community Action Group as a Political Force in the Ghetto." (unpublished Ph.D. dissertation, Yale University, 1969).

FRIEDMAN, ROBERT S., LAWRENCE B. MOHR, and ROBERT M. NORTHROP. "Innovation in State and Local Bureaucracies," Paper presented to the Annual meeting of the American Political Science Association, New York, 1966.

HALLMAN, HOWARD. "The Community Action Program: An Interpretative Analysis," in Warner Bloomberg, Jr. and Henry J. Schmandt (eds.), Power, Poverty and Urban Policy (Beverly Hills, California: Sage Publications, 1968).

HIRSCH, WERNER Z. "Cost Functions of an Urban Government Service," Review of Economics and Statistics, Vol. 47 (February, 1965), pp. 87-92.

HIRSCH, WERNER Z. "Quality of Government Services", in Howard G. Schaller (ed.), Public Expenditure Decisions in the Urban Community (Baltimore: Johns Hopkins Press for Resources for the Future, 1963).

HIRSCH, WERNER Z. "The Supply of Urban Services," in H. S. Perloff and L. Wingo, Jr. (eds.), Issues in Urban Economics (Baltimore: Johns Hopkins Press for Resources for the Future, 1968).

Institute of Public Administration. Developing New York City's Human Resources, A report of a study group of the Institute of Public Administration to Mayor John V. Lindsay, Vol. 1 (June, 1966).

JANOWITZ, MORRIS, DEIL WRIGHT, and WILLIAM DELANEY. Public Administration and the Public-Perspectives Toward Government in a Metropolitan Community (Ann Arbor: Bureau of Government, University of Michigan, 1958).

KAIN, JOHN F. "Urban Form and the Costs of Urban Services." Harvard Program on Regional and Urban Economics, Discussion Paper No. 6, October 1966.

LAMPMANN, R. J. "How Much Does the American System of Transfers Benefit the Poor," in L. H. Goodman (ed.), Economic Progress and Social Welfare (New York: Columbia University Press, 1966), pp. 125-157.

MOYNIHAN, DANIEL P. Maximum Feasible Misunderstanding (New York: The Free Press, 1969).

MOYNIHAN, DANIEL P. "Three Problems in Combatting Poverty," in Margaret S. Gordon (ed.), Poverty in America (San Francisco: Chandler, 1965).

MOYNIHAN, DANIEL P. "What is 'Community Action'?" The Public Interest, Vol. 6 (Fall, 1966).

MUELLER, EVA. "Public Attitudes Toward Fiscal Programs," The Quarterly Journal of Economics, Vol. 77 (May, 1963), pp. 210-235.

NETZER, DICK, RALPH KAMINSKY and KATHERINE W. STRAUSS. Public Service in Older Cities (New York: Regional Plan Association, 1968).

PIVEN, FRANCES F. "Participation of Residents in Neighborhood Community Action Programs," Social Work (January, 1966).

REINER, JANET S., EVERETT REIMER, and THOMAS A. REINER. "Client Analysis and the Planning of Public Programs," Journal of the American Institute of Planners, 29 (November 1963), pp. 270-282.

SHAPIRO, H. "Measuring Local Government Output", National Tax Journal, Vol. 14 (December 1961).

TEMPLETON, FREDERIC. "Alienation and Political Participation," Public Opinion Quarterly, Vol. 30, No. 2 (1966), pp. 249-261.

U.S. Advisory Commission on Intergovernmental Relations, Performance of Urban Functions: Local and Areawide (Washington: Government Printing Office, September 1963).

THE AUTHORS

The Authors...

JOEL D. ABERBACH is Assistant Professor of Political Science and Research Associate, Institute of Public Policy Studies, University of Michigan. He is currently working on a comparative administration study under a Ford Foundation grant, and a longitudinal study of race relations in Detroit under a grant from the National Institutes of Mental Health. He is co-author with Jack L. Walker of the forthcoming book, *Race and the Urban Community* (Little, Brown).

DONALD J. BORUT is Assistant to the City Administrator for Program Planning and Development, Ann Arbor, Michigan. He was that city's Acting Human Relations Director in 1969. In addition, he is an Instructor, Bureau of Industrial Relations, University of Michigan.

JAMES M. BUCHANAN is Professor of Economics and General Director, Center for the Study of Public Choice at Virginia Polytechnic Institute. He has been a Fulbright Research Scholar in Italy, Fulbright Visiting Professor at Cambridge University, and a Visiting Professor at the London School of Economics. In 1958, he was a member of the Council of Economic Advisors and is a member of the Advisory Task Force on Urban Problems. His recent books include *Academia in Anarchy: An Economic Analysis* (Basic, 1970); *Cost and Choice: An Inquiry in Economic Theory* (Markham, 1969); and *Demand and Supply of Public Goods* (Rand McNally, 1968).

JOHN P. CRECINE is Director, Institute of Public Policy Studies, and Associate Professor of Political Science and Sociology at the University of Michigan. He has served as an Economist with the RAND Corporation, the U.S. Department of Commerce, and as a Consultant with the U.S. Bureau of the Budget. He is President of B.P.T., Inc., a consulting firm. In 1967, he won the L. D. White Award of the American Political Science Association. His publications include *Governmental Problem Solving: A Computer Simulation of Municipal Budgeting* (Rand McNally, 1969); *A Dynamic Model of Urban Structure* (RAND Corporation, 1968); articles in *Public Administration Review, Urban Affairs Quarterly, Management Science,* and other professional journals. He is a member of the City Planning Commission, Ann Arbor, Michigan.

OTTO A. DAVIS is Professor of Political Economy and Associate Dean, School of Urban and Public Affairs, Carnegie-Mellon University. His articles have appeared in many published books and professional journals in economics and political science.

PETER K. EISINGER is Assistant Professor of Political Science at the University of Wisconsin. His field of specialization is urban politics, and he is currently doing research on decentralization and control-sharing in cities, and on urban political protest behavior.

LYLE C. FITCH received his Ph.D. from Columbia University. He has been President of the Institute of Public Administration, New York, since 1961. In the period 1957-1961, he was successively First Deputy Administrator and Administrator of the City of New York. Prior to that, he was Senior Management Consultant in charge of fiscal and economic research, Office of the Mayor of New York City, and held other government and university posts. He has served as a Consultant to numerous federal, state, and government agencies, among them the federal Departments of Housing and Urban Development; Transportation; Health, Education and Welfare; and Treasury, the states of New York, Connecticut, Delaware, Rhode Island, and Pennsylvania, and the cities of New York, Philadelphia, Washington, D.C., and Atlanta. He is the author or co-author of several books and numerous articles on urban administration, economics, and public finance, including *Urban Transportation and Public Policy* (Chandler, 1964).

DANIEL R. FUSFELD is Professor of Economics at the University of Michigan. He has been an Economic Consultant to the Department of Labor and in 1963-1964, served as Director of the Manpower Evaluation Staff, U.S. Office of Education. In Michigan, he has served on the Governor's Committee on State Economic Problems and the Governor's Committee on Manpower and Training. His books include *The Age of the Economist* (Scott, Foresman, 1966; Morrow, 1968) and *The Economic Thought of Franklin D. Roosevelt and the Origins of the New Deal* (Columbia, 1956). His articles have appeared in *American Economic Review, Challenge, New Republic, Journal of Economic Education,* and other journals.

ARNOLD J. MELTSNER is Assistant Professor, Graduate School of Public Affairs, University of California at Berkeley. From 1958 to 1963, he was a Cost Analyst with the RAND Corporation and worked on the installation of a PPB system in the Department of Defense. During this period, he also conducted studies of the Air Force accounting system and developed cost models for systems analysis.

SELMA J. MUSHKIN is Economist, Senior Research Staff, the Urban Institute. She specializes in studies of state-local finance. From 1963 to 1968, she was Director of the State-Local Finance Project, George Washington University. Other experience includes Economic Consultant OECD, Paris; Consultant, U.S. Office of Education; Economist, Public Health Service; and Chief, Division of Financial Studies, Social Security Administration.

RICHARD MUTH is Professor of Economics at Washington University, St. Louis. He is Associate Editor of the Southern Economic Journal and, in 1969, was a member of the President's Task Force on Urban Renewal. He has taught at the Johns Hopkins University, University of Chicago, and Vanderbilt University. His articles have appeared in numerous books and journals including *Management Science, Econometrics, Journal of Finance,* and *The International Encyclopedia of the Social Sciences* (Macmillan, 1968). He is author of *Cities and Housing* (University of Chicago Press, 1969).

DICK NETZER is Dean, Graduate School of Public Administration, New York University. He has been a Consultant to the governments of Colombia and Puerto Rico, the Ford Foundation, the Committee for Economic Development, and the Advisory Commission on Intergovernmental Relations, among others. He is the author of *Economics of the Property Tax* (Brookings, 1966); co-author of *Public Service in Older Cities* (Regional Plan Association, 1968); and co-author of *Economic Aspects of Suburban Development* (State University of New York, 1969).

ROGER NOLL is Senior Fellow with the Brookings Institution. He was Senior Staff Economist with the President's Council of Economic Advisors from 1967 to 1969. He previously taught at the California Institute of Technology and Harvard University. His published papers include "Central City—Suburban Employment Distribution" published as part of the report from the President's Task Force on Urban Growth.

ANTHONY H. PASCAL is Project Director, Human Resources Studies, RAND Corporation and lecturer at the University of California at Los Angeles. In 1966-1967, he was Director of Research, Economic Development Administration, U.S. Department of Commerce. He has served as Consultant to the Bureau of the Budget, Commerce Department, Office of Economic Opportunity, and to the Urban Coalition. Books that he has edited include *The American Economy in Black and White: Essays in Racial Discrimination in Economic Life* (forthcoming, 1970) and *Thinking About Cities: New Perspectives on Urban Problems* (Dickenson, 1970). He has authored a number of books, the most recent of which is *The Economics of Housing Segregation* (RAND Corporation, 1967).

ALBERT J. REISS, JR. is Professor of Sociology and Chairman of the department at the University of Michigan where he is also Director of the Center for Research on Social Organization. He has taught at the University of Wisconsin, the State University of Iowa, Vanderbilt University, and the University of Chicago. He has been a Consultant to the National Advisory Commission on Civil Disorders and to the President's Commission on Law Enforcement and the Administration of Justice. His books include *Occupations and Social Status* (Free Press, 1961) and *Schools in a Changing Society* (Free Press, 1965). His articles have appeared in numerous books and journals including the *International Encyclopedia of the Social Sciences* (Macmillan, 1968).

JOHN RIEW is Associate Professor of Economics and member of the Institute of Public Administration, Pennsylvania State University. His published articles have appeared in *Land Economics, National Tax Journal, Review of Economics and Statistics,* and the *American Journal of Economics and Statistics.*

TERRANCE SANDALOW is Professor of Law at the University of Michigan. He was previously Professor of Law at the University of Minnesota. He has contributed articles to professional legal journals and is co-author of *Government in Urban Areas* (forthcoming, 1970).

WILBUR THOMPSON is Professor of Economics at Wayne State University. He has been a member of the Committee on Urban Economics of Resources for the Future, Inc., the Committee on Urbanization of the Social Science Research Council, and a Research Associate with RFF. His research papers have appeared in many leading professional journals and published books and he is the author of *Preface to Urban Economics* (Johns Hopkins, 1965).

JACK L. WALKER is Associate Professor of Political Science and Research Associate, Institute of Public Policy Studies, University of Michigan. He has conducted studies of innovation diffusion among state governments under a grant from the Social Science Research Council, and is currently doing research on race relations and policy-making in Detroit under a grant from the National Institute of Mental Health. His articles have appeared in numerous political science journals.

W. ED WHITELAW is Assistant Professor, Department of Economics and Research Associate, Bureau of Governmental Research, University of Oregon. Beginning in June, 1970, he will spend a year as Visiting Research Fellow, Social Science Division of the Institute for Development Studies, University College Nairobi (Kenya). He previously taught at MIT and has been a consultant with the New England Economic Research Foundation in Boston. His monograph, *The Determinants of Municipal Finance: A Time Series Analysis,* was published by the Harvard University Program on Urban and Regional Economics (Discussion Paper Number 57, November, 1969).

AARON WILDAVSKY is Dean, Graduate School of Public Affairs, University of California at Berkeley. He is co-author with Nelson Polsby of *Presidential Elections* (Scribner, 1968), and author of *The Politics of the Budgetary Process* (Little, Brown, 1964), among other books. His many articles have appeared in the leading social science journals.

GAIL WILENSKY is Executive Director of the Governor's Council of Economic Advisors in Maryland. In 1968-1969, she served as Staff Economist with the President's Commission on Income Maintenance Programs. Her book *State Aid and Educational Opportunity* was published by Sage Publications in the spring of 1970.

AN INTERNATIONAL JOURNAL OF THEORY, DESIGN AND RESEARCH

Editor-in-Chief

MICHAEL INBAR
Hebrew University and Johns Hopkins University

Managing Editor

CLARICE STOLL
*Center for Study of Social Organization of Schools,
Johns Hopkins University*

Editorial Board: Sarane S. Boocock, University of Southern
California / James S. Coleman, Johns Hopkins University/ Richard
Duke, University of Michigan / William Gamson, University of
Michigan / Peter House, Envirometrics / Allen Mazur, Stanford
University / E. O. Schild, Hebrew University / James Sullivan,
University of California, Santa Barbara / John Taylor,
University of Sheffield / Roland Werner, United States
International University / Gerald Zaltman, Northwestern University

Simulation Review Editors: Layman Allen and Fred Goodman,
University of Michigan

Simulation and Games is intended to provide a forum
for theoretical and empirical papers related to man, man-machine, and
machine simulations of social processes. The journal will publish
theoretical papers about simulations in research and teaching,
empirical studies, and technical papers about new gaming techniques.
Each issue will include book reviews, listings of newly available
simulations, and short "simulation reviews."

Frequency and price: *Simulation and Games* will be published
quarterly in March, June, September and December—commencing in 1970.
Subscription rates will be $15.00 annually—with a one-third
discount to faculty and members of professional organizations who
wish to purchase personal subscriptions, and a one-half discount
to (full-time) students. Special two- and three-year rates are
also available. Please add $1.00 per year additional for
postage outside the U.S. and Canada.

SAGE PUBLICATIONS / 275 S. Beverly Dr. / Beverly Hills, Calif. 90212

COMPREHENSIVE URBAN PLANNING

A Selective Annotated Bibliography with Related Materials

by MELVILLE C. BRANCH, Professor of
Planning, University of Southern California,
and a member of the Los Angeles City
Planning Commission.

This reference volume includes 1500 annotated bibliographic
citations to planning literature—ranging from general and
theoretical works to methodology; administration; decision-
making; specific in-puts from other disciplines (such as geol-
ogy, engineering, law, the social sciences); subsystems such as
housing, environment, land use, transportation, utilities and
services, and bibliographies.

Among the users for which these references are designed:
practitioners of city planning and their colleagues in professional and academic fields—including researchers, historians,
students of urban planning, and teachers preparing reading
lists for instructional purposes. The selected references also
constitute an excellent list of initial acquisitions for a new
city planning library; would support a graduate educational
program; and form an excellent nucleus and general content
guide for a large library on city planning.

Also included in the volume is a chapter on colleges and
universities in the U.S. and Canada with programs in Urban
and Regional Planning. In addition to author, title, and subject indexes, sources are given (including addresses) from
which items cited in the bibliography may be obtained.

SBN 8039-0041-4   484 pp.   February, 1970

SAGE Publications / 275 South Beverly Drive / Beverly Hills, Calif. 90212

The Metropolitan Community

ITS PEOPLE AND GOVERNMENT

AMOS H. HAWLEY
University of North Carolina

BASIL G. ZIMMER
Brown University

Focussing on attitudes toward consolidation of suburban with central city governmental units and services, THE METROPOLITAN COMMUNITY, is a thorough and rigorous analysis of the economic, political and social characteristics of metropolitan residents. More than 50 tables summarize such factors as: education, age, occupation, income of household heads; church attendance; voter registration; attitudes toward present government, schools, public services, and toward proposed changes; and attitudes toward officials and their competence.

CONTENTS

(Available Clothbound and Paper)

L.C. 77-92358   160 pp.   February, 1970

SAGE PUBLICATIONS, INC. / 275 South Beverly Drive / Beverly Hills, Calif. 90212

COMPARATIVE URBAN RESEARCH: The Administration and Politics of Cities

Edited by ROBERT T. DALAND, *University of North Carolina*

This volume is sponsored by the Committee on Urban Administration and Politics of the Comparative Administration Group (ASPA)

CONTENTS

L.C. 69-18751   362 pp.   SBN 8039-0012-0   June, 1969

SAGE PUBLICATIONS, INC. / 275 South Beverly Drive / Beverly Hills, Calif. 90212

LOCAL GOVERNMENT AND STRATEGIC CHOICE

An Operational Research Approach to the Process of Public Planning

by J. K. FRIEND and W. N. JESSOP, *Institute for Operational Research (at the Tavistock Institute)*

This book is concerned with the processes of policy-making and planning in local government and explores aspects of the planning process which impinge on any local community—with particular reference to the planning of housing, transport, education, and shopping, of land use and local government finance. The book is the outcome of an intensive case study of the city of Coventry, which has important implications for urban governments throughout the world. Through a series of linked case examples in local and regional planning, the authors offer a range of suggestions (many of them original) regarding the possibilities of developing more effective aids to choice in complex strategic situations.

$12.50   SBN 8039-0025-2   320 pp.   L.C. 75-83398

REGIONAL AND URBAN STUDIES

A Social Science Approach

Edited by S. C. ORR, *Senior Lecturer in Political Economy, Glasgow University, and* J. B. CULLINGWORTH, *Director of the Centre for Urban and Regional Studies, Birmingham.*

The theme of this book is the contribution of the social scientist to the analysis of planning problems. The editors trace the evolution of planning thought, in both physical planning and social science, resulting in the present tentative steps towards collaboration. Within this perspective the authors describe their approach to specific planning problems such as: projecting industry employment and population, the labor market, housing, and transport. The uses and limitations of techniques such as the regional multiplier and regional input-output analysis are discussed, as is the methodology of the new approach to planning. The concluding chapter is a case study of urban renewal.

CONTRIBUTORS: K. J. Allen, G. C. Cameron, J. B. Cullingworth, T. H. Hollingsworth, J. T. Hughes, L. C. Hunter, K. M. Johnson, R. M. Kirwan, Sarah C. Orr and Edith M. F. Thorne.

$8.50   SBN 8039-0040-6   280 pp.   L.C. 71-84221

SAGE Publications / 275 South Beverly Drive / Beverly Hills, Calif. 90212